ERAU-PRESCOTT LIBRARY

Aviation Food Safety

Aviation Food Safety

Erica Sheward
Technical Director
Castle Kitchens Ltd

Blackwell Publishing

© 2006 by Erica Sheward

Blackwell Publishing
Editorial Offices:
Blackwell Publishing Ltd, 9600 Garsington Road, Oxford OX4 2DQ, UK
　Tel: +44 (0)1865 776868
Blackwell Publishing Professional, 2121 State Avenue, Ames, Iowa 50014-8300, USA
　Tel: +1 515 292 0140
Blackwell Publishing Asia, 550 Swanston Street, Carlton, Victoria 3053, Australia
　Tel: +61 (0)3 8359 1011

The right of the Author to be identified as the Author of this Work has been asserted in accordance with the Copyright, Designs and Patents Act 1988.

All rights reserved. No part of this publication may be reproduced, stored in a retrieval system, or transmitted, in any form or by any means, electronic, mechanical, photocopying, recording or otherwise, except as permitted by the UK Copyright, Designs and Patents Act 1988, without the prior permission of the publisher.

First published 2006 by Blackwell Publishing Ltd

ISBN-13: 978–1–4051–1581–0
ISBN-10: 1–4051–1581–5

Library of Congress Cataloging-in-Publication Data
Sheward, Erica.
　Aviation food safety / Erica Sheward.
　　p.　cm.
　Includes bibliographical references and index.
　ISBN-10: 1–4051–1581–5 (hardback : alk. paper)
　ISBN-13: 978–1–4051–1581–0 (hardback : alk. paper)
　1. Food handling.　2. Food contamination—Prevention.
　3. Food—Safety measures.　4. Airlines—Food service.　I. Title.

TX537.S485 2005
363.72′96—dc22
　　　　　　　　　　　　　　　　　　　　　　　2005007959

A catalogue record for this title is available from the British Library

Set in 11/13 pt Bembo
by Integra Software Services Pvt. Ltd, Pondicherry, India
Printed and bound in India
by Replika Press Pvt, Ltd, Kundli

The publisher's policy is to use permanent paper from mills that operate a sustainable forestry policy, and which has been manufactured from pulp processed using acid-free and elementary chlorine-free practices. Furthermore, the publisher ensures that the text paper and cover board used have met acceptable environmental accreditation standards.

For further information on Blackwell Publishing, visit our website:
www.blackwellpublishing.com

Dedications

I would like to dedicate this book to all those who gave me the help and support I needed to achieve it.

- To Sara Mortimore who will remain forever my champion and an inspiration.
- To Nigel Balmforth and Laura Price at Blackwell Publishing for being so supportive.
- To my staff and colleagues at Castle Kitchens, especially John Holdaway, Rob Prowse and Vince Bellenot, who made my sabbatical and therefore completion of this book possible.
- To my partner and best friend Martin, for his unflinching support, encouragement and love always.
- To the late Al Cordes, a pilot of extraordinary skill, who taught me to think outside the box and forced the issue of aviation food safety onto the pilot agenda, my thanks forever and love to him and his wife Bobbie.
- To all my industry friends and colleagues who helped with my research, and my cabin crew friends who encouraged me to be controversial so they could be better equipped to do the job they love without compromising safety.
- To my great friend Alison Hardie, the best British crew member there is.
- To Rohn Dubler for offering an international crew perspective and being a source of endless inspiration.
- To Susan Friedenberg, the best corporate flight attendant in the world and my colleague in all things controversial, whose insights into the world of business aviation have been invaluable.
- To my lovely friends Lucy and Alistair Drummond for their air traffic and pilot perspectives.
- To the industry I love for providing me with a passion other than red wine and Italy!
- To Tony Cooper, the first man of aviation and the one who took on all the research, encouraged my obsession and taught me all I know.
- To my late father Eric and my late grandparents Ivy and Tom, for sharing with me their wisdom, intelligence and lifelong desire to make a difference.
- To my beloved brother Lee, my lifelong collaborator in courting controversy.
- Above all I dedicate this book to my mother, Jan Sheward, who brought me up to believe that I could achieve anything I wanted. Thanks to her I have.

Contents

Dedications	v
About the author	xiv
Foreword	xv
Preface	xvii
Acknowledgements	xix
Abbreviations	xx
Introduction	xxiv

1 Aviation safety and its impact on the global economy	**1**
Aviation statistics – the crash effect and potential economic impact	2
What constitutes aviation safety?	4
The pilot factor	7
Is food safety a major aviation safety issue?	10
Aviation food safety and the global food chain link	13
2 Consumer perceptions – fact or fiction?	**16**
Airline catering overview	17
The consumer view and how it drives the airline product	19
Buy on board – the battle of the brands	21
How safe is airline food?	24
Causative factors of food poisoning outbreaks associated with meals on aircraft	24
Chain of events	30
3 Current codes of practice	**32**
The International Health Regulations (IHR)	34
Comparison of roles of WTO, WHO and CAC	35
Impact of IHR on airlines' food safety policies	40
IATA and ICAO guidelines	42
Delivery and acceptance of catering supplies	43
Meal and beverage service to the flight crew	44
Food safety and hygiene – risks and prevention	44
Crew personal hygiene	45
Delayed flights	45

Suspected food poisoning	46
Special meals	46
Galley and equipment hygiene	46
Potable water and ice	47
Insects	47
Impact of non-regulatory format of industry directives	48
Food safety legislation	49
Catering standards versus food manufacturing protocols	50
Food labelling legislation	54
Manufacturing-style approaches to airline catering labelling	55
Special-meal labelling	56
IFCA and IFSA World Food Safety Guidelines	58
WHO guidelines	60
Terrorist Threats To Food	60
Guide To Hygiene and Sanitation in Aviation	62
4 Have Airlines Considered Crisis Prevention?	**64**
Management programmes required to facilitate HACCP implementation	65
Education and training programmes essential to HACCP implementation	69
People and process analysis	71
Training requirements	72
Developing a manufacturing-based HACCP study	74
Product and process evaluation	76
Defining operational procedures to comply with the seven principles of HACCP	80
Principle 1: Conduct a hazard analysis	80
Principle 2: Determine the critical control points (CCPs)	82
Principle 3: Establish critical limits	84
Principle 4: Establish a system to monitor control of the CCPs	84
Principle 5: Establish corrective actions to be taken when monitoring indicates that a CCP is not under control	85
Principle 6: Establish procedures for verification to confirm that the HACCP system is working effectively	86
Principle 7: Establish appropriate documentation concerning all procedures and records appropriate to HACCP principles and their application	88
Aviation catering HACCP versus manufacturing HACCP	89

5 Implementing manufacturing SOPs to achieve aviation food safety utopia	**93**
Product development	94
Product-generic issues for consideration	95
Product-specific issues for consideration	95
Development issues	98
Supplier outsourcing	101
Raw material procurement	104
End product specifications	117
Goods receipt	120
Production protocols	127
Assembly protocols	139
Labelling and shelf-life attribution	145
Despatch protocols	147
Verification microbiology and product recall	148
In-flight documentation	150
6 Liability issues – protecting the airline brand	**154**
Aviation liability	155
Establishment of the Warsaw Convention/Montreal Convention	155
Liability under the Warsaw Convention	158
Liability under the Montreal Convention	160
Food safety as an aviation liability issue	162
Case study	164
Brand liability in the aircraft environment	166
7 The airline catering supply chain	**168**
Airline catering and airline caterers	168
Defining an airline caterer	169
Factors governing product development	171
Operational catering issues and safety	173
Food brokers	175
Verifying safety systems	177
Logistics and catering operations	179
To buy in or make in-house – the manufacture of airline catering products	182
Traceability – the critical issues	184
8 Fitness to fly	**190**
Pilot incapacitation and its link to in-flight food safety	190
Are pilots predisposed to food-borne illness?	195

x   Contents

The benefits of food safety training for tech crew	197
Developing aviation industry SOPs to assure food safety fitness to fly	199

9 Cabin crew – the missing link   201

Cabin crew food service role explained – chefs, or merely waiters in the sky?	204
Cabin crew as in-flight auditors	207
Cabin crew health and the risks posed to food safety	211
Cabin crew sickness reporting	211
Cabin crew role in monitoring food standards	213
Food hygiene training protocols for cabin crew	214
Long haul – scheduled	216
Short haul scheduled – domestic only	217
Short haul scheduled – non-domestic	218
Long haul – charter	219
Short haul – charter	220
Low cost – short haul	221
Cabin crew role in ensuring effective food safety management	223
Catering supply goods receipt	224
Catering supply reheat and service	225
Catering supply prerequisite issues	226

10 Managing aircraft water   228

Current standards	228
Public health authorities	231
Local authorities	232
Aircraft operator responsibilities in water safety management	233
Risk factors in aircraft water supply	234
Small animals in the water supply system	236
Controlling microbial risks in aircraft water supply	239
Water safety plan	240
Disinfection	242
Aircraft water systems	242
Aircraft water sampling	243
Aircraft water risk assessment inclusions	244
Implementing water safety standards on a global scale	245
Water consumption in-flight and its critical link to cabin health issues	249
How safe is aircraft water?	249

11 Aircraft disinsection and pest management	**253**
Vector-borne disease – the case for disinsection	253
What is disinsection?	256
The threat posed to global health by vectors	256
The history of disinsection	258
Methods of disinsection defined	260
The blocks away method	260
The top of descent method	262
The on arrival method	263
The pre-embarkation method	263
In-flight vapour disinsection	265
The residual disinsection method	265
Disinfestation	267
Fumigation	267
Chemical safety and public health	268
Disinsection and disinfestation techniques and their link to food safety	269
12 Special meals – special hazards	**271**
What are special meals?	272
Preference meals	273
Religious meals	274
Medical meals	275
Special meal menu planning and development	277
Factors influencing special meal development strategies	281
Summary of factors to be considered in special meal development	283
Menu development	285
Labelling as a menu development consideration	288
Special meal manufacturing protocols and specifications	289
Producing special meals using manufacturing protocols	291
Developing technical specifications	293
Special meal labelling	296
Applying international labelling standards	296
Role of technical specifications in product labelling	300
Nutritional analysis and data	301
Healthy options – healthy claims	303
EU legislative impact on airline SPML labelling claims	304
13 Aviation food safety versus aviation food security	**306**
Aviation food safety and its relationship with bioterrorism	307
Aviation catering security rules, regulation and guidance strategies	310

International Civil Aviation Organization (ICAO)	311
International Air Transport Association (IATA)	312
National Aviation Security Programme (NASP)	312
National Aviation Security Authority	313
World Health Organization (WHO)	314
European Civil Aviation Conference (ECAC)	314
European Food Safety Authority (EFSA)	315
Tokyo Convention 1963	316
Hague Convention 1970	316
Montreal Convention 1971 and Montreal Protocol 1988	316
National ratification of international legislation	316
European Parliament and Council of the European Union	317
Aviation and Transportation Security Act 2001	317
Homeland Security Act 2002	317
Impact of legislative directives	318
EU Food Directives 2000	319
Public Health Security and Bioterrorism Preparedness and Response Act 2002	322
Industry response to legislative directives and guidance	325
Security assurance in the food manufacturing sector	327
Operational security directives in the supply of aviation catering	331
General security considerations	331
Security measures at the catering premises	332
Preparation of in-flight catering supplies and stores	332
Transportation and delivery	333
Application of operational security measures	333
Crew food protocols	334
14 Food safety in the business aviation environment	**337**
What is business aviation?	337
History of business aviation catering	338
Who are business aviation caterers?	343
Catering supplied to general aviation aircraft	344
The general aviation food chain explained	347
Flight crew impact on the general aviation food chain	350
General aviation food safety management utopia	354
Catering providers – general factors influencing product safety	355
The product	355
Premises	355
Personnel	356
Transportation	356
Food preparation	356

Food storage	357
Refuse disposal and dish-wash	357
Conclusions	358
Aircraft operators – general factors influencing product safety	358
The product	358
The product provider	358
The crew	359
Transportation	359
In-flight storage	360
In-flight service/preparation	360
Refuse disposal and dish-wash	361
Conclusions	361
Catering HACCP in the general aviation environment	362
Prerequisite issues	362
Suitability of HACCP approaches to general aviation catering	365
Operational HACCP in the general aviation environment	368
References	371
Index	377

About the author

Erica Sheward is a director of Castle Kitchens Limited, a UK-based food manufacturing company specialising in the provision of chilled and frozen prepared meals to the aviation and retail sectors and specialist catering services to the General Aviation sector.

Erica is also a director of food safety for Jet Academy, which specialises in delivering food safety and security training and auditing initiatives to the aviation industry, and a director of Global Food Standards, a food safety auditing company specialising in the development and implementation of food safety management systems based on manufacturing standards.

Her previous publications include *Your Passport to Food Safety*, a food safety training booklet for cabin crew and *In-Flight Food Safety – Passenger Health and You*, a cabin crew food safety training programme certified by the Royal Institute of Public Health.

Erica is a qualified British Retail Consortium third party auditor and also acts as an expert witness in cases of personal injury involving food safety liability. She is a serving member of the National Business Aviation Association of America Flight Attendant Committee and trains and lectures extensively throughout Europe and the USA.

Foreword

As someone who works globally in food safety I was delighted to be asked to write this foreword for Erica Sheward's first book on airline food safety.

We are at a time when consumer perception of the airline becomes less about the obvious safety factors like its crash record but more about service factors such as comfort and food; then the quality of the food and inherently its safety becomes a critical success factor in building the brand. All passengers, and especially frequent flyers such as me, have an expectation that we will arrive at our destinations fit for work or play. Until recently I had not paused to reflect on the hazards of eating on the plane and how this might affect not just my ability to perform on arrival but even whether I might arrive at all!

When I first met Erica Sheward I was struck by her devoted passion for ensuring safety in the airline catering business. This much needed book makes a compelling case for better management of food safety for all aspects of the aircraft food supply chain. It draws our attention not just to the obvious aspects of food preparation and service but also to the implications of issues such as water supply or pest control and the regulatory framework, or current lack of it, that surrounds the industry.

The dangers of sickness in flight are easily imagined and case studies have been well documented, not just of passenger sickness but also of crew incapacitation and thereby issues of fitness to fly. In this context the issue of food safety for the flight deck crew in the hours and days prior to crew service is often neglected and is certainly unregulated.

As one of the components of general aviation safety, food service is infrequently discussed, yet the wider world of 'terrestrial' food manufacture for retail and ground-based catering has much experience and many tools to offer. There are also a global regulatory framework and global practices learned by multi-national food companies that can serve as a starting point for the airline industry. In this book we are reminded of the transnational nature of airline food provisioning with global sourcing, procurement, manufacture and storage of food and packaging. Beyond the provision of the food, we are made to consider the cultural habits of final preparation, serving and consumption of the food by consumers from any country or culture on an aircraft from any other country or culture. It is a fascinating and unique operating environment that deserves our attention.

At the heart of this important book is the contention of whether airline food service should be treated as a catering operation or a food manufacturing operation, with all the inherent regulatory and best practice procedures and controls that surround such processes. Perhaps more important than regulations is the consideration of the vital importance of training and education of the food handlers at every stage of the supply chain, not least of whom are the cabin crew.

In considering a process to establish food safety we are drawn to two aspects that are well covered in this book. Product development, as in any other food industry, is critical at the start of the process and the relationship between the 'chef' and the manufacturing food technologist needs to be considered to ensure that food safety is designed into the product. Secondly we confront HACCP (Hazard Analysis and Critical Control Points) and how this should be used, not in a generic manner but to understand and control product and process specific risks. Throughout this discussion the prerequisites such as hygiene, environment, best practice and education are well covered – which is as it should be.

Beyond the insurmountable case that is made for better food safety in the airline food supply chain we are finally, in these times, forced to consider the vulnerability of this food chain to malicious attack. The simplicity with which deliberately contaminated food can jeopardise aircraft safety and security is worrying. Whether it is from a deliberate attack of bio-terrorism or an act of neglect on the part of the airline industry, let us hope that it does not take a disaster or crisis to drive change, as it so often has in other arenas. It is time for the airlines to work for themselves with the support of national governments and transnational regulatory bodies to set better standards and change behaviours to ensure that safe food and drink is served at all times on every airline.

They are most likely doing exactly this but the discussions that this book will no doubt stimulate must surely help to raise awareness and hasten progress. The author is to be congratulated in helping us to understand this area of the food industry and I fervently hope that her enthusiasm and commitment to improve the standards will be successful.

Sara Mortimore
Qualtiy & Regulatory Director
General Mills International
Minneapolis, MN, USA

Preface

The aviation industry is a multibillion pound amalgam of emerging technologies, manufacturing processes and service profiles, which combine to create the end product. Most of us experience this to a greater or lesser degree several times during the average lifetime.

Considering all the contributory factors which combine to assure aviation safety, it is ironic that the perceived risks to aviation safety posed by the in-flight meal solutions on offer are rendered negligible.

As both a product provider and consultant to the industry I have grown increasingly convinced that the attributable risks to aviation safety posed by the catering chain of logistics and supply have been underestimated dramatically.

The 21st century technology evidenced in every other area of the industry is both necessary and obvious if the industry is to continue to meet the global increase in demand for advanced and eclectic travel opportunities. State of the art avionics and service technologies, streamlined and aesthetically driven cabin interior solutions, spellbinding in-flight communication and entertainment options: all continue to develop and evolve in tandem with an increasingly fast-paced and modern world.

The challenges are driven by customer expectations. The customer expects the prototypes to become reality as fast as their predecessors have embarked on their maiden flight. The safety expectations move just as swiftly.

It is my belief that in the 21st century the attributable risks posed to aviation safety by the catering product, rival any other in the mechanical, operational or security arenas. Any suggestion that the customer expectations which drive aviation safety standards generally are not equivocal in the foodservice area, are unfounded. The status of customer expectations with regard to the catering product is directly linked to historical deficiencies in aesthetic appeal and quality shortfalls. In this regard the customer perception of quality is not linked to safety expectations in the same way as non-food service products are.

The historical customer fascination with the quality, i.e. aesthetics, ergonomics, etc. of in-flight food products, has lulled the industry into a false sense of security over the requirement to develop products and protocols assuring all-round product safety assuming quality and safety as one in the way that food manufacturers do. Everything that appertains to product safety needs to be placed under scrutiny: date marking and labelling, nutritional data and raw material outsourcing, above all manufacturing standard GHPs, GMPs and HACCP.

Recent attempts from within to develop generic industry standards did little to sway my firmly held belief that much that the aviation industry attempts to do escapes scrutiny and avoids the primary issue which faces it. As long as the aviation community refuses to accept that every aspect of the logistical and operational food service supply chain should be governed by manufacturing standard food safety management and quality assurance protocols, it will remain a risk to aviation safety. It is only by the adoption of a manufacturers' mentality as opposed to the cosy caterers' ethics, that the emerging safety and security issues which face the aviation catering community can be effectively dealt with. How can issues of bio-terrorism be tackled when systems management protocols are so inappropriate? How can the industry react effectively to the worldwide overhaul of food labelling requirements when their specialist meal provision has no cohesive structure or basis for analysis? How can full traceability be assured when the supply chain is so haphazard and broker driven? How can menu development occur in tandem with product safety considerations, when celebrity chefs determine the critical source of supply?

The component nature of the product and the requirement for global replication should be the issues which drive forward the adoption of food manufacturing standards, not the smoke screen behind which both the aircraft operators and catering providers hide. In this book I will attempt to unravel the issues which appertain to product safety in the aviation catering environments, and suggest ways in which standards can be reviewed to assure the same advancements in food service products as one would expect to find in the aviation product itself.

The lessons learned by food manufacturers over the past 30 years can serve as a great incentive to get it right first time. The assumption that catering safety issues will be hidden from scrutiny by a flying public preoccupied with quality is a dangerous and potentially costly assumption to make.

As the numbers of those flying swell year on year, a new generation of supermarket psyched, product safety aware litigants take to our skies. Unless a proactive approach to their quality and safety demands is undertaken, instead of the historically reactive methods still employed, the aviation industry may well find itself left too far behind to recover.

Erica Sheward 2005

Acknowledgements

I would like to thank the following people for all their help and professional support in the completion of this book:

George Banks, Menu Development, British Airways, London.
Scott Chapman, Senior Partner, Medical Litigation, Tress, Cox and Maddox, Sydney, Australia.
Neil Chitty, Director, Advance Fumigation and Pest Control, Horsham, West Sussex.
Bill Cossins, Consultant, Product Authentication International, Harrogate, North Yorkshire.
Ricky Cuss, Operations, Virgin Atlantic, London.
Anne Green, Catering Representative, Qantas In-flight Services, London.
Paul Hobbs, Environmental Health Officer, Horsham District Council, West Sussex.
Caroline Kelly, Food Technologist, Waitrose plc, Bracknell, Berks.
Simon Levy, Food Safety and QA, Airfayre Ltd, London Heathrow Airport.
Sara Mortimore, General Mills, Minneapolis, USA.
Dr Sanj Raut, Principal Scientist, National Institute for Biological Standards and Control, Potters Bar, Herts.

Abbreviations

ALPA	Air Line Pilots Association
APHA	Association of Port Health Authorities
APHO	Association of Public Health Observatories
AQIS	Australian Quarantine and Inspection Service
ATSA	Aviation and Transportation Security Act
AVML	Asian vegetarian meal
BAA	British Airports Authority
BALPA	British Air Line Pilots Association
BLML	bland meal
BOB	buy on board
BRC	British Retail Consortium
BSE	bovine spongiform encephalopathy
CAA	Civil Aviation Authority
CAC	Codex Alimentarius Commission
CBP	Customs and Border Protection
CCP	critical control point
CDC	Centers for Disease Control and Prevention
CHML	child meal
CITEJA	Comité International Technique d'Experts Juridiques Aériens
COSHH	Control of Substances Hazardous to Health
CP	control point
CRM	crew resource management
DBML	diabetic meal
DfT	Department for Transport
DVT	deep vein thrombosis
EASA	European Aviation Safety Agency
EC	European Community
ECAC	European Civil Aviation Conference
EEC	European Economic Community
EFSA	European Food Safety Authority
EPA	Environmental Protection Agency
ETEC	enterotoxic *E. coli*
EU	European Union
FAA	Federal Aviation Administration

FAO	Food and Agriculture Organization
FAR	Federal Aviation Regulations
FBO	fixed base operator
FDA	Food and Drug Administration
FISA	Food Industry Security Assurance Scheme
FLTS	flights
FP	food poisoning
FPML	fruit plate meal
GA	general aviation
GATT	General Agreement on Tariffs and Trade
GC	General Council
GDP	gross domestic product
GDWQ	Guidelines for Drinking Water Quality
GFML	gluten free meal
GHP	good hygiene practice
GM	genetically modified
GMO	genetically modified organism
GMP	good manufacturing practice
GNP	gross national product
HA	hazard analysis
HACCP	hazard analysis and critical control points
HFML	high fibre meal
HNML	Hindu meal
HPC	heterotrophic plate counts
HSE	Health and Safety Executive
IATA	International Air Transport Association
ICA	intercarrier agreement
ICAO	International Civil Aviation Organization
IFALPA	International Federation of Air Line Pilots Association
IFCA	International Flight Catering Association★
IFSA	International Inflight Food Service Association
IHR	International Health Regulations
IMF	International Monetary Fund
ISR	Information Systems Research
JAML	jain meal
KSML	Kosher meal
LCML	low calorie meal
LFML	low fat/low cholesterol meal
LPML	low protein meal
LSML	low salt meal
MCS	multiple chemical sensitivity

MOML	Moslem meal
MOR	Mandatory Occurrence Report
MQS	New Zealand MAFF Quarantine Service
MSG	monosodium glutamate
NASA	National Aviation Security Authority
NASP	National Aviation Security Programme
NBAA	National Business Aviation Association
NLML	non-lactose meal
ORML	Oriental meal
PARNUTS	Particular Nutritional Uses
pax	passengers
P+	production day plus
pH	potenz hydrogen
PHEIC	public health emergencies of international concern
PHLS	Public Health Laboratory Service
PI	pilot incapacitation
PIC	pilot in charge
PIL	passenger information list
PPE	personal protective equipment
PRML	low purine meal
QA	quality assurance
QMS	quality management systems
QO	Quarantine Officer
QUID	Quantitative Ingredient Declaration
R&D	research and development
RVML	raw vegetarian meal
SARS	severe acute respiratory syndrome
SDR	special drawing right
SEP	standard emergency procedures
SFML	seafood meal
SOP	standard operating procedure
spec	specification
SPML	special meal
SPS	sanitary and phytosanitary measures
TRANSEC	Transport Security Directorate
TSA	Transportation Security Administration
TSE	transmissable spongiform encephalopathy
TSU	tray set up
ULV	ultra low voltage
UN	United Nations
VGML	vegan meal

VLML	lacto ovo vegetarian meal
WHA	World Health Assembly
WHO	World Health Organization
WQTs	water quality targets
WSP	water safety plan
WTO	World Tourism Organization

*The IFCA has changed its name (March 2005) to ITCA (International Travel Catering Association).

Introduction

Just over 90 years ago, on 1 January 1914, a gentleman named Abe Pheil became the world's first airline passenger. A phosphate miner, he sat on an open bench in the cockpit of the seaplane nicknamed Limping Lucy. The pilot in charge on board this 21 minute flight from St Petersburg to Tampa in Florida USA was a 25 year old named Tony Jannes. He wore no uniform and did not work for an airline.

Since that eventful day, the industry has made great progress; the accessibility of air travel to the ordinary man in the 21st century is such that many people commute as effortlessly on board aircraft as they do by bus or train.

Back in 1914, the success of the trip or otherwise would no doubt have been determined by the fact that the aircraft survived the flight without falling from the skies. These days, however, the average airline passenger expects a lot more by way of in-flight comforts: a comfortable seat, an in-flight meal that meets the same attributable standards of quality as for a meal served in a restaurant on the ground, a fully flushing toilet and hot and cold running water, in-flight movies and music at the touch of a button.

This expectation is today a startling reality. All credit to the ingenuity of modern technology in tandem with the vast amount of work that goes on behind the scenes at every major airport in the world, thus providing the infrastructure that brings the passenger not only movies and music at the touch of a button but also ensures the safety of the product and thereby the safety of the passenger.

Whether or not Abe Pheil was given an in-flight meal was not recorded by the newspapers of the day. The first record of in-flight catering appeared in *Flight* magazine on 14 December 1922, with the following short paragraph:

> *'Mr Lloyd, the manager of the Trust House has now arranged for luncheon boxes for any passenger who feels that he or she would like to relieve the monotony of an air journey by taking a meal.'*

It is highly likely that anyone who took up the offer of an in-flight meal was less worried about the safety of the contents of the luncheon box than the possibility that the aircraft might not survive the trip. Aviation catering has come a long way since the early pioneering days and is now a multibillion pound industry worldwide.

Introduction

In the 21st century, how safe is the food and drink available for both passenger and crew consumption on board the aircraft that we travel on to transit the globe? The reality is that, whilst in all other areas of aviation safety technology has forged ahead at a rapid rate, delivering state of the art aircraft design and manufacture, the world of aviation food safety has been dramatically left behind.

The aim of this book is to focus on every aspect of the in-flight product, its evolution and current method and level of production and attempt, to uncover why, in terms of food safety, it is a prehistoric industry.

With an in-depth knowledge of the airline catering industry, the product requirement, logistics and every aspect of the supply chain and manufacturing process are brought under scrutiny, with a best practice scenario documented and discussed in each case.

Aviation food safety management systems and current codes of practice are analysed in terms of existing practices. Airline climate and culture changes, which would facilitate the evolution of a best practice utopia, are scrutinised. Parallels are drawn between current mainstream food manufacturing systems and how they might be introduced into airline catering operations.

The major issue is to highlight a situation where, as a direct result of the refracted evolution of the industry, food safety management systems are devised and implemented along catering guidelines instead of mainstream food manufacture. This is indeed a startling reality bearing in mind the volumes involved and the requirement for global provision and replication of the product.

Analysis of the relative failure of the current codes of practice within the industry and a direct comparison with the manufacturing sector are included to highlight the gravity of the failure of current GMPs and GHPs throughout the extended supply chain. We discuss why the unique nature of the airline catering product renders this the case.

An examination of the extended and refracted supply chain and the component nature of the product highlights the difficulties inherent in airline catering supply and why every aspect of the operation is a potential food safety management nightmare. In the industry many crucial aspects of the supply chain are overlooked or ignored and these are focused on in detail in the book. The burgeoning requirement to prepare, cook and serve an eclectic array of menus from scratch actually on board is a major example; the impact of flying food handlers' fitness to work is another.

If one takes into account the diverse nature, specification and evolution of the product, coupled with the airlines' obligation to supply a safe product all over the world, regulatory compliance issues are rendered something of an irrelevance in the devising of food safety supply, manufacture and management systems.

Best practices and a new perspective are outlined with accompanying examples of systems documentation, in order to encourage and support the requirement to restructure the processes, improving levels of food safety and security globally.

Aligning food safety requirements and GMPs with aviation safety, security and aircraft design issues poses a major headache. Traditionally, extended consultation and liaison with sources of food safety management expertise from outside the industry have never been an option. The historical climate of self-regulation renders the industry susceptible to criticism and failure. The result of the industry's shortcomings is examined in terms of its potential impact not only on food safety but also on aviation safety.

A fundamental reassessment and restructuring of every product and process appertaining to the product delivery has to occur so that aviation food safety standards can fall into line with the rest of the industry's emerging technologies.

1 Aviation safety and its impact on the global economy

Whilst it may seem somewhat strange to begin a technical book on aviation food safety by discussing aviation safety statistics in general, I feel it is crucial from the outset that one understands what constitutes aviation safety, the effect that safety breaches may have on the global economy and how aviation food safety issues play a pivotal role in what has historically been regarded as the wider, unrelated picture.

As a supplier of catered food products to the industry as well as an industry consultant and advisor, I have never lost sight of the relevance of the part that I have to play in assuring not just food safety but aviation safety also. Too often it seems, the connection between catering provision and the capacity to compromise the personal safety of both passengers and crew and the economic integrity of nations worldwide is readily overlooked. Food and water safety issues, whilst viewed as major bio-terrorist threats on the ground, are not perceived as such in the air. The flight-deck crew and their food safety fitness to fly are also a factor ill considered and not trained for, despite the well guarded industry data to suggest that a large proportion of pilot incapacitation occurs as a direct result of food-related illnesses.

The aviation industry provides the ideal vehicle for the transmission of a whole host of food, water and vector-borne diseases, yet it escapes unified regulatory scrutiny and is governed instead by 'best practice' ideals with gulf-wide clauses open to all manner of sensory interpretation. Those who really understand how it works generally have a pecuniary interest in keeping those secrets to themselves and many who work in the industry have no experience or connection with the nature and standards of food manufacturing provision elsewhere. The world of aviation food provision is complex and difficult but it requires a cohesive and proactive approach to safety, taking account of the bigger picture and the other safety issues it has the capacity to impinge upon.

Shortly after the 9/11 World Trade Center terrorist attack, one of the many crew members I had trained in food safety told me that following my course she had returned to work with many new perspectives, the overriding one being that the easiest way to bring an aircraft down would be to poison the crew. She told me that at the next airline security training she attended she had regaled those attending with this notion and the basis for it, and had been vilified by both the trainer and her fellow attendees.

To underestimate the impact that aviation food provision has on the wider issues of aviation safety and the global economy is to show a fundamental misunderstanding of the precise and intricate nature of the application. The multimillion unit replication, the global outsourcing requirement and the 'export' nature of the food product are all key to understanding the true potential that airline food has to permeate aspects of wider economic and international significance.

Throughout this first chapter we set the scene as to why and how it is absolutely critical that aviation food safety systems are scrutinised, standardised and regulated effectively in line with the highest possible advances in food manufacturing protocols. The inadequacy of the current situation is reflected in the prevalence of non-regulated, ad hoc, mass catering standards, which provide no assurance that each and every aviation food product is verified, consistent and above all traceable.

Aviation statistics – the crash effect and potential economic impact

The aviation industry generates a whole range of direct economic benefits to the global economy. It is a major employer, it provides vital trade links and above all it supports the movement of goods. For many countries also, it provides the opportunity for a thriving aerospace technology industry to evolve.

There are also indirect economic benefits from associated jobs created as a result of airport activity. The British Airports Authority (BAA) estimated in 2000[1] that the number of jobs in the UK indirectly supported by airport activity, such as hotels and food provision, was around 380 000. In San Francisco the airport provided 20 000 jobs, with a further 200 000 in the associated visitor industry depending on the region's airports for their customers. In 1998 the aerospace industry in the UK employed 154 000 people and contributed around £6.1 billion to the UK economy, equivalent to 0.8% of the gross domestic product (GDP).

Aviation transport is vital for business, with business traffic accounting for approximately 24% of all passenger traffic, and it is expected to become more important over the next 30 years. Air transport is also an important means of transporting freight and currently accounts for 20% of all UK exports by value. Traditionally, air freight has been used for high value commodities, fragile goods, emergency items and spare parts for production line breakdowns. However in recent years the range of goods carried by air has widened significantly to encompass luxury foods, exotic fruits and flowers, chilled meat and fish, newspapers and fashion items.

The presence of air connections is of paramount importance to the location decisions of many overseas investors and the importance of easy air access to any nation looking for regional development cannot be underestimated. In 2000 the World Tourism Organization (WTO)[2] estimated that world tourism grew by an estimated 7.4% in 2000, its highest growth rate in the last ten years and almost double the increase of 1999. Europe accounts for 58% of international tourism and grew by 6.1% in the year 2000 to 403 million arrivals – an increase of nearly 25 million on the previous year. Whilst the WTO reported that all regions of the world hosted more tourists in 2000, the fastest developing region was East Asia and the Pacific with a growth rate of 14.7% and 14 million more tourists than in 1999. The WTO predicted that international arrivals were expected to reach over 1.56 billion by the year 2020. Of these 1.18 billion would be intraregional and 377 million would be long haul travellers.

The WTO also noted in 2000 that the receipts registered for international tourism exceeded US$1 billion per year in 59 countries and suggested that:

> *'the tourism industry is one of the biggest industries in terms of employment and contribution to GNP (Gross National Product). It is often an essential component of sustainable development.'*[2]

So having established the bare bones of exactly how critical are the presence of successful and effective aviation travel opportunities to the global economy, we need now to turn our minds back to the context of where safety breaches in the industry may impact on that situation. In the context of the aviation model there are numerous opportunities for safety to be impacted upon. Generally most consideration and public consciousness is focused directly on the 'crash effect' and the direct connections made between potential breaches in mechanical and technical aviation safety and aircraft disasters.

However, the wider picture constitutes a host of other factors that all have a part to play in the aviation safety debate but are less prevalent on the public stage. The historically reactive nature of aviation safety mechanisms dictates that many of the topics on the safety agenda are there as a result of their direct implication in an incident or accident. In the 21st century also, the profound media interest in aviation health matters, such as deep vein thrombosis (DVT), severe acute respiratory syndrome (SARS), cabin air quality and cosmic radiation, has driven previously ill-considered cabin safety issues onto the mainstream aviation safety agenda, at a time when traditional crash effect issues were becoming less of a concern.

Many of the potential safety factors that have the capacity to impact on passengers, crew and the global economies of nations that sustain an aviation infrastructure, are food and water related, but despite this they are widely ignored and excluded from the mainstream aviation safety debate.

Recently I attended an annual industry conference on 'cabin health' organised jointly by the International Air Transport Association (IATA) and the WHO. The agenda was packed with all of the previously mentioned legitimate cabin health issues, most of which had been the focus of intense media interest over the past few months. I suggested that the debate in relation to all was perhaps a little academic unless the wider connection of the issues could be made in terms of their relationship to appropriate food and water provision also.

What is the point of a debate on DVT and taking steps to minimise its impact when one of the prescribed proactive remedies is increased water consumption on board and the integrity of the tanked water on board is not called into question?

What is the point of a debate on whether 'obviously sick' people should be prevented from travelling when the industry proffers a whole host of meal solutions dedicated exclusively to the 'obviously sick'?

So long as food and water safety on board aircraft remains an isolated and non-integrated aviation safety debate, its true capacity to impact on aviation safety will remain undetected. Whilst the industry is very quick to champion its safety record in the food safety arena, the reality for anyone who dares to challenge it is a wall of silence, a denial out of hand and a laying of the blame at anyone else's door. Only those on the inside, the intricately well informed and those who are prepared to challenge that assertion to the bitter end, will discover the true picture of a reactive compensation culture cowering behind 30-year-old food safety systems and an irrelevant industry best practice which is neither adopted nor enforced but provides a useful smokescreen for the industry, designed to keep the regulators at bay.

Harsh words maybe, but designed to throw down the gauntlet and make a genuine contribution to the world of aviation food safety with a full and given knowledge of what is required to achieve aviation food safety utopia.

What constitutes aviation safety?

I am not, nor would pretend to be, any great expert on technical and mechanical aviation safety matters. That is not what this book is about. However I do feel entitled to proffer an opinion as to what I believe constitutes aviation safety, in the context of the very real connection that food and water provision issues have with the wider implications of the subject matter.

I believe that anything that has the capacity to impact on the effective functioning or intended work activity of either the technological infrastructure or physiological capabilities of the person or persons engaged in a safety function on board an aircraft, needs to be considered an issue of aviation safety importance. That being said, we must now look at all the possible food and water-related provisions that may fall into that category:

- Food and water quality provision for flight-deck and cabin crew in-flight
- Food and water quality provision for flight-deck and cabin crew when down route
- Passenger and crew exposure to air-borne, water-borne or food-borne disease in-flight
- Passenger and crew exposure to air-borne, water-borne or food-borne disease down route
- Passenger and crew exposure to foods containing undeclared allergens in-flight
- Passenger and crew exposure to foods containing undeclared allergens down route
- Passenger and crew exposure to disease carrying vectors in-flight
- Passenger and crew exposure to disease carrying vectors down route
- Passenger and crew exposure to terrorist activity via food and water provision in-flight
- International food chain exposure to non-traceable food waste and products.

As we can see from the above list, there is immense capacity for a whole concentration of safety issues to be focused around the perceivably simple, logistical matter of providing catering services to an aircraft. The wider picture of safety issues appertaining to the quality and integrity of aviation food products is focused on their 'exported' food potential.

In 2001 the UK was ravaged by the worst outbreak of foot and mouth disease for over 50 years. The economic impact was devastating, not just for those involved in the primary production and farming industries but also for those involved in the travel and tourism sectors, as millions of would-be visitors ceased to come. Protecting the safety and integrity of the national food chain is a matter of international significance and any travel vector involved in the transportation of finished food products or raw materials, has to come under the strictest possible scrutiny so that the potential contamination of the food chain does not remain a viable possibility. If one ignores the obvious 'export' status of airline catered food products, and therefore the potential opportunity for the international food chain to be compromised by inappropriate and inadequate food production standards, then a major loophole in the transit of foodstuffs with the capacity to impact on the safety and integrity of the food chain has been dramatically overlooked.

If one is to take account of the broader aviation safety picture a dedicated focus of attention must be placed on the potential interrelationships that the provision of food and drink products have with other aviation safety models. Integrated safety perspectives are abundant in the aviation sector and form the basis for the development of training solutions and the advancement of technological systems management. However, the same integration of approaches to aviation food provision and the acknowledgement of its significance as a major aviation safety issue, are not evident in the same way. It is essential that the provision of aviation food products is placed into the appropriate safety context and a logical and systematic assessment of their safety function is undertaken in order that the interrelated issues are acknowledged.

Unless the broadest possible view is taken of the potential application that food and drink provision may have in the aviation safety arena, then the logical chain of events and protocols that needs to be established in order to ensure supply chain integrity will also not be effectively established. The food safety and security issues that face the industry are immense, affording a unique opportunity to impact on an industry which if associated with safety or security shortfalls, in turn has the capacity to impact on the stability of the global economy like no other.

Throughout the following chapters we will examine each interrelated safety issue in turn and attempt to place them all in the true context of their potential impact on aviation safety matters. We will acknowledge the complex and intricate connections between those involved in the aviation supply chain, and how deviations from food manufacturing standard operating procedures (SOPs) have the capacity to create havoc in what is traditionally considered a 'catered' environment. The factors of proportionality of scale of production which form the basis upon which all defined food standards are ultimately interpreted and implemented, are also examined and the crucial link is explained between safety and quality systems management operating in tandem in these environments.

The requirement for global replication and outsourcing of aviation food products demands that a standardisation and acknowledgement of amalgamated food production protocols is made and enforced by regulatory compliance rather than industry 'best practice'. The industry is currently open to all sorts of interpretive methodology in the food safety sector and guidance standards are set against a backdrop of mass catering ethics and fluctuating food standards based on international deviation. It is for the industry to have the parameters of appropriate food standards dictated and defined for it, not by it, in the same way as manufactured food products designed for international export. As long as the unique safety aspects of aviation food provision are defined and determined by the industry itself, then the real safety factors will remain hidden and unconsidered. Safety costs and, in an industry so driven by fiscal constraints, the profile of food safety issues will continue to be driven away from the mainstream aviation safety debate, only to be played out in front of an invited audience of industry insiders dedicated to keeping the true picture of aviation food safety failure obscured.

The pilot factor

One of the most indelible factors to link food safety matters with aviation safety matters is the direct impact that the consumption of in-flight food and drink products potentially has on those charged with the ultimate safety responsibility in flight – the flight-deck crew.

Pilot incapacitation (PI) is defined in aviation safety terms as an incident or accident that affects the pilot at the controls of an aircraft. It must be sudden and total and take place during a critical stage of flight.

There is immense potential for the flight-deck crew to have their capacity to carry out their duties effectively compromised by the consumption, either in-flight or down route, of food and drink products that are unfit. Pilot incapacitation statistics are closely guarded and kept under wraps by aviation organisations such as the International Civic Aviation Organization (ICAO) and the Federal Aviation Administration/Civil Aviation Authority (FAA/CAA) and are designed to be used to determine the evolving itinerary of airlines' emergency training procedures. Whilst I have often heard the statistics debated at cabin safety conferences attended by the aviation inner circle, and whilst I had a brief insight into the causative factors whilst I was at ICAO in 2002, they are generally not for public consumption. What they illustrate time and time again is that a vast percentage of reported incidences of PI are due to the flight-deck having been affected by food poisoning or food and water-related illness.

In Chapter 8, 'Fitness to fly', and Chapter 9, 'Cabin crew – the missing link', we look at the issues surrounding the consumption of in-flight food products not just by the flight-deck but also by the cabin crew, and suggest that the best auditors of the safety and quality of the in-flight food products are the crew themselves, those who are consistently eating a variety of in-flight food products from a variety of outstations throughout their working lives.

Whilst PI situations are trained for relentlessly to ensure that every conceivable safety malfunction has a policy and procedure afforded to it, the wider issues affecting the flight-deck crew and their food safety fitness to fly are not redressed. To do so would mean having to confront the reality and enormity of aviation food safety issues as they really are and placing aviation food safety on an agenda alongside the *real* aviation safety issues as the industry is happy to have them portrayed. As soon as connections are made, and interrelationships defined between the mainstream aviation safety agenda and aviation food safety a worrying nebula of concern descends over the whole picture, and the historical ease with which aviation food safety and security issues are sidelined onto an alternative agenda becomes a little more difficult.

Whilst the pilot incapacitation factor remains a sensitive and secretive one as far as the industry is concerned, there are some exceptional examples in recent aviation history that made the headlines. The following two examples illustrate all too clearly the crucial link between aviation food safety and the pilot factor, and the ultimate link to aviation safety itself.

In November 1989 a flight was en route from Sydney to London. It made a fuel stop and crew change in Singapore and then continued on its journey to London, stopping once again in Abu Dhabi. Once more there was a crew change and extra fuel was loaded as the news came to the captain in charge that the weather condition upon arrival into London was likely to be mist.

Several hours into the sector from the Middle East to London the first officer and flight engineer became extremely unwell with suspected food poisoning. Whilst they had not consumed any of the same food on board, they had eaten together the night before. They were so incapacitated that the crew took the decision to place them in the bunks. The captain, Glen Stewart, an extremely experienced pilot of 35 years took steps to contact the ground in London. Anxious not to incur all of the associated inconvenience and costs inherent in an aircraft divert, and anxious to ensure that his fellow crew members got some medical attention as soon as possible, he made the decision to proceed to London flying the aircraft alone.

Upon arrival into London the weather condition was not mist, it was fog, and by virtue of the extra fuel that had been loaded in the Middle East, Captain Stewart was able to circle around for an hour waiting for permission to land. He was becoming more and more concerned for his fellow crew whose condition had worsened. He began his approach without them beside him, doing his checks and balances in the cockpit, only to discover in the nick of time that during the poor visibility he had mistaken the A4 road carriageway for the runway.

The aircraft missed crashing into the top of a local hotel by barely 100 feet, and those on the ground who watched the belly of the aircraft descend from the fog laden sky that crisp November morning, gasped in horror.

On board the aircraft, the passengers were none the wiser. The extreme professionalism of the crew ensured that the passengers were oblivious to the near danger they had been placed in. Indeed, Captain Stewart himself made an immediate announcement to the effect that they had merely decided to abort the landing for safety reasons and assured them that they would be back on the ground shortly.

Ultimately, had the flight-deck crew members not been incapacitated they both may have been in a position to detect the near fatal error. Captain Stewart was immediately suspended from flying, accused of gross misconduct and committed to trial. In May 1990, at Isleworth Crown Court, he was found guilty after an immensely technical and high profile case. The crew involved in the incapacitation incident were never called to give evidence and there was no reference to the incident having been a potentially causative factor in Captain Stewart's alleged error of judgement. Captain Glen Stewart never flew again and later that year he tragically took his own life.

In a letter to one of the passengers on the aircraft that day, who had written to him whilst he was on trial expressing her sympathy and gratitude for ultimately landing the aircraft safely, he said:

> *'I don't know why we found ourselves in that situation that day, all I know is that I have since been diagnosed with a myopic condition that they didn't test pilots for (they do now) and that may have affected my vision in restricted visibility conditions. I was without my checks and balances on the flight-deck that day but ultimately I just wanted to get that aircraft down for the sake of my colleagues who were sick. The fatal mistake I made was loading extra fuel in Abu Dhabi, if I hadn't we would have had to have diverted when we reached London.'*[3]

Whilst this story takes its place in the aviation safety hall of 'near miss' incidents that litter the archives of aviation history, the ensuing human tragedy could so easily have been avoided. Whilst the issues of certain types of myopia, having been identified, were immediately redressed by the industry, the issue of pilot incapacitation due to food poisoning wasn't even discussed in connection with this case, let alone reacted to and acted upon. Is it unreasonable to expect pilots to take care of what they eat and where they eat down route when they are on company time and have passenger safety to consider? Is it unreasonable to incorporate mandatory crew training to highlight food safety issues and how they have the capacity to impact on flight-deck incapacitation and crew fitness to fly? Is it unreasonable to look at the real picture laid bare by an examination of pilot incapacitation statistics and look at the essential requirement for safe crew food provision above all else?

The second tale forms part of even more recent history. In June 2002 an aircraft was en route between South Africa and London. In the first class section were a group of several key performers who were on their way to take part in the Queen's Golden Jubilee celebrations. Four hours into the flight the female first officer became extremely ill with all the signs of food-borne illness. Having already suffered several bouts of such illness in her flying career she took extreme care when eating and drinking down route especially. Having been incapacitated so severely and having been placed in the bunks, this lady began to hallucinate. The pilot in charge was only too aware of the cost and inconvenience that would be caused if he made the decision to divert the aircraft to seek medical help for his colleague rather than continue to London alone on the flight-deck. But so grave were his concerns for his colleague that he made the decision to divert into Barcelona.

The ensuing hours and days that followed saw all manner of complicated issues having to be redressed, not least having to charter an aircraft to get the VIP guests to London in time to form part of such an historic celebration. Then there were the problems of crew logistics, hotel accommodation and transfers for all the other passengers, and so it went on.

The lady first officer was diagnosed with a severe bout of campylobacter enteritis and was hospitalised for several days in Barcelona until fit enough to passenger home. There was no ensuing evaluation of exactly how and why this could have happened to this officer yet again but it transpired afterwards that on her way to the airport in Johannesburg on the morning of the flight, she had stopped to buy in the shopping mall a healthy breakfast of freshly squeezed orange juice. She had no food safety training and had no idea that something so perceivably healthy could pose such a dire threat to her health if not handled and stored correctly.

Yet again this is an example of the direct connection between aviation food safety, the pilot factor and impact on the wider issues of aviation safety generally. It will become clear as the book progresses that there are so many interrelated issues in aviation that are dealt with in isolation and form aspects of unconnected agendas. The requirement for all crew members to be trained in the food safety issues commensurate with their personal safety responsibility as well as the relative risk factors inherent in their career, for those who travel to the far flung corners of the globe, cannot be underestimated, as the debate over the most prevalent causes of PI rages on.

Is food safety a major aviation safety issue?

Having already established the connection between the pilot factor and aviation food safety, we can begin to see the fashion in which the true picture of aviation safety should be defined. To disconnect the peripheral service-related aspects of the aviation product from the technical and safety dominated ones is to negate the potential impact that they may have on any predetermined safety agenda. In the same way, the cabin health issues that dominate another sphere of the aviation safety debate remain isolated by their context and not amalgamated by their common denominators.

A list of the mainstream perceived aviation safety issues may well look something like that shown in Table 1.1.

An equally mainstream and defined list of cabin health and safety issues may well look like that in Table 1.2.

Indisputably, many of the generalised issues that form aspects of the above two mainstream aviation safety agendas have a direct link with the safe, secure supply of food and water both in-flight and, as far as the crew are concerned, down route also. To suggest that aviation food safety issues do not belong or have no major part to play in ensuring that the other safety agendas are fully considered and all attributable risks accounted for, is to have a fundamental misunderstanding of the nature, ergonomics, logistics, prerequisites, extent and potential impact of the aviation catering supply chain.

There is no doubt in my mind that the safety and security of food and drink products supplied to aircraft should be of paramount importance and given the credence they deserve on the mainstream aviation safety and security debating stage. Throughout this book we will look at all of the inter-related issues and how they each have an impact on the success or failure of the aviation food chain.

If one looks at the 'catered' airline food product in terms of passenger perception, it is the one thing most likely to engender some kind of subjective response from those who fly. Whilst the aesthetic quality and appeal of the products is constantly called into question by an ever more demanding flying public, it is not for the public consciousness to draw attention to the provision of food and drink as an aviation safety issue. It is for the industry itself to focus on the connections and to assess the risk factors inherent in the interrelationships between the food and drink provision on board aircraft and the safety and security issues that face it in the 21st century.

Table 1.1 Perceived mainstream aviation safety issues

1. Mechanical and technical failure
2. Pilot error/incapacitation
3. Terrorism and bioterrorism
4. Unruly passengers

Table 1.2 Perceived mainstream aviation cabin health issues

1. DVT – passengers and crew
2. Cosmic radiation
3. Spread of communicable disease – SARS
4. Spread of vector-borne disease (malaria, dengue-fever, etc.)
5. Cabin air quality

If one looks at the nature of aviation food provision in detail, as we do throughout this book, it becomes clear that the safety and security issues that have always faced the aviation sector in terms of food manufacture and provision, were once issues they faced in unique isolation. Nowadays the requirement for mass produced, ready to eat foods, manufactured to satisfy an export market requirement, is a challenge embraced with huge success by the multinational supermarket food manufacturers on a daily basis.

Since the mid 1970s food manufacturing and food technology protocols have been driven by the need to secure the export successes of the products made, with little opportunity for sentimentality in terms of product development and packaging. The major difference is that all food manufacturing protocols have been developed proactively, to ensure the success of a 'getting it right first time' ethic. The large manufacturers and the supermarket chains they supply cannot afford to have a national or international food safety or security scandal linked to their lack of proactive systems management protocols. Meanwhile the airlines have stood back, refused to acknowledge the impact that aviation food safety has on the mainstream aviation safety agenda and cowered behind its catering not manufacturing roots in the hope of avoiding detection. The traditionally reactive approaches to situations of food safety crisis management have meant that the evolution in food safety management protocols has never taken place.

How can this be and why has it remained so in such a litigation conscious society? Whilst the supermarkets remain static and obvious targets, the food service and travelled nature of the end products supplied on board aircraft, provide no such static and obvious shop window. Instead the ever evolving, travelling, secular world of airline food products, whilst it faces the same challenges as those manufactured products found on supermarket shelves, does not become subject to the same level of subjective, legislative scrutiny.

In any situation where levels of safety consistency are critical to the successful operation of an industry, it is essential that the safety practices are regulated and mandated specifically to meet the requirement by external agencies with no pecuniary interest. In all other areas of aviation safety this is positively the status quo. In the world of aviation food safety, who scrutinises and audits the safety attributes of the supply chain? The reality is that the airline caterers audit the suppliers (sometimes!) and the airlines or their representatives audit the airline caterers. Thus, the industry self audits and is open to no objective review whatsoever.

So as we can see, the emerging picture is one of secrecy and self-regulation with innovation being stalled by the financial necessity to suppress costs at all costs! I am acutely aware that my colleagues in the industry would challenge my perspective vehemently, unfortunately, for many of them are not exposed, nor ever have been, to the bigger picture in terms of the types of standards of production and systems management protocols that are required to shore up the manufactured food chain. They are fortunate that they do not have to face the legislative reality of labelling claims and responsible supplier outsourcing. What they cannot comprehend, however, is that in an ever evolving, demanding and fast paced world, the reactive Russian roulette approach to food safety management cannot continue. When the evolution of airline catering product development is happening so fast, there is no way that the present standards of food safety management are robust enough, nor detailed and risk-based enough, to ever keep up.

In the next chapter we look at consumer perceptions and how a potential food safety crisis can affect an already distorted customer view of what, how, when, where and by whom product development, production and safety ergonomics are handled.

Aviation food safety and the global food chain link

Throughout this book we will examine the link between the provision of catered food products on board aircraft and the potential impact any compromise in product integrity may have on the wider global food chain.

In any manufacturing environment, it is essential that the highest possible attributable standards of product safety are not only ensured at the point of production by way of the implementation of advanced food safety systems management protocols, but assured by external, operational and institutional forces that have a vested interest in preserving the wider aspects and integrity of the global food chain.

Any food product that is export in nature, either by virtue of its mainstream export potential or its inclusion on a menu of airline-catered food items, should become part of the wider picture in terms of product safety, quality and legality assurance. Never has there been a greater media and public awareness of the threats to global economies posed by inappropriate scrutiny of the food chain leading to national and international epidemics of human and animal disease. It seems bizarre then that the most obvious vectors for such products, i.e. aircraft that fly internationally, are not subject to the same scrutiny as mainstream food exports.

It is essential that all exported food products remain fully traceable to raw material source, a factor long since recognised as critical to the assurance of the food chain generally yet by no means apparent in the production activity of many airline food products. The non-food technology lead procedures and protocols employed and the non-standardisation of production protocols when global replication and outsourcing are a major issue, renders the aviation food product susceptible to being implicated in the worst kind of food safety disaster with ramifications on a global scale.

In Chapter 7, 'The airline catering supply chain', we examine all of the potential outsourcing possibilities. It becomes clear in that chapter just how non-cohesive the approaches to aviation food provision really are and therefore how difficult it has become to issue industry guidelines on product safety when the outsourcing protocols are so fragmented and mismanaged.

To most of the outside world the evident systems management procedures in many airline catering units appear akin to those in mass catered environments. Unfortunately the prepared meal, dietary claims and export nature aspects of the application, render the requirement a little more technically demanding. The unseen picture is one of brokered finished product and components from a whole host of untraceable sources and no throughput translation of documentation to ensure component-by-component traceability. The potential impact on the food chain as a whole is disturbing to say the least.

Food handlers who travel and handle food for a living have long been recognised as having an impact on the spread of global disease[4]. These groups of travellers are mainly confined to the aviation and cruise ship industries but yet again the aviation industry escapes the kind of scrutiny given to its cruise ship counterpart. The requirement to have all cabin crew food handlers trained in food hygiene matters commensurate to their work activity is an issue glossed over and diluted by the airlines in an effort to keep training budgets to a minimum, and in order to avoid the catering issues that crew empowerment in this area would inevitably bring.

There is no doubt that the human and airline food product link between aviation and the safe guarding of the global food chain is a tenuous one. With the massive upsurge in recent years of low cost 'buy on board' food services comes the added issue of allowing passengers to bring their own personal food exports on board rather than incur the inflated costs of a catered meal product in-flight. If the only travel sectors where this is possible are domestic ones, then that is less of an issue; however, with consumer habits dictating the catering climate of the airline augmentations of tomorrow, the more concerning aspect of personal food imports via 'snacks on board' brings its own brand of difficulty.

The overriding factors inherent in minimising the impact that aviation food products, whatever they may be, ultimately have on the wider concerns of global food chain preservation are that all food products carried on board are ultimately traceable to source and manufactured to the highest levels of food technology protocol.

Every chapter in this book will pose some kind of dilemma as regards the link that aviation food products have with the assurance of the global food chain. It is for the industry to adopt an integrated and cohesive approach to matters of aviation food safety, to place it into the context of an amalgamated food safety and aviation safety debate and as a result develop systems management protocols that assure any incidences whereby aviation food safety and mainstream aviation safety matters meet.

2 Consumer perceptions – fact or fiction?

Since the first regular airline passenger service began in 1919 in Europe, between England and France, the service of food and beverages has become an integral part of any flying experience[5]. Initially the only products available for consumption in-flight were snack items such as sandwiches, whilst tea and coffee was served at intervals throughout the flight. However in the 1930s the in-flight food and drink service revolution came of age with the advent of a hot meal service on several routes.

With the introduction of jet engine passenger services in the 1960s the airline catering evolution really came of age, as package tour travel became available to the masses and air travel no longer remained the exclusive domain of the wealthy. With any downward spiralling of ticket costs, the inevitable issue is adherence to fiscal constraints, which can only be attained through the suppression of non-fixed costs. Inevitably throughout their lifetime, airline catered food and drink products have become an intrinsic part of the airline budget battle. No other non-fixed costs have the capacity to cost so much. No other airline service variable can ultimately be so variable.

From the outset of the growth of mass tourism in the 1950s and its resounding influence on the way that lives in the 20th century were transformed, to the vital role that air travel has to play in the global economies of the 21st century, the significance and importance to the global economy of securing and assuring the entire aviation product, cannot be underestimated. In the 1950s there were 25 million tourist arrivals worldwide, in 1960 69 million, in 1970 160 million and in the 1990s 400–600 million tourist arrivals recorded worldwide annually[2,5].

It is obvious that these figures reflect the burgeoning of a global industry on an unprecedented scale over a very short space of time. It is inevitable that with any industry boom will come inherent difficulties in ensuring that associated products and services keep ahead of the game in order to fulfil the requirement, satisfy customer expectations and support the brand successfully. It is in that last statement that the industry dilemma begins and ends. Customer perceptions and delivering in line with airline brand expectation are of paramount importance to any airline business if the sustainable future of the aviation product and global economy is to be secured.

Airline catering overview

Throughout this book there are consistent references to the claim that the necessary safety evolution in aviation food safety has not taken place, as a direct result of the industry's entrenchment in catering not manufacturing ethics. The basis for this theory and the evidence to substantiate it become more apparent as one reads on; however, the reasons behind the status of this situation and why it is significant can be found in the following paragraphs.

A review of the rapidity with which the airline passenger travel industry has grown over the past 50 years is enough to give some kind of indication of how and in what way the aviation catering industry has burgeoned in tandem. The ever-changing, eclectic nature of the requirement has been as rapid as the sheer increase in volume. As such the aviation catering industry can be forgiven for having had no real opportunity to plan its safety evolution effectively.

The early days of aviation food provision provided a golden age of high end, top quality, small volume menu and service delivery ethics without the modern day constraints of global replication and logistics. The design of the galley space and the ergonomics of the in-flight meal delivery were not driven by the requirement to serve hundreds of low cost, in-flight meals within a very short space of time and the focus was on a five star restaurant-style experience in the sky.

With the ever-changing nature of aircraft capability came the requirement to take more people to further reaching destinations more often and as such the nature of the food service products changed also. The lowering of airline ticket prices and the mass market appeal of the 1950s and 1960s brought about another airline meal revolution with the requirement to keep meal costs as low as possible whilst still delivering some level of product quality. As aircraft grew in size and long-range capability, the galley areas shrank and the weight distribution to fuel ratios became an intrinsic consideration in airline meal design. Logistics, ergonomics, fiscal constraints and weight considerations, above anything else, drive the modern day airline meal concept.

For an industry so driven by everything but the product itself, it is easy to see how safety considerations have not become an integral aspect of the airline meal evolution in the same way as other considerations. The requirement to get the job done in the most aviation-appropriate and cost-efficient fashion has left something of an hiatus in the manner in which food safety technology has failed to become integrated, and the wider safety impact of every new product revolution has not been given consideration in the widest possible context.

In the early pioneering days of airline catering development there was no similar industry revolution in food manufacturing. The ready meal phenomenon had not yet begun and the associated technologies in product manufacture were not yet developed. The mass catered nature of the products on board was the only known method of large-scale manufacture and the shelf-life attributes and export impact of such products had not been brought under the type of scrutiny now given to the modern day prepared meal industry.

It is clear from any mass-catered safety code of practice that the food safety parameters by which the products are produced bears no resemblance to the food safety mechanisms employed in large-scale food manufacture[6]. The major differences can be witnessed in the high-risk application of manufacturing prepared meals versus the lower-risk applications of mass cooking and bulk serving. The airline food industry has always been about the assembly of prepared meals whether they be ready-to-eat cold products such as salads and desserts or prepared meals for reheat on board. It is ironic then that in terms of safety protocol the industry persists in sustaining its mass catering credentials rather than adopting the infinitely more appropriate food manufacturing protocols inherent in the safe production of mass-produced prepared foods worldwide.

The reasons for this are many and complex and are explored chapter by chapter throughout the book, but I believe the critical issues lie in the extremely close relationships that still remain between the airlines themselves and the companies that produce and handle the logistics associated with the provision of catered food products, on their behalf.

Historically, at the outset of the evolution of airline catering many of the major airlines owned and operated their own catering operations, which were a natural and necessary extension of their service operation. In many cases this was the norm until relatively recently when, with the fluctuating economic fortunes of many airlines, their catering chattels were sold off. Since the mid 1970s this new chapter of owner not airline operated units in airline meal production has seen a global explosion in the expansion of a handful of airline catering companies dedicated exclusively to this task.

In terms of what this has meant to the aviation industry, there has been a significant shift in the nature of the airline/caterer relationship that has had to be contractually addressed. Whilst the issue of food safety liability once lay firmly in the hands of the airline who owned and operated the catering unit itself, there now came a requirement to redefine the liability issues between independent catering operators who manufactured and provided the food and the airline who stored, reheated and served it on board.

Ultimately it is the airline that has the greater brand liability to consider so it is in the airline's interest to bond its liability with the catering provider. This type of operational bonding is made all the more effective if the perception remains that both the food production environment and the end product service environment are governed by the same safety parameters. It is therefore in the aviation industry's interest to perpetuate the myth that 21st century airline food provision is governed by mass catered and food service safety parameters in the same way as the service environment on board.

The reality, as we will see, is about as far from that as one could get, with the operational, logistical and supply chain gulf between food production and food service growing wider year on year. The drive for innovation and cost reduction, without all due consideration being given to the wider aspects of product safety definition, will ultimately be what determines brand vulnerability. It is essential that the wider issues appertaining to product safety are determined in the context of the wider safety picture and are amalgamated with other factions of the aviation safety debate.

The greatest benefit for the airline catering industry, were it to acknowledge its food manufacturer status, would be to learn from the lessons of the large food manufacturer as they have forged ahead in developing systems management protocols that are capable of delivering brand consistency, reliability and extended shelf-life on a global scale. In Chapter 5 we examine exactly what is entailed in achieving this in terms of food safety systems management overhaul and implementation.

Invariably with the complexity and variable economic factors that determine the airline food product profile, the industry continues to be faced with all sorts of operational, logistical and political problems that it has to solve. My assertion is, however, that unless these are tackled in tandem with safety issues also being allowed to drive the debate, then the consumer view and customer perceptions stand the best chance of being the only factors that really matter as the extent of brand vulnerability becomes a startling reality.

The consumer view and how it drives the airline product

I have often experienced mixed feelings about my association with the airline catering industry, as you can be assured that airline catering is a subject guaranteed to spark a debate in any company. I have had complete strangers form an immediate opinion of me, based on even a vague suggestion that I may have some part to play in the production or delivery of the dreaded in-flight meal!

Whether the industry likes it or not, it has, since the advent of the low budget tray set-up meal, fallen victim to the worse type of product association. The millions and masses of the global population who fly are all defying the industry to deliver something satisfying, innovative and tasty to relieve the monotony of the in-flight experience and to reinstate their faith in the quality and appropriateness of in-flight meal offerings.

The strange dichotomy is that, despite the fact that airline food products are responsible for engendering so much vitriol in the flying public, they are also the most evident product service variable upon which consumers choose to fly with one airline over another. The challenge for the industry then is to attempt to surf the wave of consumer opinion and employ as many obvious food service comforters as they can to soothe the mood of those who fly. All in-flight food service offers are developed with consumer views firmly in mind. The problems inherent in developing products that fit the cost and ergonomic requirements are the first challenges followed by the restrictions on product availability port to port.

Much time is dedicated to evaluating the fiscal impact of equipment design and usage on board, and for the larger international airlines their ability to circulate equipment resources between ports is critical to the cost efficiency of any meal offer. Whilst the introduction of any new piece of equipment may cost the airline millions, it is believed to have a significant link to improved passenger perceptions of product quality. The introduction of disposable equipment on package and charter routes in the early 1960s brought with it an aviation food product equipment revolution whereby the requirement for airline caterers to wash and return catering equipment was negated and costs to the airlines plummeted. The result was a massive dumbing down of the perceived quality of aviation food products by the consumers. This factor, together with the environmental issues inherent in so much packaged food waste being dumped, resulted in the introduction of the aviation industry's happy medium: rotable food service equipment. One of the best examples of this type of reaction was the removal of all metal cutlery from all aircraft after the events of 9/11. For business and first class travellers the pressure was on the airlines to find a disposable cutlery product that would engender the same kind of quality aspiration in the passenger as silver cutlery had done.

So it would seem that consumer perceptions have the ability to transform many aspects of the airline food product, whether it be the equipment models that carry the products or the products themselves. Issues such as route demographics will also drive the menu choices of aviation menu developers, with the percentage passenger loads defining the profile of what types of food products are likely to be served. In this way multicultural meal offerings are often available and in many cases are predominant if the passenger profile reveals that this is the apparent and consistent demographic even if the home domicile is Western-based. To this end one can expect to find a distinct Asian food influence on many of the most travelled Far Eastern routes out of Europe and the same is true of European airlines leaving out of the United States.

There are many other typical customer associations that are capable of driving the definition and menu profiles of aviation food products. Anything that is representative of the home destination engenders route association with home domicile demographics and products will be deliberately chosen on the basis of their capacity to evoke a familiar brand message in the consumer. The skill in brand marriages where aviation food products are concerned rests with the capacity of a branded food product to underwrite traditional airline food brand deficiencies – quality, safety and luxury.

In recent years the airlines have relied upon the branded product supply chain to enhance the perceived quality, safety and luxury aspects of their food offer. The advantage is that the consumer focus becomes the branded components and not the catered meal aspects and these can range from cookies and crisps to bread products, salad dressings and ice cream. Unlike single unit food manufacturers who have to make a hit with a single product, the food service, multicomponent nature of the in-flight food offer allows for a situation whereby the airlines have several opportunities to make an impact.

The traditional airline focus is on consumer perceptions of quality and luxury. Whilst it remains clear that consumer aspiration is also driven by perceptions of product safety attributes, the aviation industry will allow customer assumptions in this area to carry them through.

The assumptions I am referring to are those made by consumers when confronted with strong brand marriage messages on a single tray set-up. The assumption is that all products on the tray have been manufactured to the same quality, safety and legality parameters and that risk of brand jeopardy would not allow any product safety shortfalls in the airline catered products. As will become evident later, the reliance on branded associates to enhance the safety aspects of any given airline food product is big business and is designed to ensure that product safety attributes remain one area of concern that does not become driven by the consumer.

Buy on board – the battle of the brands

Having established how important it is for airlines to commit themselves to successful brand partnerships or marriages with mainstream food products, it is interesting to look now at recent innovations in airline operations that have led to a massive upturn in the requirement for brand partnerships between the airlines and food manufacturers.

In the late 1990s came an explosion in the number of low cost, no frills airlines who were establishing themselves in not just the European but the US and Australasian marketplaces. Both the low cost and no frills tags had a direct association with the fact that neither food nor drink offers were an integral aspect of the ticket price. To this end the consumer expectation was low in terms of service, which is one of the reasons why the buy on board concept was born. Hidden behind the PR drive to add a little service to a deliberately no service airline concept, came an innovative method of underwriting some of the low fares on offer, with the introduction of an additional in-flight revenue stream. Previously, on-board sales had been limited on the scheduled carriers to duty-free sales and on their charter partners to alcoholic drinks. Here now was an opportunity to extract some considerable revenue from on-board sales of food items.

At the time no one could have imagined what an impact this entire concept would have on the future of mainstream aviation food products, but less than a decade later even full service carriers are having to admit that buy on board concepts may well be a significant part of the future landscape for airline food provision. With such concepts come a whole new set of inherent problems, not least of which is how such concepts would work on long haul, multisector routes. What is fascinating, however, is the changing dimension of brand relationships in this area of the industry compared to the traditional management of brand relationships in the full service area of the industry.

Previously the airlines were in the driving seat as far as dictating the benefits of brand listings, and the branded manufacturers paid handsomely to have their products featured as part of the on-board food offer by many of the major international airlines. The assumption had always been that the airline shop window provided a captive and emotive vehicle for brand exposure and such benefits had to be compensated to the airlines with listing fees or product cost waivers.

With the advent of the low cost concept has come something of a role reversal. Having no on-board service frills to offer and now reliant on on-board food sales as an additional revenue stream, the battle of the brands between the low cost carriers was waged on a radical basis. The obvious associations began with the quality and profile of the beverage (particularly coffee) offers, where major high street coffee shop brands became the biggest prize, whilst the vend associated versions fell by the wayside. Here was a fantastic illustration of the branded product adding value to the airline and taking precedence over the airline brand itself in terms of its food service status. Totally driven by consumer perception and demand, the low cost airline financial gain was now derived from the success and profitability of on-board food and beverage sales, rather than the listing fee benefits of their full service airline partners.

It has always been a source of huge fascination to me that some of the biggest food product brand names in the business have sought so desperately to be associated with airline brands. Everything from coffees, sodas and salad dressings to gin, chocolate bars and ice cream are clearly visible as global food and drink brands on every major airline.

So what benefit does the airline shop window have? The answer is the massive exposure to the transiting population of the world, and the association with and exposure to a captive cross-section of the wealthy and status conscious in premium cabins. All scenarios lead to the potentially glorious symbiosis that is aviation and food product brand marriage.

Inevitably the flip side of such harmonious brand unions lies in the not so favourable aspects of large food brands being implicated in any airline food safety scandal. In the case of high-risk foods, particularly, ice cream and yoghurt, sandwiches and fruit juices, the potential for brand damage should the said products become embroiled in a food safety issue is immense. I have often witnessed the poor integrity of such products at point of service in-flight and cringed on behalf of the manufacturer, who is oblivious to the obvious pre-flight and in-flight abuses of their products.

A practical example I can cite was a company which was supplying a chilled fruit smoothie product to one of the airlines. It became subject to a whole wave of passenger and crew quality complaints. The bottles showed evidence of having blown and the drink was reportedly rancid at point of service. The investigation that followed showed no evidence of traceability to batch at point of loading, as no information attributable to the product had been recorded on the flight loading documentation. Further investigation showed that despite the product's scientifically verified eight-day life, and despite the daily loading requirement, three days either side of the reported problem none had been ordered from the manufacturer who directly supplied the products.

In the absence of any loading records and any in-flight temperature monitoring records to verify the nature of the on-board chill chain, the trail went dead. Whilst my brand is insignificant in the global scheme of things, it is significant to me, and having a full and given knowledge of the extent of aviation food safety shortcomings, I have always to consider as a manufacturer the potential cost to business reputation even if ultimately vindicated.

So, as we can see, the issues for the brand leaders are not at all straightforward and despite the companies' obvious technical capacity to defend any claims against them, the issues inherent in placing such perceivably safe brands in public awareness jeopardy need to be considered, particularly if one is paying for the privilege of 'franchising' the airline shop window.

The use of branded products in the airline food service environment looks set to increase over the next decade in both the low cost and full service airline environments, driven by both consumer demand and airline aspiration. The airlines have come to rely more and more on brand marriage to underwrite the integrity of their multicomponent tray set offerings, whilst the brands have seen a positive benefit to utilising the captive, transiting market of the aviation passenger demographic.

How safe is airline food?

This chapter is dedicated to a focus of attention being placed on the consumer view and how it impacts and predetermines the action or inaction of the aviation industry.

In terms of food safety strategies it is difficult to gauge how effective the consumer view is in driving enhanced performance and systems management forward. Often it is difficult to identify a food safety crisis in the aviation environment, due to the non-captive nature of the flying public and the fact that if a causative agent has a longer incubation period than the duration of the flight, passengers become ill after disembarkation. This results in a cluster of food-borne illness amongst airline travellers from many different nations that are difficult to recognise and almost impossible to trace to origin. It is understandable that airlines do not feel inclined to publish any data on food-borne outbreaks that they have been associated with, as it gives rise to a whole host of bad publicity and loss of consumer confidence and therefore revenue[7].

The lack of statistical data in this regard leaves the more obvious consumer view to be one of general poor quality and not specific in terms of attributable safety aspects. This has led to a huge focus by the industry on improving perceived standards of quality without considering safety and quality attributes as one in the drive to assure total product safety. Throughout this book we will look at numerous aspects of the product development and delivery supply chain that support this assertion, but for now our focus is on examining the nature of the little data available, in order to draw some conclusions and to set the tone for the rest of the book.

The first reported food poisoning outbreak associated with a meal served on an aircraft, occurred in 1947 and implicated sandwiches[8]. The most up-to-date data that are widely available show the reporting of 41 outbreaks since 1947. It is important not to take such a relatively low figure too seriously as it does not take account of the internal complaints data of each individual airline which cannot be accessed without permission from the airline itself. My experience is that the airlines are understandably unwilling to sanction their food safety complaints data being published and therefore offer up the documented safety failures of their in-flight food products willingly and for the benefit of media consumption.

Causative factors of food poisoning outbreaks associated with meals on aircraft

Tables 2.1–2.5 illustrate the nature and causative factors of the entire 41 outbreaks of food poisoning. From the table we can see that *Salmonella* spp.,

Table 2.1 Reported outbreaks implicating *Salmonella* on board aircraft 1947–1997

No.	Year	Country of origin	Causative organism	Affected no. of pax	References
1	1947	Anchorage	*Salmonella typhi*	4	Williams et al.[8]
2	1966	Adelaide	*Salmonella, Staphylococcus*	3	Munce[9]
3	1967	Vienna	*Salmonella enteritidis*	380	Munce[10]
4	1973	Denver	*Salmonella thompson*	17	Tauxe et al.[11]
5	1975	Rome	*Salmonella oranienburg*	23	Munce[9]
6	1976	Las Palmas	*Salmonella typhimurium*	1800	Svensson[12]
7	1976	Paris	*Salmonella brandenburg*	232	Bottiger and Romanus[13]
8	1976	New Delhi	*Salmonella typhi*	13	Tauxe et al.[11]
9	1983	New York	*Salmonella enteritidis*	12	Tauxe et al.[11]
10	1984	London	*Salmonella enteritidis*	631	Tauxe et al.[11]/ Burslem et al.[14]
11	1985	Faro	*Salmonella enteritidis*	30	WHO[15]
12	1986	Vantaa Fin	*Salmonella infantis*	91	Hatakka[16]
13	1989	Palma de Mallorca	*Salmonella enteritidis*	80	Jahkola[17]
14	1990	Bangkok	*Salmonella ohio*	5	Jahkola[17]
15	1991	Greek islands	*Salmonella*	415	Lambiri et al.[18]
16	1997	Canary Islands	*Salmonella enteritidis FTI*	455	De Jong[19]

Table 2.2 Reported outbreaks implicating *Staphylococcus aureus* on board aircraft 1947–1997

No.	Year	Country of origin	Causative organism	Affected no. of pax	References
17	1947	Vancouver	*Staphylococcus aureus*	13	CDC[20]
18	1965	Adelaide	*Staphylococcus aureus*	4	Munce[9]
19	1966	New Delhi	*Staphylococcus aureus*	15	Munce[9]
20	1973	Lisbon	*Staphylococcus aureus*	247	CDC[21]
21	1975	Anchorage	*Staphylococcus aureus*	196	Eisenberg et al.[22]
22	1976	Rio de Janeiro	*Staphylococcus aureus*	28	CDC[23]
23	1982	Lisbon	*Staphylococcus aureus*	6	Svensson[12]
24	1991	Los Angeles	*Staphylococcus aureus*	25	Socket et al.[24]

Table 2.3 Reported outbreaks implicating *Vibrio cholerae* and *Vibrio parahaemolyticus* on board aircraft 1972–1992

No.	Year	Country of origin	Causative organism	Affected no. of pax	References
25	1972	Bangkok	*Vibrio parahaemolyticus*	9	Tauxe et al.[11]
26	1973	Bahrain	*Vibrio cholerae*	47	Sutton[25]
27	1973	Bahrain	*Vibrio cholerae non 01*	64	Dakin et al.[26]
28	1976	Bombay	*Vibrio parahaemolyticus*	28	Desmarchelier[27]
29	1978	Dubai	*Vibrio cholerae non 01*	61	Desmarchelier[27]
30	1992	Lima	*Vibrio cholerae 01*	80	Eberhart-Phillips et al.[28]

Table 2.4 Reported outbreaks implicating *Shigella sonnei* on board aircraft 1947–1991

No.	Year	Country of origin	Causative organism	Affected no. of pax	References
31	1971	Gran Canaria	*Shigella sonnei*	219	Oden-Johanson & Bottiger[29]
32	1971	Bermuda	*Shigella sonnei*	78	CDC[30]
33	1983	Acapulco	*Shigella*	42	Tauxe et al.[11]
34	1988	Twin cities	*Shigella sonnei*	240	Hedberg et al.[31]

Table 2.5 Reported outbreaks implicating various other organisms on board aircraft 1947–1991

No.	Year	Country of origin	Causative organism	Affected no. of pax	References
35	1969	Hong Kong	Multiple	21	Tauxe et al.[11]
36	1969	Hong Kong	Multiple	24	CDC[32]
37	1991	Melbourne	Norwalk-like agent	3053	Lester[33]
38	1993	Charlotte, USA	ETEC	56	CDC[34]
39	1970	Atlanta	*Clostridium perfringens*	3	CDC[30]
40	1967	London	*Escherichia coli*	1	Preston[35]
41	1971	Bangkok	N/A	23	Mossel & Hoogendoorn[36]

Staphylococcus aureus and *Vibrio* spp. are the most commonly reported causative organisms. Thousands of flights have been affected and over 9000 passengers and crew have been reported to have suffered from food poisoning; the number of reported deaths involved in these tables stands at 11. A *Salmonella enterica* serovar typhimurium infection via cold salads served on charter flights from Las Palmas was implicated in the six deaths that occurred in 1976 (Table 2.1, outbreak 6).

Salmonella enterica serovar enteritidis was the cause of two deaths in a major outbreak in 1984 in which 3103 flights and over 600 passengers and crew were affected (Table 2.1, outbreak 10). *Vibrio cholerae* caused two deaths in 1972 and again in 1992; both implicated cold appetisers as the infected food item (Table 2.3, outbreaks 25 and 30). In 1971 another death was implicated in a shrimp and crab salad-related incident from which the causative organism remained unknown (Table 2.5, outbreak 41).

The information contained in Tables 2.1–2.5 is documented in detail in the following paragraphs.

Salmonella spp.

Salmonella spp. are the most commonly implicated pathogens in aircraft food outbreaks. *Salmonella* has been reported as responsible for 15 food poisoning outbreaks, affecting over 4000 people. Eight different serotypes have been identified, with *S. enteritidis* being the most common, causing six outbreaks. *Salmonella enterica* serovar typhi was the cause of two outbreaks. What appears to have been typical of the *Salmonella*-caused outbreaks in most cases was that the dissemination of contaminated food continued for several days, implicating many flights in every outbreak.

The first widespread *Salmonella*-caused outbreak connected with airline meals occurred during the early days of mass tourism on intercontinental flights from Australia to the UK via Europe in 1967 (Table 2.1, outbreak 3) when over 400 people were affected. During the investigation that followed it was discovered that contaminated mayonnaise from a Vienna flight kitchen was to blame. The irony of this situation is that three decades later a similar incident occurred on an Australian airline between Sydney and the Far East when a dessert was found to have been contaminated with *Salmonella*.

The largest *Salmonella*-caused outbreak occurred in 1976 when approximately 1800 people from all over Europe fell ill as a result of eating infected food on several charter flights (Table 2.1, outbreak 6). The investigation that followed revealed that cold salads with mayonnaise prepared in Las Palmas, Spain, were the source of infection.

In 1984 an incredibly widely spread outbreak of *S. enteritidis* occurred when a vast number of aspic-glazed appetisers were served on over 3100 flights, affecting 631 first class and business class passengers and 135 crew (Table 2.1, outbreak 10)[14].

In the 1990s there were two further outbreaks affecting over 400 people in each case (Table 2.1, outbreaks 15 and 16). Both outbreaks involved several charter flights, one catered from a flight kitchen in the Greek Islands and the second from a flight kitchen in the Canary Islands.

Staphylococcus aureus

The major difference between the outbreaks implicating *Staphylococcus aureus* and *Salmonella* in Table 2.2 has less to do with the frequency and more to do with the restricted number of flights affected. In five of the eight outbreaks chilled desserts were seen to be the likely vehicles of infection, whilst in the three other cases it was hot meals that were implicated.

The 1970s saw two major outbreaks occur, the first on three flights from Italy to the USA via Portugal in 1973 (Table 2.2, outbreak 20) and the second on a flight from Japan to France via Alaska and Scandinavia. During the first a custard-style dessert was found to have excessively high counts of *S. aureus*, and an antibiogram showed the same results in patients as in the dessert. In the second incident, ham included in the breakfast loaded in Alaska was shown to be contaminated with the same phage type and enterotoxin-producing strain as was isolated in the patients and linked to an inflamed finger lesion of one member of the catering staff. In the second incident, 142 passengers and 1 crew member were hospitalised upon landing in Copenhagen, following the very swift onset (30 minutes–2.5 hours) of symptoms in-flight.

During investigations of the two smaller outbreaks also, high levels of *S. aureus* were found in the suspected foodstuffs, with the same strains being present in the patients. However, in neither case was the possible role of food handlers as the source of infection investigated.

Vibrio spp.

Vibrio spp., *V. cholerae* 01, *V. cholerae* non 01 and *V. parahaemolyticus* were reported as causing six outbreaks via aircraft food (Table 2.2). The endemic occurrence of cholera in some Asian countries caused the seventh cholera pandemic and was linked to *V. cholerae* outbreaks registered on long haul flights from Europe to Australia in the 1970s. The gastrointestinal illness of passengers was traced to cold food loaded in Bahrain (Table 2.3, outbreaks 26 and 27); Bahrain was experiencing an outbreak of cholera at the time. The products implicated in the outbreaks were cold plated foods but it was also suggested that ice on board may have been a vehicle because of the capacity of *Vibrio* to survive for prolonged periods of time in iced water.

The largest airline associated outbreak of cholera took place relatively recently, in 1992 (Table 2.3, outbreak 30). Seventy-five of the 336 passengers who had flown from Peru to Los Angeles were infected, resulting in the death of one of them. An epidemiological study undertaken at the time realised a strong connection between the consumption of a cold seafood salad and illness[28]. This outbreak was extremely significant as it displayed for the first time indisputably the capacity of aviation food products to be implicated in the spread of disease from endemic to non-endemic areas and highlighted the risks inherent in consuming foods produced in cholera-infected areas. It is also interesting to note that in 1998 the WHO reported an increase of nearly 100% of cholera cases worldwide on all continents[15]. South America had not seen cholera incidence before 1991 but by 1992 the epidemic had spread to 20 countries in Latin America, causing over 5000 deaths from 600 000 cases[37,38]. The intrinsic link between the quality of aviation food and water and the spread of disease had at last been established in the most vibrant and terrifying fashion.

Shigella spp.

Four food poisoning outbreaks caused by *Shigella* via aircraft meals have been reported (Table 2.4). The first in 1971 was traced to in-flight meals served to charter passengers on several flights from the Canary Islands to Sweden

(Table 2.4, outbreak 31). The meals, prepared in Las Palmas, were reported to have infected 219 passengers. In another incident, a seafood cocktail was linked to the illness evident in 19 passengers following consumption on board a flight from Bermuda (Table 2.4, outbreak 32).

More recently, in 1998, a wide reaching outbreak caused by *Shigella* was associated with meals served on 219 flights to 24 different US states and also to cities in Europe and South America (Table 2.4, outbreak 33). Once again the outbreaks were connected with cold foods, emanating from Minnesota. The fascinating aspects of this outbreak are that, due to the inherent difficulties associated with tackling an outbreak, i.e. a long incubation period (1–4 days), relatively low attack rates (4%) and the dispersion of ill passengers over a wide geographic area, this particular one may have gone undetected had it not been for the fact that a professional football team travelling together were involved[31].

Clostridium perfringens

Clostridium perfringens has been implicated in one outbreak, with a hot meal containing turkey involved (Table 2.5, outbreak 39). A total of 394 passengers and crew over eight flights were exposed to the contaminated product, which displayed a mean incubation period of approximately 11 hours with diarrhoea being the main symptom.

Escherichia coli

Oysters contaminated by *E. coli* were the cause of incapacitation of 22 crew members over a period of four days in 1967 (Table 2.5, outbreak 40). It is probable that faecal contamination was involved; however, the incubation period and symptoms were similar to Norwalk-like virus. At the time there were no methods of virus detection but since then the development for the recovery of viruses from bivalve molluscs has proved that raw or cooked shellfish contaminated by viruses was documented as being the cause of numerous outbreaks in the 1990s[39,40].

One outbreak caused by enterotoxic *E. coli* (ETEC) was described in the USA in 1993 (Table 2.5, outbreak 40). The outbreak affected 47 passengers on one flight and was associated with raw carrots in a salad. A further nine passengers reported gastrointestinal illness on a different flight where the same salad had been served. Epidemiological investigation of a local outbreak at the same time also revealed ETEC.

Norwalk-like virus (Norovirus)

In 1991 more than 3000 passengers and crew were affected on several flights from Melbourne, Australia. A supplier of fresh orange juice was common to all caterers involved. Surveys revealed 100% attack rates in the orange juice drinkers and 0% in the non-orange juice drinkers. The incubation period was variable between one and three days whilst the symptoms were consistent with the typical clinical picture for this type of viral disease and the presence of a Norwalk-like agent in faecal samples taken from the passengers. In this case, therefore, despite the fact that detection of the agents in the orange juice failed, there was strong epidemiological evidence to link the incidence of gastrointestinal illness with the consumption of orange juice.

In October 2004 an ongoing investigation into a suspected viral outbreak affecting over 30 passengers and crew on board a flight from London to Bermuda was reported[41]. The resulting impact was a series of flight delays and cancellations. At the time of writing, no conclusions had been drawn about the precise nature or cause.

Chain of events

The contributing factors associated with all of the recorded outbreaks in Tables 2.1–2.5 varied hugely but were generally consistent with a chain of events combining to cause ill effects. The most frequent factor leading to an outbreak via airline food was insufficient refrigeration. Following behind was contamination via infected food handlers which combined in many cases with insufficient refrigeration to cause a problem. In four outbreaks infected food handlers were implicated either by ignorance or negligence. The misuse of high-risk food items such as mayonnaise and aspic glaze were causative in three outbreaks, whilst inadequate hygiene, toilet and hand-washing standards were detected in three cases. Finally, cross-contamination was linked to two cases.

It is interesting to witness in cold statistical detail the potential enormity of what constitutes just one recorded outbreak. It is true to say also that the outbreaks that make it into the statistical league tables, such as those discussed within the parameters of this chapter, normally have had some kind of widespread and global implications and therefore cannot avoid external investigation. Meanwhile, the perceivably isolated incidents involving less than a critical mass of five passengers and crew will historically be dealt with by the airlines' internal mechanisms and will remain under the detection threshold for statistical analysis. It is ironic that in the case of all major outbreaks of disease, whether they be food borne or otherwise, early reporting and the sharing of information are critical in establishing the cause and isolating the problem in order to contain the spread. The shroud of secrecy that veils airline food poisoning reporting procedures and data is instrumental in the true picture of aviation food safety being distorted, whilst the industry's capacity to share information that may prove critical to curb an outbreak is hampered.

It is for consumer awareness and perception to drive the quality and safety standards of the airline food product forward in the same way that passenger perception drives everything else.

It is for the industry to understand the consequences of food safety liability and the impact on consumer confidence that any implication in a food safety crisis may bring. The consumer drives the brand distinctions and in turn drives the brand evolution. If the quality and safety of food products are not considered in the wider context of brand protection then all attributable product enhancement messages will be lost, falling on the deaf ears of an increasingly food-safety aware flying public who demand commensurate levels of safety and service in the air as they do on the ground.

3 Current codes of practice

In this chapter we examine the current codes of food safety management practice that relate either directly, generally or specifically to the aviation industry. In broad terms the adopted codes fall into two main categories: those dictated by regulatory compliance or those designed by the industry as best practice ideals. The assumption is that regulatory compliance standards are more generalised in their approach, whilst industry best practice codes are more peculiar to the nature of aviation food provision.

I would challenge that view. My theory is based on a fundamental belief that the 21st century application of the food provision requirement falls firmly in the food manufacturing, not catering, standards sector and therefore all prevailing food standards should reflect this. My assertions are linked to several key factors appertaining to the modern aviation product requirement:

- Volume of food provision.
- Global replication and standardisation requirement of products.
- Prepared meal requirement.
- Extended life requirement.
- Special meal (SPML) requirement and labelling claims attribution.
- Component compilation of the meal.
- Food security and bioterrorist issues.
- Potential impact of the product on the global food chain.
- Capacity for the product to be implicated in a disease outbreak of urgent international health importance.

Later it will become clear that globally devised, regulatory compliance, food standards documentation does actually mandate manufacturing protocols, whilst the industry persists in cowering behind its catering-based best practice.

I am also of the firmly held belief that much of the evolution of food standards regulation in this area has been overlooked, by virtue of the fact that the requirement and the methods of provision have emerged from their catering roots and failed to embrace the emerging technological requirements of prepared meal manufacture. This is understandable given the burgeoning pace at which food systems management has advanced during the past 30 years in particular; however, that is not to negate the requirement to apply the appropriate protocols and compliance directives to the product, in consideration of all of the above pervading factors.

The following list shows the regulatory compliance and best practice directives associated with the provision of airline food and associated products. We will examine them each in turn in order to understand their influence on the safe production, distribution, service and disposal of airline food and where regulatory compliance ends and best practice begins:

- The International Health Regulations, including the role of:
    - WHO (World Health Organization)
    - FAO (Food and Agriculture Organization)
    - WTO (World Trade Organization)
    - SPS Committee (Committee on Sanitary and Phytosanitary Measures)
    - CAC (Codex Alimentarius Commission).
- Port Health Regulations.
- Food Safety Legislation (variable nation to nation).
- Food Labelling Legislation (variable nation to nation but formalised under Codex).
- IATA (International Air Transport Association) and ICAO (International Civic Aviation Organization) guidelines.
- IFCA (International Flight Catering Association) and IFSA (International Inflight Food Service Association) World Food Safety Guidelines.
- WHO guidelines *Terrorist Threats To Food* (2002).

It is important to understand that in order to embrace a standardised set of SOPs that would encompass all regulatory compliance directives from across the globe, any industry best practice code has to establish the highest possible standards. In the light of the export nature of the product, its potential impact on the global food chain and its possible association with outbreaks of communicable disease, it is even more critical that the industry is cohesive in its approach to food safety management and adopts the broadest possible view of the food chain so that all appropriate systems management protocols are considered and incorporated.

To adopt an isolated and refracted approach is to undermine the effectiveness of what has been considered and to overlook the potential emergence of new issues that may prove hazardous to the overall integrity of the food chain itself. By bringing together in one chapter many of the regulations and recommendations that have a bearing on aviation food provision, I hope to assist the process of determining what protocols really need to be established to protect the industry from associations with food poisoning outbreaks or with the transmission of food and water-borne disease.

The International Health Regulations (IHR)

The IHR, agreed by the international community and adopted by the WHO in 1969, represent the only regulatory framework for global public health safety and security. The IHR can prevent the international spread of infectious disease by requiring national public health measures that are applicable to travellers and products at the point of entry.

The current IHR have been in force since 1971, when they replaced the International Sanitary Regulations which were originally adopted by the World Health Assembly in 1951. The IHR are legally binding on member states and their core obligations currently require member states to:

- Notify the WHO secretariat of certain diseases and epidemiological evidence.
- Provide health services at ports and airports.
- Issue international health documents and apply health measures to international traffic, these being no more stringent than those described in the regulations.
- Make supplementary recommendations during urgent international public health events, including the despatch of WHO teams to assist member states to deal with these events.

However, at the time of writing the IHR are being revised and modernised to reflect changes in disease epidemiology and control and as a reaction to substantial increases in the volume of international traffic. Emerging epidemics specifically associated with air travel, such as the severe acute respiratory syndrome (SARS) outbreak of 2003, have prompted an overhaul of the current directives.

Earlier, in 1998, as part of the IHR revision process, the WHO secretariat approached the members of the Committee on Sanitary and Phytosanitary Measures (SPS) of the WTO with the key objective of determining if it was possible to minimise the effect of any conflict in the application of measures under the SPS Agreement and the IHR, since both organisations share almost the same membership. The revisions to the IHR have been promised since 1998. The Intergovernmental Working Group on the Revision of the IHR met in Geneva in February 2005 to progress the amendments further towards acceptance, but there is still no time-frame for completion. When the revisions happen they will impact on both WHO and WTO members, and it is important to consider the key changes proposed. They include the following.

Notification modifications

A change in notification is proposed, from three diseases – cholera, plague, yellow fever – to 'any disease outbreak or event of urgent international health importance'. These changes are likely to result in more frequent application of the IHR, which in turn will increase potential conflict with the SPS

Agreement. The IHR requirements are legally binding on member states, as is the SPS Agreement. The SPS Agreement covers a wide range of trade concerns, including those related to human, animal and plant health.

Structural changes

The IHR will move away from an integrated text to a document containing 'core text' obligations with annexes giving specific and current technical recommendations. The core text will retain the rights and obligations, prohibitions and permissions (55 articles), whilst the Annexes will still be regulatory but will provide greater technical detail and will also be subject to change and amendment (10 annexes)[42]:

Annex 1. Core capacity requirements, for surveillance, response and designated points of entry.
Annex 2. Notifications decision instrument.
Annex 3. Determination of PHEIC (public health emergency of international concern) and temporary recommendations.
Annex 4. Technical requirements for conveyances and operators.
Annex 5. Measures for vector-borne disease.
Annex 6. Certificates of vaccination and prophylaxis.
Annex 7. Requirements for vaccination or prophylaxis.
Annex 8. Maritime Declaration of Health.
Annex 9. General Aircraft Declaration.
Annex 10. The Review Committee.

Comparison of roles of WTO, WHO and CAC

Since the Codex Alimentarius Commission (CAC) is jointly operated by the FAO and the WHO and is directly cited in the SPS agreement as the standard-setting body for food safety, it has been included in the following comparison of the roles of the WTO and WHO. The WTO/WHO discussions raised several questions related to the respective roles and functions of WHO, WTO and CAC. In simple terms the three organisations could be characterised as follows.

WTO

The World Trade Organization is the principal international body concerned with negotiating trade liberalising agreements and with solving trade problems between countries. WTO is not part of the United Nations (UN) systems. Accepting the requirements of the SPS Agreement is one of the obligations implicit in WTO membership.

WHO

The World Health Organization is a specialised agency of the UN. The WHO acts as the directing and coordinating authority on international public health. It promotes technical cooperation for health among nations, carries out programmes to control and eradicate disease, sets out international health standards and strives to improve the quality of human life. Under its constitution, WHO may create regulations such as the IHR.

CAC

The Codex Alimentarius (food code) is a collection of internationally adopted food standards presented in a uniform manner. The CAC implements the joint FAO/WHO Food Standards Programme, the principal purpose of which is to protect consumer health and ensure fair practices in the trade of food.

In order to fully understand the role each of these organisations has to play in the assurance of global food safety, it is essential to compare their respective roles and functions and identify where potential conflict may exist between them, and conversely areas of potential functional synergy.

Purpose

WTO

- Helps trade to flow as freely as possible.
- Serves as a forum for trade negotiations.
- Settles trade disputes.

SPS

- Recognises the right of governments to take sanitary and phytosanitary measures.
- Maintains they should only be applied to the extent necessary to protect human, animal or plant health.
- Maintains they should not be misused for protectionist purposes and should not result in unnecessary barriers to trade.

WHO

- Provides worldwide guidance in the field of public health.
- Promotes technical cooperation.
- Carries out programmes to control and eradicate disease.

IHR

- To ensure maximum security against the spread of diseases.
- To ensure minimum interference with world traffic.

CAC

- Implements the Food Standards Programme by:
  - preparing food standards
  - publishing food standards in the Codex Alimentarius
  - promoting the coordination of all food standards work undertaken by intergovernmental and non-governmental organisations.

Legally binding

WTO

- Established by a legally binding treaty which has since signed 135 members.
- Replaced the General Agreement on Tariffs and Trade (GATT) organisation in 1995 at the end of the Uruguay Round negotiations 1986–1994.
- Membership of the WTO entails acceptance of all the results of the Uruguay Round, except for two plurilateral agreements.

SPS

- All WTO members are also members of the SPS Committee.
- The SPS Agreement came into force with the establishment of the WTO.

WHO

- Established as a specialised agency of the UN in 1948, has 191 members.
- Issues non-legally binding guidance and directives. However, the WHO constitution makes the IHR legally binding on WHO member states.

IHR

- The current IHR were adopted by the 22nd World Health Assembly (WHA) in 1969.
- WHO member states have the right to reject or make reservations to the IHR.

CAC

- The standards, codes of practice, guidelines and other recommendations of the CAC (or contained in the Codex Alimentarius) are not legally binding.
- The SPS Agreement recognises CAC standards as the reference for food safety requirements when they affect health and international trade.

The legal aspect of the interrelationships between all these organisations and agreements is extremely difficult to explain, but a simple summary would be:

- WTO makes SPS legally binding on members.
- WHO constitution makes the IHR legally binding on WHO members but they can opt out or make reservations.
- CAC is not legally binding but its standards are adopted by WTO under the SPS agreement to deal with disputes.

Core principles

WTO/SPS

- Trade must be conducted on the basis of non-discrimination.
- Members to provide equal treatment for tariffs and trade with all members.
- Domestic and imported products treated in the same way.
- Trade rules must be transparent.
- Measures must be based on scientific evidence.

WHO/IHR

- Strengthen the use of epidemiological principles to detect, reduce or eliminate the sources of infection.
- Improve sanitation in and around ports and airports.
- Prevent the dissemination of vectors.
- Improve national and international activities to help prevent the establishment of outside infection.

CAC

- Protect consumers from unsafe food and fraudulent practices.
- Codex Alimentarius intended to guide and promote the definitions and requirements for foods to assist harmonisation and facilitate trade while protecting consumer health.

Governing bodies

WTO

- Headed by ministerial conference, meets every two years.
- General Council oversees the operation of agreements on a near monthly basis.
- General Council also acts as dispute settlement body and trade policy review mechanism.

SPS

- The committee operates on a consensus basis. Overseen by the General Council.

WHO/IHR

- The World Health Assembly (WHA) is the body which determines policy direction for the WHO.

CAC

- Meets annually with representation on a country basis. An executive committee meets between sessions and acts as the executive organ of the Commission.
- The Secretary is appointed by the Directors-General of the FAO and WHO.

Key functions

WTO

- Facilitates through committees the implementation of all agreements (goods, services, intellectual property) and legal instruments in connection with the Uruguay Round.
- Resolves difficulties related to the implementation of agreements.

SPS

- Protects human life from risks in food and beverages caused by additives, contaminants, toxins and disease-causing organisms in food.
- Protects health from pests, or diseases carried by pests and animals.
- Encourages harmonisation of national measures based on international standards, guidelines and recommendations.

WHO

- Sets global standards for health.
- Cooperates with governments in strengthening national health programmes.
- Develops and transfers appropriate health technology, information and standards.
- Issues global guidance in the field of health.

IHR

- Provide a global legally binding framework of international reference for health measures.
- Regulations are an international code of practice to ensure maximum security against the global spread of disease.

CAC

- Determines priorities and initiates preparation of draft standards.
- Acts as an international forum for dialogue and focal point for all aspects of food quality and safety.
- Publishes and adopts standards.
- Reviews and amend published standards.

The principal member rights and obligations of these organisations are illustrated in Table 3.1.

Impact of IHR on airlines' food safety policies

It is important to remember that there is potential for conflict between the application of measures considered necessary by a state under the SPS Agreement, versus the maximum measures contained in the IHR.

So, having unravelled the intricacies of the cultural and structural relationships between these key international bodies, all of which have a significant role to play in either developing global health policy or implementing it, it is important to look at how this impacts on food safety protocols adopted (or not) by the airlines.

In essence, the mandates of the IHR impact on the airlines in several ways. The fact that they are legally binding on WHO member states means that in the case of most of the world's nations their home airlines have an obligation to comply, the focus being in terms of the transit of vectors and in the spread of communicable disease. In both these, the issues of food and water provision are critically implicated. In terms of both also, the quality of food and water and the vector control measures at airports as well as on board aircraft must also be considered.

The requirement under the IHR to identify and notify the WHO of a potential outbreak or spread of a communicable disease of urgent international health importance is a significant burden for the aviation industry to

Table 3.1 Principal member rights and obligations under the IHR

WTO/SPS	WTO/IHR	CAC
Right to (unilaterally) restrain trade to protect health Right to exceed international standards if measures are based on scientific evidence	Measures in IHR are the maximum measures applicable to international traffic	Standards are not mandatory

carry. The capacity for the aviation industry to be directly implicated in an incident of disease spread is inevitable, bearing in mind the burgeoning increase in the use of air travel both for human and animal cargo. All the more reason then why systems management protocols must be established and work proactively to minimise the risks, rather than reactively to deal with an epidemic. In terms of food and water the following factors are key:

- Source of food supply and full product traceability.
- Disinsection methods and records.
- Cabin crew training.
- Crew health screening.
- Controlled disposal of food waste.
- Sanitary controls over potable water supply and approved country-specific water supply plans (WSPs).

Some of these critical issues are covered by mandatory national regulations, i.e. disposal of food waste and Port Health Regulations. However, in the case of those non-mandated aspects or those governed by best practice directives, it is essential that full consideration is given to the reality of the burden of responsibility in this area before the adoption of specific protocols.

In terms of the requirement to implicate food and food waste in epidemics, one has to consider protocols that are robust enough to control everything from Norovirus to foot and mouth disease. Full product traceability is critical and in this context that has to mean full component traceability. The application of provision must also be considered in terms of the passenger's capacity to bring their own food on board, or whether return catering systems are brought into play. In the event of a disease outbreak, full product traceability will be critical to establish and isolate the cause and recall other product or isolate any other implicated flights. Detailed, specific, cabin crew training to manage the systems will also prove critical.

In order for the revisions to the IHR, that remove the barriers from specific diseases to a more general requirement for all disease of international health importance, to prove totally effective, there needs to be some synergy with the creation of mandatory food safety management obligations being made on airlines in line with the appropriate CAC guidelines. Too much is still left to the industry's discretion where the capacity and efficiency of the IHR to identify and control, let alone prevent, such outbreaks of disease where food is implicated, are sorely compromised. Having established that product traceability, crew training and health screening procedures are essential to both prevent disease spread and control it at the point of epidemic, then it is equally essential that minimum compliance standards in these areas are mandated not suggested.

Articles 14.2[43] and 14.3[44] of the IHR state the following only in relation to the quality and safety standards of food and water provided to airports and aircraft.

Article 14.2

> *'Every port and airport shall be provided with pure drinking water and wholesome food supplied from sources approved by the "health administration", for public use and consumption on the premises or on board ships or aircraft. The drinking water and food shall be stored and handled in such a manner as to ensure their protection against contamination. The "health authority" shall conduct periodic inspections of equipment, installations and premises, and shall collect samples of water and food for laboratory examinations to verify the observance of this Article. For this purpose and for other sanitary measures, the principles and recommendations set forth in the guides on these subjects published by the Organisation shall be applied as far as practicable in fulfilling the requirements of these Regulations.'*

Article 14.3

> *'Every port and airport shall also be provided with an effective system for the removal and safe disposal or excrement, refuse, waste water, condemned food, and other matters dangerous to health.'*

In the absence of any formal qualification within these Articles as to the specifics of the nature, detail and frequency of inspections by the health authority and/or health administration and in the absence of any firm definition of what constitutes 'as far as practicable' with regard to CAC compliance, the level of sensory interpretation allowed within the context of Article 14 potentially leaves the door wide open for aviation food products to be implicated in a public health emergency of international concern (PHEIC).

IATA and ICAO guidelines

Both the IATA (International Air Transport Association) and the ICAO (International Civil Aviation Authority) are industry-related bodies that provide industry-specific guidelines in all areas of aviation operational safety and security. Members of either or both organisations are under no obligation to adopt the recommendations of these organisations but generally it is expected that members will notify them if they are not going to comply. Member companies and nations are represented on the policy formulation committees of both organisations, and both organisations will have an advisory role in any regulatory activity under consideration or due for implementation by international governments.

Guidance on the catering standards of aircraft food both provision and service can be found in section 7 of the IATA *In-flight Management Manual*[45]. The manual was developed by the IATA member airlines under the authority of the IATA in-flight board. It is intended as a best practice, not a regulatory, guide and is designed to be used as a benchmark standard for airline management when establishing in-flight catering policies, procedures and also training programmes for cabin crew in this area.

Issues appertaining to safe airline food service, as opposed to standards and methods of safe food manufacture, are more predominant in this manual and that has much to do with the fact that these guidelines are primarily designed as a tool for the airlines themselves and not necessarily their catering providers. Nonetheless these standards do have a significant role to play in identifying and acknowledging that the safe continuation of the food chain on board is essential in the overall assurance of aviation food safety. Issues covered in this document include:

- Delivery and acceptance of catering supplies on aircraft.
- Meal and beverage service to the flight crew.
- Food safety and hygiene – risks and prevention.
- Crew personal hygiene.
- Delayed flights.
- Suspected food poisoning.
- Special meals.
- Galley and equipment hygiene.
- Potable water and ice.
- Insects.

Throughout section 7 of the document most of the pertinent issues appertaining to food safety and integrity are mentioned in some format or other; however, some of the best practice initiatives, in my opinion, do not go far enough.

This situation is symptomatic perhaps of the fact that, whilst this section of the document has been developed by a respected industry guidance body, its specific expertise resources in the field of food safety management may have been limited. To evaluate the effectiveness of the IATA directives and where they do or don't go far enough, we will look at them each in turn in broad terms.

Delivery and acceptance of catering supplies

Required

- Effective interaction between caterer and cabin crew.
- Cabin crew aware of type of catering uplift and check against requisition.

- Cabin crew check correct stowage location, with food sealed and protected from heat, dust and insects during loading.
- Time interval between food taken out of refrigerator and time loaded in aircraft remains within acceptable limits.
- In the event of a delay appropriate measures have been taken to prevent spoilage of food.

Not required

- Product specifications available to crew or product quality and safety parameter training.
- Temperature checks, labelling and date marking checks.
- Clarification of acceptable limits.
- Clarification of appropriate measures.

Meal and beverage service to the flight crew

Required

- Airline policy for flight-deck meal service must meet aviation regulations.
- Crew do not eat the same meal as the passengers (pax).
- Avoidance of certain types of high risk foods.
- No alcohol.

Not required

- Compliance with named regulations.
- Flight-deck meals to be generated from different suppliers.
- Flight-deck to consume different meals 24 hours prior to down route departure.
- Exclusion of specific foods from flight-deck consumption whilst in-flight or down route.

Food safety and hygiene – risks and prevention

Required

- Airlines guided by Hazard Analysis and Critical Control Point (HACCP).
- Compliance with IHR.
- Crew properly trained in food handling-company regulations and procedures, essentials of food hygiene, cabin crew health requirements, use of personal protective equipment (PPE), food handling code of practice, personal hygiene, special meals (SPMLs), airline catering orders.

Not required

- In-flight HACCP.
- Temperature monitoring during receipt storage and reheating.
- Defined standards of galley hygiene.
- Crew trained to understand the rudimentary aspects of food poisoning, recognising the symptoms and dealing with the impact.
- Crew to understand allergen sensitivity of SPMLs.

Crew personal hygiene

Required

- Cabin crew food handlers with same responsibilities as food handlers on the ground.
- General guidance as to hand-washing protocols – how and when.
- General guidelines governing fitness to fly in terms of food handling.
- Suitability.
- Senior crew to supervise other crew in the safe handling of food in-flight.

Not required

- Dedicated food handlers.
- Mandatory hand-washing training.
- Specific safety parameters concerning fitness to fly, as in criteria established in food premises on the ground.

Delayed flights

Required

- Responsibility for determining course of action to be taken with food once loaded in the event of a delay, rests with airline.
- Ultimate decision over safety of food on board after flight delay rests with crew.
- Determined dividing line between caterer and airline responsibility is point of crew acceptance.

Not required

- Specific time-frame parameters for rejection and recatering to occur.
- Specific crew training to equip them to make appropriate safe food decisions in the event of delayed flight status.
- Temperature logging data to monitor high risk foods throughout delay.

Suspected food poisoning

Required

- Crew trained to deal with consequences of food poisoning outbreak.
- Crew to fill out a medical incident report if food poisoning is suspected.

Not required

- Mandated specifics of crew training in this area.
- A specific process for food poisoning reporting only.
- Action to be taken if the food poisoning outbreak implicates technical or cabin crew.

Special meals

Required

- SPMLs to be available under IATA coding mechanism.
- Meals ordered at time of reservation.
- SPML requests to appear on passenger information list (PIL).
- Cabin crew familiar with meal codes and characteristics.
- SPMLs identified by label or tag that corresponds with PIL.
- Meal content to follow IATA guidelines.

Not required

- SPMLs to be labelled to reflect content, nutritional value or allergen hazards.
- Crew trained in specific hazards associated with SPML service.
- IATA SPML content guidelines to embrace labelling compliance regulations or Codex standards.

Galley and equipment hygiene

Required

- Crew to ensure that all galley equipment and utensils are kept clean during the flight.
- Galley areas and stowage kept clean during flight.
- Clean and dirty equipment kept separate.

Not required

- Definition of clean.
- Equipment hygiene parameters to be established.
- Methods of dirty galley reporting and responsibilities.
- Specific chemicals required as essential items on board, i.e. detergents, disinfectants, sanitisers, etc.
- Specify cleaning activity in the context of disinsection application.

Potable water and ice

Required

- Water on board must meet WHO standards.
- Ice must be delivered in sealed bags for use in drinks.
- Only cooling ice to come in block form.
- Ice only handled with utensils not hands.
- Amount of potable water loaded sufficient and related to flight length.

Not required

- Definition of sufficient.
- Crew training in ice handling and understanding of water safety issues.
- Remedial action parameters in event of only dirty ice being available, including incident reporting.

Insects

Required

- Airline policies to determine course of action to be taken in event of pest infestation.
- Crew to report insect activity to pilot in charge (PIC).

Not required

- Pest policy specifics.
- Risk assessment on potential port-specific pest activity.
- Crew training to understand risks to food safety caused by pests.

Impact of non-regulatory format of industry directives

It is clear from an in-depth analysis of the IATA guidelines, section 7 as above, that they fall considerably short in parts of what one would normally expect to be dictated by many food safety legal compliance standards. However, it is also true that IATA member airlines whose national food safety legislation is either non-existent or falls below what would normally be expected in developed food standards cultures, could indeed benefit from these guidelines.

The essence of the problem will become clearer as we move on in the chapter, but it hinges on the fact that the specifics of industry guidelines are not legally binding and the legal compliance directives, where they exist and where they apply, do not take great enough account of the uniqueness of the industry provision requirements.

In essence, what this book is all about is to suggest the requirement for an aviation-specific, legally binding set of mandates that account for every single aspect of food and beverage provision to the aviation industry worldwide. Often it is difficult to believe that in an industry so driven by consumer safety perceptions and expectations, there are no legally binding standards outside of what is considered in the IHR to specifically govern food and beverage application.

Having already proven the case that product-specific quality and safety parameters are essential to assure the global replication and provision of safe food standards on aircraft, in the same way that retailed global food brands do, let us not forget also the potential safety impact on the food chains of the world and the role that 'travelled' foods have the capacity to play in the attribution and spread of communicable disease.

It is critical never to lose sight of the broader picture when assessing the risk factors attributable to the entire supply chain. Where industry guidelines are vague, one must look to regulatory compliance standards to assure the procedures. This may prove more difficult in terms of the wider context of issues that have to be taken account of and how their interrelationships may vary depending on the destinations to which the airline is travelling.

A good example of this is highlighted by the issue of disinsection procedures into some countries. Despite the industry guidelines that state that there is no necessity to disinsect aircraft that herald from countries within western Europe, unless an epidemic has been identified, certain nations in Asia will insist upon it anyway just because it is required of them. This tit for tat attitude, which ultimately influences many airline relationship dynamics, would be avoided by a greater reliance culture on regulatory standards rather than best practice industry guidelines.

The International Civil Aviation Organization (ICAO) is a specialised agency of the United Nations and came into being in 1947 following the ratification of the Chicago Convention 1944. Article 44 of the Chicago Convention[46] assigns the functions of ICAO as:

- To ensure the safe and orderly growth of international civil aviation throughout the world.
- To encourage the arts of aircraft design and operation of peaceful purposes.
- To encourage the development of airways, airports, and air navigation facilities for international civil aviation.
- To meet the needs of the peoples of the world for safe, regular, efficient and economical air transport.
- To prevent economic waste caused by unreasonable competition.
- To ensure that the rights of contracting states are fully respected and that every contracting state has a fair opportunity to operate in international airlines.
- To avoid discrimination between contracting states.
- To promote safety of flight in international air navigation, and promote generally the development of all aspects of international civil aeronautics.

ICAO is governed by a sovereign body called the Assembly and a governing body called the Council. The governing body meets every three years and is convened by the Council. The Council is a permanent body responsible to the Assembly and comprises 33 contracting states elected by the Assembly.

Article 47 of the Chicago Convention[47] gives legal capacity to ICAO and states:

'the Organisation shall enjoy, in the territory of each contracting State such legal capacity as may be necessary for the performance of its functions. Full juridical personality shall be granted wherever compatible with the constitution and laws of the State concerned.'

By this statement under this Article, it is evident that the processes by which ICAO operates within each contracting state allow for what are ostensibly ICAO directives to become legally enforceable, as long as the contracting state agrees the principal objectives of such directives. Much of the ICAO general industry guidance is devolved in a practical, interpretive way through the specifics laid down in IATA guidelines and is representative of the special, symbiotic nature of the relationship enjoyed by these two industry organisations.

Food safety legislation

The extremely variable nature of food safety legislation state to state, nation to nation and country to country, is what renders best practice guidelines perceivably the only solution to issues of food standards formalisation, when the potential backdrop is global outsourcing and global replication. The aviation industry has much to learn from the food manufacturing sector in this regard. Over the past 30 years the requirement to develop global food standards has become a priority in order to prevent food safety issues from becoming a barrier to international trade, and in direct response to the need to assure comparable standards of product integrity on an international scale. Global food standards have been devised that assure not just food safety but also food quality and labelling criteria, all of which contribute to the harmonisation of standards to assure total product safety.

The potential relevance of any aspect of food safety law has to be considered in the formalisation of any global food standard. Interestingly enough, aspects of food law become less diverse the closer to manufacturing standards one gets. Whilst catering standards legislation remains highly variable across the globe, particularly with regard to minimum compliance temperatures, times, methods of production and chilling, the same cannot be said of the less specific yet more far-reaching requirements of food manufacture. Where the potential export nature of the product becomes an issue, legal compliance standards become an amalgam of the highest possible criteria for production, packaging, labelling and distribution.

Catering standards versus food manufacturing protocols

It seems to me that in any industry where the potential 'export' nature of the food is a given, then the following other factors have to be considered in the case for adherence to manufacturing rather than catering regulatory standards:

- Proportionality of scale of production.
- Requirement for prepared meals.
- Requirement for the product to achieve shelf-life parameters.
- Requirement for global standardisation of safety, quality and labelling parameters to be achieved.
- Full product traceability.
- Nutritional composition labelling.

None of the above factors have any connection with the development of catering safety standards. Even in mass production catering where life attributes may be an issue, the designated safety parameters will not account for prepared meal production or for the proportionality of scale associated with most airline food manufacturers.

In such standards as Codex CAC/RCP 39-1993[48], the Code of Hygienic Practice For Precooked and Cooked Foods In Mass Catering, many of the safety parameters quantified and qualified as acceptable are reflected in the IFCA/IFSA World Food Safety Guidelines. This is an excellent example of the industry guidelines potentially misinterpreting the appropriate purposes for which these Codex standards are designed. It clearly states in CAC/RCP 39-1993 Section 1.1[6]:

'This code is not intended for the industrial production of complete meals but may give guidance on specific points to those who are involved.'

In assessing the interrelationship between best practice industry guidelines and regulatory compliance standards, the CAC and Codex standards are an interesting point in question. Earlier in the chapter we learned of the non-regulatory role of the CAC under the WTO and WHO; however, what we also saw was the role that Codex standards have to play in dispute resolutions between members of the WTO under the SPS agreement. In these scenarios the Codex standards are brought into play to affect the legally binding aspects of the SPS on WTO members. In the same way, WHO member states are obliged to use Codex standards references in terms of formalising their food safety regulatory parameters and particular food safety and product integrity issues to be covered by legislation.

Having established that the Codex standards for mass catering clearly define prepared meal provision as manufacturing, not catering, setting about devising a best practice guidance document for the aviation industry should involve a positive reflection of the same global food standards adopted by food manufacturers. In this way the legal compliance issues nation to nation only become an issue when the best practice standard is perceivably lower than that dictated by the law. It is critical to acknowledge and embrace the uniqueness of the product requirement in the devolution of guidelines and never to lose sight of the facets of the product that render it capable of being implicated in a PHEIC.

The impact of any national or international food law on the aviation industry will only be witnessed in its true context if an allegation of food or water-related illness associated with aircraft travel culminates in litigation. Here best practice issues will be swept away in favour of a direct focus on whether the courses of action taken with regard to all aspects of the product outsourcing, preparation, production, packaging, labelling, storage and transit meet legal compliance standards. Interestingly enough, the aspects of food law cited in such cases will be hugely variable depending upon which country the case is heard in and, as we will see in Chapter 6, 'Liability Issues', will be directly affected by the application of the provisions of the Warsaw Convention.

All the more reason then to attribute minimum legal compliance mandates specific to the manufacture and provision of food products to the aviation industry as a whole, whatever the location of the home domicile.

The potential application of such a notion is not as far-fetched as it seems, if the CAC/WTO/SPS model was to be adopted. Here the industry best practice would become an industry mandate that could be signed up to or not by member airlines and could be called upon as the defining protocol as a means of settling any action or litigable dispute.

The other fascinating aspect of the impact that food law has on airline food provision can be witnessed in the contractual arrangements developed between airlines and their catering providers. Article 8 of the IATA Standard Catering Services Agreement[49] refers to the rather unique relationship experienced between the catering provider and the airline itself in terms of liability and indemnity. It states:

> 'In this article all reference to the Carrier (Airline) or the Caterer shall include their employees, servants, agents and sub-contractors.
>
> Article 8.1 The carrier shall not make any claim against the Caterer and shall indemnify it (subject as hereinafter provided) against any legal liability for claims or suits, including costs and expenses incidental thereto, in respect of:
>
> (a) delay, injury or death of persons carried or to be carried by the Carrier; and
> (b) injury or death of any employee of the Carrier; and
> (c) delay of baggage, cargo or mail carried or to be carried by the Carrier; and
> (d) damage to or loss of property owned or operated by or on behalf of the Carrier and any consequential loss or damage;
>
> arising from an act or omission of the Caterer in the performance of this Agreement unless done with intent to cause damage, death, delay, injury or loss or recklessly and with the knowledge that damage, death, delay, injury or loss would probably result.
>
> PROVIDED THAT all claims or suits arising hereunder shall be dealt with by the Carrier and;
>
> PROVIDED ALSO THAT the Caterer shall notify the Carrier of any claims or suits without undue delay and shall furnish such assistance as the Carrier may reasonably require.
>
> 8.2 Notwithstanding the provisions of Sub-article 8.1 above the Carrier shall be entitled to make such claims as it sees fit against the Caterer, and the Caterer shall indemnify the Carrier against any legal liability for claims of suits including costs and expenses incidental thereto in respect of:
>
> (a) death, injury, illness, or disease of persons carried by the Carrier; and
> (b) death, injury, illness or disease of any employee of the Carrier;
>
> arising from the caterer's failure to comply with clauses 5.9(a) and 5.9(B) herein.'

Clauses 5.9(a) and 5.9(b) are left to the carrier's discretion but these are designed to be appendixes to section 5.8 which requires the caterer's compliance with:

'the standards of hygiene specified by all applicable local and international laws, regulations and procedures and requirements.'

We have already learned that the specifics of the hygienic production of food are governed by prerequisite issues or good manufacturing practices (GMPs) that require the integration of food manufacturing standards not necessarily mandated by law. The uniqueness of the product provision will also not be taken into account by legal mandates so the airline's reliance on local food law, without more specific references being made, may result in the clauses becoming a contractual liability in themselves.

What can be witnessed from these clauses is the manner in which the airline contractually absolves the caterer from direct liability in terms of any product or service provided should they be implicated in a litigable action. This means that despite the fact that the airline is neither the producer nor provider of such products and services, merely the provider of the service environment, it is prepared to accept full legal and ultimately litigable responsibility, with the caterer only having to meet the costs of indemnifying the airline against such actions.

To make the same analogy in a similar situation on the ground, it would be like the supplier of a retail branded product not being contractually obligated to accept legal responsibility for its safety and integrity because the supermarket that sold it would do so instead. Whilst both ultimately play a part in the overall safety picture in terms of a continuation of safety protocols in store, post-production and despatch, the major focus of attention would always be the producer. Any contractual arrangement would not absolve the producer; it would set specific legal parameters for minimum compliance.

To contractually absolve the producer from legal responsibility in terms of the end consumer appears a bizarre thing for the airlines to do, particularly in light of the fact that the only further stipulations made are in terms of 'applicable local and international hygiene laws'. They do not stipulate which, or to what standard, and invariably local hygiene regulations will vary hugely in developed nations from those in the third world. It is not just hygiene regulations that need to be considered either; in the light of the component and potential export nature of the product, there would be the wider issues of supplier outsourcing, traceability and labelling compliance.

The answer to the conundrum lies in the very specific nature of aviation liabilities governed by the Warsaw, now Montreal Conventions (see Chapter 6,

'Liability issues'). Here the carrier can avoid liability far more effectively than a caterer on the ground by citing the accidental nature of the incident to achieve absolution. Food law prerequisites would be historically considered head to head with the requirements of the Warsaw Convention and the specifics of the airlines right to prove accidental status.

Modern food law defence mechanisms like the European-based due diligence defence will, in the case of accident or death, have to be considered alongside the specifics of the industry statutes such as Warsaw 1929 which actually govern all liabilities associated with air travel, not just food-related ones. However, interestingly enough, the new provisions of the Montreal Convention 1999 remove the ability of the airline to rely on 'all necessary precautions' or 'due diligence' as a defence.

So here we have to question the impact of non-specific local and international hygiene laws, not just on the industry itself but also on the industry providers. If the incentive to produce food that meets legal compliance directives on the ground remains a contractual prerequisite that is made null and void by the ultimate product liabilities to the service environment in the air, then how proactive are the airlines and their catering providers going to be in ensuring total product safety parameters throughout the extended supply chain? Not very, is the likely answer.

Food labelling legislation

Probably one of the most controversial issues to face the industry, and the one which offers the most suitable litmus test of the industry's commitment to integrated food standards, has to be the issue of food labelling.

I could have dedicated an entire chapter to this topic and how levels of industry compliance, particularly in certain areas, have a massive impact upon determining the success of total product safety parameters. Instead, I have chosen only to touch on the issues appertaining to legal compliance and best practice mandates, amalgamating the other labelling issues as they appear, chapter by chapter.

In the application of food labelling directives, we can witness the best example in aviation food provision of the catering versus manufacturing divide. The issues that face the industry are based on the service protocol divides between the indisputable prepacked status of the product in most economy/buy on board applications and the premium cabin delivery of a restaurant-style product service. This often results in a situation where the prepacked products should perceivably carry attributable labelling, whilst the plated food service products need not.

The catering status of the labelling issues dates back to the days when all airline ready meal application resulted in the end product being presented to the final consumer unwrapped or plated. Latterly, with the evolution of a new breed of service styles and applications, many airlines' tray service will comprise a variety of prepacked, manufactured components, including the hot meal assembled on a tray set. Invariably many, if not all, of them will not carry individually attributable ingredient declarations and nutritional labelling; indeed many I have seen on board will refer the end consumer to a phone or fax number for ingredient or nutritional information – not entirely practical at 38 000 feet!

So it is against this multi-faceted backdrop of airline food provision that guidance directives and compliance mandates have to be attributed. Bearing in mind that there is no mention of labelling compliance standards in any best practice documentation to date, with the exception of the attribution of 'use by' or 'best before' criteria to end product or raw materials, it is for me to suggest best practice labelling initiatives as I see them to appertain to total product safety.

Manufacturing-style approaches to airline catering labelling

I have written a lot so far about integrated approaches to food safety management and how the aviation catering sector fails to embrace them in the same way that manufacturers do. The absence of defined labelling parameters and protocols is an excellent illustration of this point. The labelling mechanisms undertaken in any food manufacturing environment demonstrate a practical link to the systems that underwrite food safety standards. The manner in which they are devised and implemented in any given business offers an insight into the level to which total product safety is likely to be assured.

If the products concerned are of a prepacked nature and are to be presented to the end consumer in that fashion with no further processing or repackaging, it is essential that the packaging carries all the attributable product safety information. This should include all of the following:

- Full nutritional data.
- Full ingredients declaration.
- Allergen information.
- Service instructions including heating times and temperatures if applicable as well as storage temperature parameters.
- 'Best before' or 'use by' durability information.
- Batch code.
- Product name.
- Name and address of supplier.

This situation will be true of both high risk and ambient foodstuffs, whether they are offered as a component or as an individual item. Whilst there is no specific legislation governing the labelling protocols of airline food in isolation, it is for the industry to acknowledge and accept the role that labelling has to play in food safety management. It is vital to embrace the impact that appropriate labelling has on perceived product safety and to cease to negate its relevance and importance in spite of its trading standards not food standards enforcement status.

When determining the appropriate labelling protocols to apply, international standards are formalised under Codex directives and are summarised and customised to be relevant to the following industry-related products:

- Prepacked products, including those carrying prepared meal status, require full product labelling to parameters defined above.
- Prepacked products that are contributory components which go to make up the overall meal, i.e. tray-set components or 'deli bag' components, should carry individually attributable labelling or may carry generic labelling attributable to the entire tray or bag contents that go to make up the meal.
- Non-prepacked products delivered in a restaurant-style format, i.e. plated and presented in packaging other than a transit container, need not carry individual, attributable product labelling information. However, centrally-held product specifications documenting all relevant quality and safety parameters should be in place as defined above.

With the growing industry trends towards buy on board food products, the labelling credentials of what have been traditionally considered food service products are impacted by their status of direct sale to the end consumer. It can be argued, therefore, that the service environment is actually a retail environment and as such the buy on board offers should be labelled and presented in the same fashion as retailstyle pre-packaged foods on the ground.

The major debate over the historical non-evidence of full product labelling on airline food products has always hinged on the industry's reluctance to accept its manufacturer status. Applying catering standards labelling protocols to tray-set components, even when the products are prepacked, assists in perpetuating the catering standards obligations myth.

Special-meal labelling

If there is one area in which labelling standards protocols are mandated by regulatory compliance issues, it is the special dietary meal arena. Chapter 13 talks in far greater detail about the safety issues appertaining to the provision and production of specialist dietary meals, and includes references to the labelling protocols. However, when one puts the labelling issues into the context of SPML provision, then particular reference to certain mandatory and non-mandatory directives has to be made.

Under certain UK and European legislation, the following directives are mandatory and take no account of the catering application of the product in terms of its direct sale credentials. The labelling regulations account only for the basis on which the dietary claims are being made, the prepacked status of the product and the marketing of these products to the end user who has a perceived idea as to their nutritional suitability and content. I am aware that regulatory issues evolve all the time and the aim is not to date this book by making extensive references to existing food law. The point is merely to illustrate the types of legislation to which SPML provision on aircraft is subject and why it is critical that it is taken into account by airlines and their catering providers, in the development, supply and labelling of SPMLs. The legislation is:

- Particular Nutritional Uses (PARNUTS) Directives 89/398 (see below) amendment to Food Labelling Regulations 1996.
- Article 9 of the Council Directive 89/398 EEC as amended by Directive 1999/41/EC of the European Parliament and of the Council.
- Notification of Marketing of Food For Particular Nutritional Uses Regulations (Northern Ireland No. 35) (Scotland No. 50) (England and Wales No. 333) 2002.
- The Medical Food Regulations (England) (Wales) (Scotland) (Northern Ireland) 2000.
- EC (Proposal 2003/0168COD) Article 3.

Under the auspices of the above regulations it is necessary for foods 'intended for particular nutritional uses' to be notified to the relevant authorities before they are offered for sale if:

'owing to their special composition or process of manufacture, they are clearly distinguishable from food intended for normal consumption'

and

'are sold in such a way as to indicate their suitability for their claimed particular nutritional purpose.'[50]

The types of meal categories currently offered by the aviation industry under IATA guidelines which may prove subject to these guidelines are:

- low calorie meal (LCML)
- diabetic meal (DBML)
- gluten-free meal (GFML)
- non-lactose meal (NLML)
- low salt meal (LSML)

58 *Aviation Food Safety*

- high fibre meal (HFML)
- low protein meal (LPML).

All other allergen sensitive meals are covered by specific allergen labelling regulations. Taking the international context of the industry into account, it is important to note that each WHO member nation represented by an airline will be subject to a variation on the above types of labelling legislation. In the absence of any clear, nationally implemented, legislative directives, the airline has an obligation under the IHR to comply with Codex directives. These are formalised under several specific CAC directives and include the following:

- Codex Standard 146-1985 – General standard for the labelling of, and claims for, pre-packaged foods for special dietary uses.
- Codex Standard 118-1981 (amended 1983) – Standard for gluten-free foods.
- Codex Standard 180-1991 – Standard for the labelling of, and claims for, foods for special medical purposes.
- Codex Standard 53-1981 – Standard for special dietary foods with low sodium content.
- Codex Guidelines for Use of Nutrition Claims CAC/GL 23-1997, page 5 – low calorie claim.

So it is clear that despite the fact that habitual labelling of airline products to retail standards is still far from being realised as an industry standard, its application in the SPML arena is unavoidable. In tandem with this, it is clear that in the case of special dietary meal production, labelling is seen as a crucial aspect in the verification of effective product safety standards.

IFCA and IFSA World Food Safety Guidelines

First published in 2003 after an extensive consultation period between those involved in their formulation and publication, the International Flight Catering Association (IFCA) and International Food Service Association (IFSA) World Food Safety Guidelines were designed to be:

> '*an effective food safety control concept applicable to airline catering establishments world wide and accepted by international airlines as the basic reference document for airline catering food safety without reservations or additions.*'

Aspects of food safety management included in this reference guide comprise the following:

- Hazard Analysis and Critical Control Points (HACCP)
- food safety process flow

- hazardous meal ingredients
- microbiological guidelines
- supplier audits
- stock rotation/date coding
- final cooking temperatures
- aircraft delay policy
- product recall procedure
- foreign objects
- employee health
- personal hygiene.

You will find references throughout this book to this particular industry best practice guide, sadly more in terms of what it fails to do rather than in terms of what it achieves. As with all best practice documentation, in order for it to be formulated without bias and self-interest and in order for it to be a true reflection of the concepts for which it is designed and the people whom it is designed to assist, it must be formulated with direct input from those with expertise on the subject matter who are willing and able to offer input without fear of commercial retribution.

Unfortunately this document, championed throughout the industry as definitive at the time of publication, leaves many critical directives out of the equation and places the industry bar significantly lower than the prescribed standards of the food manufacturing sector. The generic, non-product or process-specific nature of much that it redresses results in an extremely generalised notion of what is or indeed is not acceptable.

A step-by-step comparison of the HACCP standards advocated by this document and how they compare to good manufacturing practice is given in Chapter 4, which gives a more precise indication and illustration of where this document fails to deliver effective food safety management guidance in several key areas.

Perhaps the best lesson to be learned from this type of document is that where so may interested industry parties have had a hand in its formulation, an integrated, cohesive approach is almost certainly unobtainable. The absence in its formulation of a spectrum of non-industry determined expertise makes it difficult for it to be anything other than a diluted, non-specific account, instead of an industry generic, food safety management utopia.

Real best practice is only achieved in a constantly evolving environment of informed ideologies which remain true to the critical aspects of the agenda, i.e. total product safety and integrity throughout the extended supply chain, with all aspects of the supply chain acknowledged and given consideration and with fiscal concerns proving no barrier to safety issues. The emphasis on established critical limits, based on sound scientific advice and a real and genuine consideration of the uniqueness of the product in terms of its specific development scope and environmental attributes, have in my opinion been overlooked in the formulation of this documentation and there is also no useful gauge as to where the best practice advocated may not satisfy some established regulatory compliance standards.

WHO guidelines

Terrorist Threats To Food

There is no doubt that in the 21st century the malicious contamination of food for terrorist purposes is a real threat. Since the terrorist events of 9/11, every conceivable aspect of the aviation industry has been intrinsically linked to the potential utilisation of aircraft by terrorists in order to realise their objectives. The intricacies and specifics of aviation security and how it practicably interrelates with food safety management are discussed fully in Chapter 14; however, in terms of industry directives to deal with the threat of food terrorism, it is important to consider general codes of practice and how they may impact also.

The WHO document, *Terrorist Threats To Food*, published in 2002, seeks to offer advice on the bio-terrorist risks to the food chain in a non-aviation specific format. It advocates a proactive approach to the establishment of detection and alerting procedures for all food businesses from farm to fork. This document should be utilised in conjunction with the provisions of many of the industry specific security directives, implemented even before 9/11 as a direct response to the increase in hijackings implicating airline catering supply and catering suppliers.

The document emphasises the importance of assessing the potential risks to any food business posed by food terrorism and suggests that the basis of response mechanisms should be founded on an assessment of the likely threat posed. It identifies the most vulnerable foods and food processes as any one or combination of the following:

- The most readily accessible food processes.
- Foods that are most vulnerable to undetected tampering.
- Foods that are the most widely disseminated or spread.
- The least supervised food production areas and processes.

In terms of the risks associated with the aviation industry, a combination of all of the above factors could be considered depending on the country of origin and the level to which the applicable food safety management protocols have or have not been established. The risk-assessed possibility, for intentional and deliberate contamination by chemical, biological or radio-nuclear agents at any point in the food chain, requires this type of consideration to be an integral part of safety considerations. Typical food safety management programmes in the industry, including Good Manufacturing Practices (GMPs) and HACCP, would need to take account of the potential for terrorist threats to food in all aspects of the supply chain.

The emphasis on prevention in this document is quickly followed by the requirement for quick response identification procedures, alerting procedures and full product traceability – the key to effective product recall in the event of an incident. The guidelines offer specific advice on all of the following in an effort to inaugurate the terrorist threats to food, where identified, into existing food safety management infrastructures:

- processing and manufacture
- storage and transport
- wholesale and retail distribution
- tracing systems and market recalls
- monitoring.

Whilst these guidance directives are designed to be just that, it is essential that the aviation industry, having already been exposed and implicated in numerous terrorist activities over the years, treats the recommendations as seriously as those that mandate operational, rather than process attributable, catering security (see Chapter 14) and proactively seeks to amalgamate security risk identification systems into food safety management standard operating procedures (SOPs).

Whilst the WHO advice on issues appertaining to terrorist threats to food is based on advisory best practice, many governments around the world have formalised their own directives into new legislation, or in amendments to existing legislation. The best example of this can be seen in the USA where amendments to the Bioterrorism Act were made in 2002, to specifically include food provision. The four provisions in Title III Subtitle A of the Act were made by the Secretary of Health and Human Services through the Food and Drug Administration (FDA). They included:

- Section 303 – administrative detention.
- Section 305 – the registration of food and animal feed facilities.
- Section 306 – the establishment and maintenance of records.
- Section 307 – prior notice of imported food shipments.

The impact of this legislation on aviation food production has been piecemeal but was intended to be two-fold: the mandatory registration of food businesses (Section 305) and the establishment of mandatory traceability systems (Section 306) to determine 'the immediate previous sources and the immediate previous recipients of food'. It is somewhat ironic that the focus of good manufacturing practice in food production businesses has been traceability for over 30 years; now recognised as a vital tool in the fight against bio-terrorism in all types of food businesses, it is being mandated under this legislation. The bottom line is that if aviation catering providers had adopted GMPs and based their food safety systems management on these doctrines before, the mandate to do so under terrorist legislation would not have proved necessary.

Guide To Hygiene and Sanitation in Aviation

The *Guide To Hygiene and Sanitation in Aviation* by James Bailey was first published by the WHO in 1960, and formed the basis of all documented and cohesive strategies on any aspect of operational sanitary practice, food and water supply and pest control. In terms of the remit of the book with regard to food safety management, it was designed to lay down initiatives to cover the wider context of food provision at airports and most specifically in terms of catering provision to aircraft.

Having been revised on numerous occasions, it formed the basis for the development of many of the defined operational food safety management standards during the early 1970s and 1980s, many of which are still in evidence. However since its last revision in 1977, much has changed in terms of emerging technologies and food safety management techniques as well as food safety legislative requirements. The substantive increase in demand for the ever eclectic nature of aviation food products renders much of its application in this sector in the 21st century, outdated.

An inspired rewrite would be most welcome at this time to deal with the modern issues that face the aviation industry in terms of an integrated approach to hygiene systems management.

Throughout this chapter we have looked in detail and in turn at all the relevant codes of practice that potentially have a bearing on the safe provision of aviation food. It is interesting to notice that in many ways they are all interconnected, with the focus being on preparedness, good systems management and effective systems reporting in the event that something goes wrong. The essence of satisfying best practice initiatives as well as meeting regulatory compliance standards focuses upon the ability of the industry to achieve an integrated and cohesive approach to food safety management issues and to take the broadest possible view of all of the issues that have the capacity to affect product safety and integrity.

To satisfy one without the other is to undermine the effectiveness of what has been done. To lower the bar rather than raise it is to negate the impact that partially robust approaches will ultimately have on total product safety.

4 Have Airlines Considered Crisis Prevention?

Whilst it is essential that any technical book on food safety management would not be complete without a chapter on hazard analysis critical control point (HACCP), the purpose of this chapter is not to try to cover the whole subject; there are many other books that do that. I merely wish to examine the legitimacy and effectiveness of the traditional aviation catering attitudes to HACCP and to assess what impact the adoption of a manufacturer's approach might have on them.

Throughout this book my assertions have been that outdated catering considerations have permeated every aspect of operational and technical food safety management in the airline food sector, and as food manufacturing sector activity has emerged in tandem with burgeoning product advancements the aviation industry has been dramatically left behind.

The main issues, as I see them, are the generic attitudes to HACCP devolvement displayed particularly within the aviation industry best practice documentation, which do not allow for product and process-specific risk attribution to be considered. Whilst it is essential that an operational overview of general process control and production hazards is formulated under an initial operational risk assessment, the key factors associated with individual products or product groups as well as certain processes must be considered in isolation. This should include consideration of intrinsic factors such as product pH and water activity, key processes such as mixing, assembly, batch portioning and freezing, and specific hazards associated with individual products and product ingredients. Without this type of product and process-specific application of HACCP, it is inevitable that all of the hazards inherent in the process will not have been identified or controlled.

Crucial also to breaking down the barriers that currently exist and that prevent the adoption of manufacturing HACCP formulation in aviation catering, is an acknowledgement of the divide that exists between the production aspects of the food products and the service aspects. By this I mean that consideration needs to be given to the process flow and risk attributions of the steps controlled by the airline catering provider, segregated from the process steps and risk attributes of the on-board service provider.

A separate and defined HACCP plan needs to be formulated by those responsible for the integrity of the on-board service environment, namely the airlines themselves, as opposed to the current amalgam of HACCP plans formulated by the airline caterers that incorporate in-flight service hazards in areas over which they have neither jurisdiction nor control. In this way the airline caterer is deluded into believing that they are indeed in the business of 'food service' when in reality they are most certainly in the business of mass production and packaging of prepared meal components. This assertion is given credence in Chapter 13 when we look at the aviation catering industry's reluctance to accept its food manufacturing status and comply with a new wave of food safety and security legislation. To suggest that an airline kitchen environment in the 21st century can in any way compare with a hotel or restaurant environment in terms of production or process controls is ridiculous and illustrates a fundamental flaw in the approach to the identification and management of the associated hazards.

So, having established the basis upon which this chapter will evolve, our attention now turns to looking at the issues in some kind of chronology that will make more obvious the comparisons between what is currently in place and what needs to be in place.

We look at a step-by-step approach to HACCP formulation and implementation, and examine the prerequisite issues that need to be established and how they may have a direct impact on the successful implementation of the HACCP plan. We also look at the product and process control work that needs to be done to establish the specifics of the particular issues affecting each product or group of products, and define them in terms of general factors, intrinsic factors, key process factors, main hazards to be controlled and main control measures to be employed.

Throughout the chapter we look also at key operational criteria that affect the potential microbiological soundness of airline catered food products and consider them in the context of their operational diversity from manufactured products. Such factors include controlling of low and high risk separation, consideration of transfer as a process step with risk factors associated with it and the physical contamination issues thrown up by the through processing, handling, washing and packaging of finished products in breakable glass or ceramics. In terms of packaging we also look at the non-identification of packaging hazards generally in aviation catered food environments and the potential risk factors posed by them.

Management programmes required to facilitate HACCP implementation

It is essential that despite the huge amount of attention focused on HACCP as the panacea to all ills in terms of assuring food safety utopia, it is a system that cannot work and will not work in isolation. In order for HACCP to be effectively developed, implemented and managed, it is essential that other management systems are established and operating within the business and that they combine to provide an adequate framework upon which a HACCP system can be managed, through control of the processes involved. These management systems can be defined in the following way:

- established management practices
- prerequisite programmes/good hygiene practice (GHP)/good manufacturing practice (GMP)
- quality management system (QMS).

Whilst the focus of attention may well rest for the most part with GMPs and the QMS, it is essential to consider the general management requirements that will need to be resolved in order to facilitate the successful implementation of an organic and working HACCP plan. I mention this factor first as it is the one comparison I always make immediately when I visit airline catering units as opposed to food manufacturers. It is an established precedent in the food manufacturing sector that a company management commitment must be witnessed in an operational as well as statement of policy format. Retail food standard audit criteria issued by third parties, such as by the BRC (British Retail Consortium), recognised that the manifestation and operational impact of management commitment to food standards must be witnessed in a food business to underwrite the validity and success of the HACCP system.

The management teams controlling HACCP formulation and implementation in food manufacturing environments are multifaceted and multidisciplined teams trained in HACCP with a full and given knowledge of not only what it is all about but what factors and commitment of time and resource are required to make it effective and successful. It is essential that a strong technical team leads the way in terms of systems development and implementation, and with regard to all of the resources that will be required, there will need to be a strong company commitment to making financial as well as time resources available. A dedicated HACCP team leader will need to be assigned based on ability and time available to dedicate to the project. The HACCP team leader's key roles will be to set the agenda in terms of plan development, implementation and evolution.

Most of the normal business skills required will probably be available within any average size business and it is important not to overlook the potential involvement requirements of administration and data analysis, retrieval or collation personnel. In terms of the formulation and collation of data there will be a requirement to ensure that the paperwork is both assembled and reviewed and this may well also involve some business management skills base.

Prerequisite programmes are defined by the WHO as:

'practices and conditions needed prior to and during the implementation of HACCP and which are essential for food safety.'[51]

Prerequisite programmes are also known as Good Manufacturing Practices (GMPs) or Good Health Practices (GHPs) and have for a long time been considered essential and fundamental to successful food safety management. Prerequisite programmes provide essential support to the operational success of the HACCP system and may comprise any one of a number of the following good housekeeping issues. They are used to control the general standards in the kitchen or factory rather than the specific product and process hazards which are managed through the HACCP systems:

- Control of raw materials.
- Personnel hygiene training.
- Premises and equipment design suitability.
- Pest management.
- Waste management.
- Operational and process controls.
- Building maintenance and sanitation.
- Product traceability.
- Product recall.

The establishment and management of effective prerequisite systems will allow for the operational management of a host of other hygiene-related issues such as time and temperature control, cross-contamination and control of all incoming raw materials including packaging. Operational controls will need to be established in order for all of the identifiable risks in the HACCP system to be effectively monitored and controlled, and therefore must include systems to manage water supply, drainage and waste disposal, production protocol, staff training, cleaning, air quality and ventilation, building and equipment maintenance and pest control issues.

The QMS is dedicated to ensuring that customer expectations are met and are not the same as prerequisite issues. It will often be used to manage the prerequisite programme and HACCP systems to assure total product quality, safety and legality. The focus of these systems is the prevention of product non-conformity and a standardisation of product and process control and validation. This again is a vital systems support and management tool, based on a 'getting it right first time' approach, which is currently completely absent from aviation catering systems management. The concern here then is that the consistent production of products that meet all attributable quality, safety and legality requirements and satisfy consumers' expectations, may be compromised by the non-existence of a QMS. The ability also of the

68 Aviation Food Safety

HACCP team members to audit the QMS status of their own suppliers will be affected without a team understanding of the necessity and importance of having a QMS in place.

> *'Whilst a QMS is not a prerequisite programme in terms of good hygiene practice, it is often used to manage the prerequisite programme and HACCP systems so that any element of the operation can be effectively controlled.'*[52]

Figure 4.1 illustrates the manner in which a total quality management programme may hang together.

Having established that the support programmes necessary to ensure that the HACCP plan can be fully devised, implemented and managed fall into three categories, they need to be considered as a three-legged stool. With the absence, removal or compromising of any one of the legs, the stool becomes unbalanced and unstable and most importantly cannot be used to support the weight of the structure of the HACCP plan (see Figure 4.2).

It is essential therefore that before the commencement of the development of the HACCP programme, the current health and status of all three supporting management programmes are assessed in order to determine what is available to support the HACCP system and what is still required in terms of skill set, environment and procedures. In engaging in a process of systems management

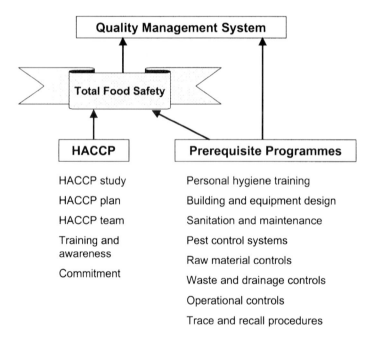

Figure 4.1 Quality management programme (from Mortimore & Wallace[53]).

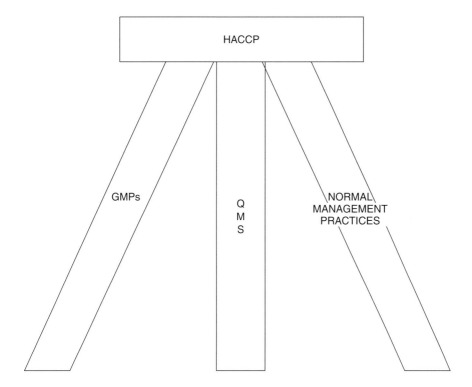

Figure 4.2 HACCP stool.

deficiencies analysis, the HACCP team can identify what exactly is still required in a systematic and structured way whilst simultaneously reaffirming the relationship dynamics between the management systems stool and the HACCP plan and control point systems.

Education and training programmes essential to HACCP implementation

Having established what management systems are in place and then the status of them in terms of what is done and what is left to be implemented or achieved, the next logical step in the process is to assess the appropriate personnel required to develop, implement and manage the HACCP system. Having selected members of the HACCP team it is essential that they are fully trained and conversant with HACCP principles and that they have a sound knowledge of the attributable products, processes and supply chain issues likely to impact on the quality, safety and legality of products supplied.

As I suggested earlier, in food manufacturing businesses the HACCP team members are largely made up of multifaceted, multidisciplined team members with a strong emphasis on operational as well as technical knowledge. My experience in looking at the development of teams in the aviation catering sector is that the teams are largely non-operationally based and top heavy with quality assurance (QA) managers rather than hands-on operations personnel. Whilst a certain level of technical expertise is most certainly essential in the profiling of HACCP team members, those individuals practically charged with the day-to-day responsibility for the safe production and through processing of the products manufactured have an invaluable contribution to make to any HACCP team in terms of hazard identification and risk management proposals.

Invariably all team members will need to have a sound understanding and working knowledge of HACCP principles and this will require some considerable investment of time and resources in training those involved from the shop floor up. It is a mistake to avoid having to make the training investment, by assigning to the team only senior members of management staff, who may well be already trained, in an effort to avoid the training investment in lower grades of personnel. The team members must be representative and reflective of every aspect of the business to be managed and not just those commonly associated with managing perceived food safety risk. Operational personnel throughout the supply, production and distribution chain will need to be considered in order for the broadest possible spectrum of risk management resource to be consulted. Maintenance and engineering personnel who can provide a working knowledge of process equipment capability and design will also provide the necessary skills required to assess equipment and process risks.

Undoubtedly there is a direct relationship between the success of the HACCP system and the level of training and education delivered to those charged with the responsibility of implementing and managing the system. It is both useful and necessary to remind all employees of their specific role and responsibilities in the successful production of safe foods, and initial and ongoing training strategies should include information on the hazards associated with all aspects of the food chain. Training requirements in all departments need to be satisfied; in order to facilitate the successful implementation and integration of HACCP throughout all production and process protocols, every member of staff must have a knowledge of HACCP commensurate with their hazard identification and risk management responsibilities.

In order to facilitate this, there needs to be a strong senior management commitment to providing adequate time for thorough education and training as well as equipping personnel with the practical tools and materials necessary to complete their tasks effectively.

Training and education strategies will vary depending on the complexity of the business. However, the nature of the product and process application in most airline catering facilities will require a firmly risk-based management and personnel training strategy with training initiatives being devolved through the team leader once gaps in the basic management protocols that underwrite the HACCP system have been identified.

People and process analysis

In order to identify what needs to be done and how it should be implemented, a simple people and process analysis can be undertaken to make an accurate assessment of where gaps in GMPs, GHPs and normal management strategies exist.

Step 1

Gaps in prerequisite programmes, GHPs, GMPs and QMS assessed and identified. This may need to be reviewed often, particularly if a new product or process is to be introduced.

Step 2

Identify potential members of the HACCP team. Members should not exceed six and can be recruited from three key areas initially:

- QA and/or technical with expertise in contamination hazards and methods of control.
- Operational and/or production with expertise in the practical application of the production processes.
- Maintenance and/or engineering with expertise in the capacity and capability of the design layout and equipment.

Decisions about the precise make-up of the HACCP team may be influenced by deficiencies identified in the normal management systems. For example, if it has been assessed that raw material traceability systems are not adequately established under the QMS to meet new legislative requirements, then it may be sensible to include a team member from goods receipt or despatch to help devise, implement and manage a practical system of product traceability.

Other areas from which it may be considered crucial to devolve HACCP team members could be research and development, distribution, purchasing or external consultants.

Step 3

Assign HACCP team leader and set out their responsibility. Assignment should be based on knowledge through training and experience and on strong leadership skills. Setting out team leader responsibility will ensure that all training requirements are identified and training programmes initiated.

Step 4

Decide on level and types of training initiatives to be undertaken and a time-frame for completion. Address the risks posed to operational control and successful HACCP implementation by outstanding training deficiencies and devise a plan to monitor this. Training initiatives to take any of the following forms:

- External training from an approved and recommended source.
- In-house training by team members and relative to the in-house processes undertaken.
- Online training via e-learning initiatives.

Step 5

Re-evaluate management systems in terms of success through training, eliminating any existing skills gaps. Test skills attainment success by assigning HACCP team tasks for evaluation prior to devising the HACCP plan or conducting a review.

Figure 4.3 illustrates the skills development required to develop and implement a HACCP system and demonstrates where the skills required are HACCP-specific.

Training requirements

By virtue of the essentially technical nature of the information required in order to conduct an effective hazard analysis, it is recommended that any training content ensures that there is a given knowledge and experience to:

- conduct a hazard analysis (HA)
- identify potential hazards
- identify and distinguish which hazards need to be controlled
- recommend control measures appropriate to the product and process application, critical limits appropriate to preserving the quality, safety and legality of the product supplied and procedures to facilitate effective monitoring and verification
- recommend appropriate corrective actions when a deviation occurs
- validate the HACCP plan.

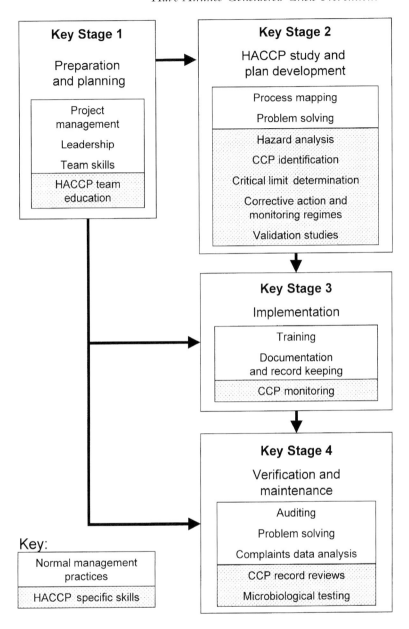

Figure 4.3 HACCP system development – skills requirements (from Mortimore & Wallace[54]).

Having established the very direct connection between the nature and level of HACCP training received by not only members of the HACCP team but all operational personnel, completion of the training initiatives needs to be adequately documented and independently verified to ensure due diligence. In essence, the success of the development, implementation and management of the HACCP plan will depend almost entirely on the successful training of the skills base employed to carry out the requirement.

Developing a manufacturing-based HACCP study

We look at the contrasts and comparisons between the establishment and implementation of a manufacturing versus catering-based HACCP system later when we analyse why it is crucial that the aviation catering industry embraces and adopts manufacturing-style hazard analysis to ensure that all attributable hazards have been effectively identified and controlled. However, we now look at the steps necessary to make manufacturing-style HACCP a reality.

Having assessed the status of the prerequisite and management programmes, evaluated the training credentials of the skills base within the business and decided how existing protocols and procedures need to be incorporated into the HACCP plan, it is now for the team leader to set out the parameters for the project and to ensure that all team members are fully conversant with how it will look in its entirety.

The complexity of the operation will determine the type of plan to be established and implemented in terms of whether it will be a linear, modular or generic approach (see Chapter 14 for definitions).

The structure of the plan will be determined by the team, based on the nature of the products and their intended uses and a time-frame planner to establish how all of the key preparatory features to HACCP implementation are going to be rolled out. At this stage, with a clear time-frame in mind, the team should have a fair idea of the make-up of the system and how and where it interrelates to existing GHPs, GMPs and the QMS.

The time-frame determinants will form the basis of building the project, whilst also providing an indication of exactly what controls will need to be established at each stage to ensure implementation is successful. The project steps will look something like this:

- Assign HACCP team and train.
- Assess the content and status of existing management systems and SOPs and identify any deficiencies.
- Complete all preparation for project formulation.
- Construct product and process-specific flow diagrams.
- Conduct a hazard analysis.
- Identify critical control points (CCPs).
- Complete charts to document and demonstrate control measures.
- Undertake training of production and process-related operatives.
- Establish monitoring systems.

- Train monitoring personnel.
- Establish equipment requirements and variables.
- Conduct a series of internal audits to verify full compliance and implementation.
- Revalidate HACCP plan.

With the HACCP team assigned and trained, the development of the HACCP study can begin and will comprise the development of several key pieces of documentation, all of which come together to make up the HACCP system. Whilst the HACCP plan remains the key piece of reference documentation, it in itself comprises other documents such as a process flow diagram and HACCP control chart. Figure 4.4 illustrates the principal processes in the HACCP study.

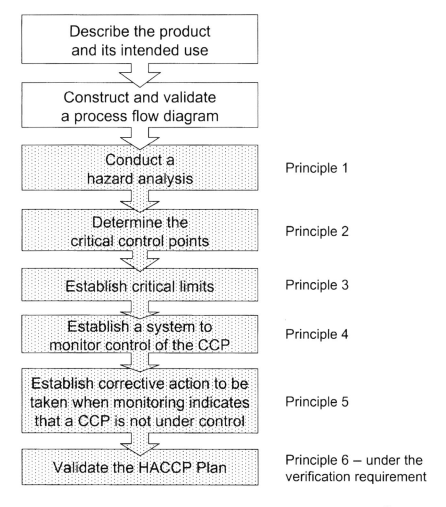

Figure 4.4 Key stage 2: the HACCP study (adapted from Mortimore & Wallace[55]).

76   *Aviation Food Safety*

The initial work of the team, in preparation for the development of the HACCP study, will need to have determined several key factors:

- Scope of the HA: are all three hazard groups to be considered simultaneously?
- Status of the prerequisite programmes.
- Structure of the plan to be adopted: linear, modular, generic, etc.
- Start and end points of the study to ensure that all attributable hazards are considered.

Product and process evaluation

The next step is to describe the product and/or group of products to which the HACCP study is going to relate. In aviation catering environments a modular approach will work best to represent a situation whereby several basic processes are used to produce a number of different products. The HACCP principles will need to be applied separately to each basic process, a module, and then these modules combined to make up the complete HACCP system. Figure 4.5 illustrates how this might look for an airline catering unit that combines product manufacture and assembly with finished product throughput and tray-set assembly.

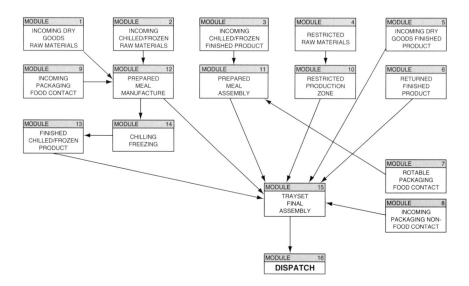

Figure 4.5 Modular process flow for a generalised airline catering production unit.

Once the product or group of products has been identified, the production process needs to be studied and understood by all the team members, and all of the pervading factors associated with the product, the process and the potential uses of the product will be evaluated in terms of the general control measures required. The findings of this evaluation of product, process and intended uses need to be recorded in a brief document, which will need to be referred to at various stages of the system development.

As part of the same assessment of product and process evaluation, the product formulation will need to be examined in order to gain an understanding of how the raw material and product formulation may affect finished product safety. Intrinsic factors are described by Mortimore and Wallace[56] as:

'the compositional elements of the product which can affect microbial growth and therefore product safety. The main intrinsic factors to be considered in foodstuffs are pH, and acidity, preservatives, water activity and the ingredients themselves.'

pH and acidity

- pH levels can be used to prevent the growth of food poisoning or food spoilage organisms.
- The optimum level for growth of micro-organisms is pH7 but they have the capacity to grow at levels as high as pH8 and as low as pH4.
- Prolific growth of micro-organisms at low pH levels has the capacity to raise the pH level of the food to ones which provide optimum conditions for growth.
- Consideration should be given to the capacity for micro-organisms to grow given the natural pH of the food product.
- Yeasts and moulds will grow at pH<4.
- Bacterial spores present in low pH raw materials may grow when mixed with other raw materials of higher pH, to contaminate the finished product.

Preservatives

- Chemical preservatives added to foods need to be subject to legal limit considerations.
- Additives and preservatives may induce allergic reactions and may need to be declared on a label even if the product is for food service consumption.

Water activity

- Reducing water content by using dried or cured products in recipe compositions inhibits microbial growth where the chill chain cannot be assured.
- Adding sugar or salt to food will reduce water activity.

Ingredients

- The intrinsic properties of ingredients should be assessed both individually as raw materials and then after interaction with each other as the finished product.
- Ingredient specifications will illustrate any inherent hazards in the finished product, such as allergens inclusion and presence of additives or preservatives. If allergens are present they will need to be controlled throughout the storage, transfer and production processes to avoid cross-contamination.
- Hazards associated with finished products should be identified as well as those associated with raw materials. In the manufacture of special diet meals, excluded raw materials and entire food groups will need to be considered.

The next step that needs to occur in tandem with product evaluation, is process evaluation. This will include an examination of the production processes inherent in the manufacture of the product and may well include any of the following:

- cooking
- mixing
- cold assembly
- tray-set assembly
- blast chilling
- freezing
- packaging: flow wrapping, hand wrapping, air exclusion, foiling, use of china or ceramic as a packaging material.

Any of the above processes can have an impact on the product safety attributes and must be considered as part of the process evaluation. The construction of a process flow diagram will generate the real basis for the hazard analysis; it will follow each step in the process from raw material receipt right the way through to end product delivery and will attribute as much technical detail as required to accurately assess the potential hazards throughout the processes. Data should include all of the following information[57]:

- all raw materials and packaging
- all production processes and activities

- storage conditions throughout the process
- time and temperature profiling
- transfers of product both within production areas and between production areas
- equipment and design features.

The process flow diagram should contain as much information as possible if it is to prove a valuable tool in the assessment of hazards and identification of CCPs. In the aviation catering environment, adoption of the modular approach will allow the inclusion of more detail within each module, to avoid any steps being missed out. The process flow diagrams for each module will need to be placed together to gain an in-depth picture of the whole operation. Figure 4.6 illustrates the process flow diagram for one of the modules illustrated in Figure 4.5.

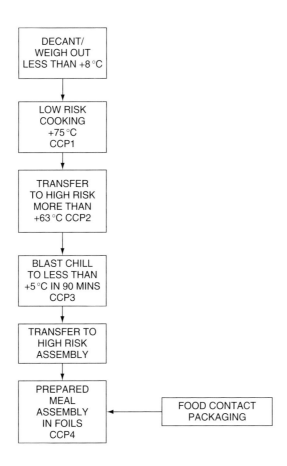

Figure 4.6 Process flow chart module 12 – prepared meal manufacture.

On completion of the process flow diagram, it must be validated by the HACCP team to ensure accuracy in a real life environment against the diagram; at this point gaps in its validity can be identified and the flow either amended or the deviations corrected.

In such a manually-based operation as aviation catering, ensuring that the operational staff are not deviating from the documented flows will be a sizeable task. The more labour intensive a function the more open to systems abuse it may be. Whilst an operative may have a perfectly legitimate rationale behind performing a task in contravention of the documented flow, it may pose serious hazards to the product and process if it remains undetected by the HACCP team. The importance of regimented practices from all members of the operational staff cannot be stressed enough and all staff must be trained and supervised to ensure that all tasks are being carried out in the same way on every shift.

The benefits not only to product quality and safety but also to consistency, legality and meeting consumer expectations cannot be underestimated when choosing to adopt a manufacturer's approach to HACCP planning and development. Whilst the complexity of aviation catering environments in terms of multiproduct and production requirements makes the prospect seem a little daunting, if one accepts that the inherent complexities result in a direct connection with multiple risk management issues, the detail and the total product safety assessments delivered by manufacturing-style HACCP development seem rather reassuring. The only preclusion to full and successful implementation is a willingness to change and to embrace the benefits of a well-documented, well-established system already proven in manufacturing environments all over the world.

Defining operational procedures to comply with the seven principles of HACCP

As mentioned at the beginning of this chapter I am not attempting to cover the whole subject of HACCP formulation, implementation and management, as it would take up the entire book. The following sections, relating to the seven HACCP principles as defined under Codex Alimentarius, bring those principles into focus in the context of manufacturing-style ethics.

Principle 1: Conduct a hazard analysis

The essential aspect of conducting a hazard analysis is that safety concerns attributable to the product or production process are differentiated from safety and quality concerns that will be covered ultimately by the QMS. In the absence of a defined QMS, it is easy for an assumption to be made that quality as well as safety attributes will be dealt with by HACCP. Whilst a consideration of both quality and safety as one leads to a culture of total product safety, the QMS has to operate side by side with the HACCP systems and not in isolation.

A hazard is considered to be significant if it is likely to cause harm to the consumer unless it is properly controlled. All significant hazards are managed through HACCP, whilst non-significant hazards are managed through prerequisite programmes or the QMS, as discussed already. It is essential not only that all significant hazards are controlled but also that the HACCP team are capable of recognising and identifying them in the first place. Often this is difficult for those team members who are close to the process, as products and protocols that may appear hazardous to an outsider may not appear so to those who are consistently carrying out that particular process or function. If the HACCP team members have not adequately considered all attributable product and process hazards, then the resulting HACCP plan will remain unsound and ineffective, even if it is followed to the letter.

In my experience, the generic status of catering-style HACCP systems results in a fundamental misconception of the full spectrum of attributable product and process-step hazards. Hazards attributable to recontamination of foods postproduction, unscientific shelf-life attribution, allergen control during special meal production and non-documented transfer of product as a process step with associated hazards, are all examples.

The other major hazard historically ill considered by caterers is the risk posed to the finished product by the associated physical and biological hazards attributable to food contact packaging. I have yet to witness the documented process flow and hazards identification of packaging materials through a catering facility, on either a generic plan or flow diagram, even when the prerequisite programmes make reference to it or when it constitutes an obvious physical hazard by being fabricated from glass, china or ceramic.

The key issues are summarised as:

- Hazard analysis has to involve all team members in the systematic evaluation of both raw material and finished product alongside all attributable process steps. At each stage identification and analysis of all potential hazards has to be made based on sound scientific advice.
- Hazards are defined as biological, physical or chemical.
- Assessment of the likelihood of occurrence is needed.
- Assessment of the severity of the outcome is needed.

- External sources of information should be consulted in the assessment and evaluation of potential hazards. These may include outside consultants, specialists or research facilities and publications.
- Control measures must be specific to each hazard.

Once hazards significant to the product and process have been identified by the HACCP team, consideration needs to be given to the required control measures that need to be established in order to eliminate the hazard or reduce it to an acceptable level. These may well already be in place, for example many chemical and indeed physical hazards can be effectively controlled through the robust nature of the prerequisite programmes, but evaluation of hazard control at every step still needs to be undertaken.

More than one control measure may need to be employed to control an identified hazard, just as more than one hazard identified at a particular process step may be controlled by a single measure.

Principle 2: Determine the critical control points (CCPs)

There are several factors involved in the identification of CCPs but most of them are based on the HACCP team having a full and given knowledge of the product and processes and being competent to identify not only the hazards but the measures required to control them. It is the information accumulated during the process flow and hazard analysis exercises that will be used to determine the appropriate CCPs.

We have already discussed the fact that control measures designed to assure product quality and legality specifically are not CCPs and are not managed by HACCP. The team's comprehension of this concept is pivotal to their effective determination of CCPs correctly. Too many CCPs may result in the credibility and effectiveness of the system being undermined, whilst too few nay result in food safety being compromised.

The decisions made about whether a process step or hazard constitutes a CCP can be influenced by posing several key questions[58].

Q1: Do control measures exist?

Make sure that control measures really are in place and operational within the business (If the answer to Q1 is yes then move to Q2. If the answer is no move to Q1a.)

Q1a: Is control at this step necessary for safety?

If the answer is no (for example control measures may be in place further along in the process that will control the hazard) move to the next process step or hazard.

If the answer is yes, a modification in the product or process must be implemented to ensure that control measures can be built in. When a suitable control measure has been identified, go back to Q1 and progress.

Q2: Is the step specifically designed to eliminate or reduce the likely occurrence to an acceptable level?

The question refers to the process step not the control measure. If the control measure is considered here instead, the answer will always be yes and will result in the process step being wrongly labelled as a CCP. If the answer to Q2 is yes, then the process step is a CCP.

Q3: Could contamination with identified hazards occur in excess of acceptable levels, or could these increase to unacceptable levels?

To answer this question the team members should use the information recorded on the HA chart together with their expert knowledge of the process and its environment. The issues to be considered here include:

- time and temperature conditions
- the production environment (design, hygiene, manufacture)
- cross-contamination from personnel, another product or raw material
- acceptable levels for significant hazards.

If the answer to Q3 is yes, proceed to Q4. If the answer is no, move to the next process step or hazard.

Q4: Will a subsequent step eliminate identified hazards or reduce the likely occurrence of a hazard to acceptable levels?

This question is designed to acknowledge the presence of any hazards that will be removed by subsequent steps in the process or by the consumer.
    The identification of CCPs is also subject to a fair degree of common sense judgement and can be considered inappropriate at any process step where:

- a hazard cannot be controlled
- there is no guarantee of establishing a scientifically-based critical limit
- the step cannot be monitored.

Having identified all of the CCPs it is useful to document them alongside the CPs in order to verify which process steps are being controlled by the HACCP and which are being managed by other systems. It is critical that CCP identification is carried out on a product and process-specific basis and that the specific HACCP and process flow documenting all of the CCPs is attached to the end product specification. It is crucial that operational devolvement of the CCPs can be witnessed in the production environments and documented on work instructions and production records as a reminder to the operatives engaged in the control of the operational hazards.

Principle 3: Establish critical limits

In order for the HACCP team to differentiate between what constitutes safe and what constitutes 'unsafe', for each CCP, defined parameters need to be attributed, whereby when the product is seen to fall outside of these defined 'critical limits', it may constitute a safety hazard.

Codex[59] defines critical limits as:

> *'a criterion which separates acceptability from unacceptability.'*

Critical limits are in effect the safety limits that must be met by each control measure at a CCP to assure food safety and can be defined by legal compliance standards, best practice guidelines and scientifically proven values. In defining the nature of the critical limits, consideration must be given by the HACCP team to the type of hazards to be controlled and their ability to be measured and monitored through scientific evaluation or observation. Each CCP will have one or more control measure applied to the product or process step and each control measure will display one or more critical limit.

Critical limits may be based on several product intrinsic factors as well as process controls such as time, temperature, moisture level, water activity, pH, presence of preservatives, etc.

In food manufacturing environments the employment of operational limits may be set in excess of the designated critical limits in order to build additional safety features into the product and/or process step. This is a typical example of the proactive nature of manufacturing-based HACCP and the desire to adopt a 'get it right first time' approach.

Principle 4: Establish a system to monitor control of the CCPs

The Codex definition of monitoring is:

> *'The act of conducting a planned sequence of observations or measurements of control parameters to assess whether a CCP is under control.'*[59]

It is necessary to monitor the CCPs to ensure that they are under control and that the manufacture of safe food products is not being compromised. The methods of CCP monitoring are defined by the nature of the control measures employed to control the hazards. It is essential that the procedure of monitoring is not confused with the control measures themselves.

The monitoring process serves three main purposes:

- It tracks the operational trends with regard to food safety before a deviation from critical limits occurs.
- It is used to determine loss of control and product or process deviation at a CCP, i.e. exceeding a critical limit. At the point of deviation corrective action must be applied.
- It provides written documentation for use in verification steps.

Monitoring procedures, whilst they may constitute either observations or physical measurements, must be a planned not haphazard or accidental sequence of activities. The monitoring results need to be continuous and consistent and recorded on a scheduled batch basis.

It is important also that the personnel responsible for undertaking the monitoring functions at each process step that constitutes a CCP are suitably trained to be familiar with the process, are trained in the techniques, and are aware of the corrective actions to employ should a CCP fall outside of critical limits and of how to document the outcome of the monitoring procedures to ensure that the effective monitoring results of CCP control measures are recorded on the appropriate documentation.

Depending on the nature and frequency of the monitoring activity the types of documentation may vary. All CCPs in the process can be monitored on dedicated data collation paperwork, or monitoring activities can be amalgamated into the production paperwork and recorded in that way. What is certain, however, is that the data resulting from the monitoring procedures will be essential in ensuring when and how deviations from critical limits may have occurred and at what point in the process product safety may have been at risk.

Principle 5: Establish corrective actions to be taken when monitoring indicates that a CCP is not under control

Should the monitoring procedures undertaken as part of HACCP principle 4 show that a deviation from critical limits has occurred, then corrective action needs to be taken to deal initially with the product or process deviation and then to review the process to ensure that further deviations do not occur. The nature and frequency of monitoring activities should allow for the process to be brought under control as soon as possible after the point of deviation has occurred, and allow also for the associated affected products to be identified.

Having established which products or raw materials may have been affected by the process deviation, it is then necessary to take steps to isolate them and assess the impact of the process deviation on product safety. Whilst responsibility for the types of corrective action employed lie ultimately with the HACCP team and will be documented on the HACCP plan, both production supervisors as well as operational staff will be involved in establishing how and why the deviation has occurred and what corrective actions need to be implemented. For the most part corrective actions will need to be undertaken with immediate effect and involve those charged with the responsibility of controlling the CCP in the initial stages of corrective action application and reporting.

Having established that a process deviation has occurred and resulted in the requirement to employ corrective action, the HACCP team will need to look at any changes that may need to be made to the process, and therefore the HACCP system, to prevent similar deviations from recurring. These will need to be formally agreed by the team and their implementation documented; also, staff may need to be retrained and controlled process documentation used to monitor the CCPs altered.

Principle 6: Establish procedures for verification to confirm that the HACCP system is working effectively

The Codex definitions of verification and validation are[59]:

> *'Validation – obtaining evidence that the elements of the HACCP are effective. Verification – application of methods, procedures, tests and other evaluations, in addition to monitoring, to determine compliance with the HACCP plan.'*

Having developed and implemented the HACCP plan it is essential that its effectiveness, if controlling food safety hazards, is both validated and verified.

Validation involves ensuring that all of the relevant hazards have been identified and can be controlled. This type of validation assessment must be undertaken for every CCP, with the focus on ensuring that both the control measure and the critical limits established will be effective in controlling the hazard. In my experience it is often useful to gain an external opinion of the validity of the HACCP at the point at which the plan is established but has yet to be implemented. It is all too easy to be so close to a project that hazards can be missed or CCPs misinterpreted. Whilst validation procedures are about internal auditing in essence, it never hurts the maintenance of the process to engage an outsider's view, assuming of course that they are HACCP conversant and have some degree of knowledge of the processes and products. It is only after the HACCP team members have completed the internal HACCP validation process and satisfied themselves that all hazards have been appropriately identified and the correct critical limits set, that the HACCP plan can be implemented.

Verification differs from validation in that it is usually undertaken at the point at which the HACCP plan has been implemented and can be seen to be working. Verification techniques need to be designed to demonstrate that the control measures have been met during the process steps and need to be considered as part of an ongoing activity. They may include review of the system as well as the systems data and need to span the entire spectrum of the HACCP plan, not just focus on the control measures monitoring data. This will involve auditing all of the attributable systems that support the HACCP, including the QMS and prerequisite programmes.

The frequency of HACCP validation needs to be determined from the outset as part of the initial HACCP team planning, but may need to be increased if complaints trends or microbiological data show that problems with the system have occurred (see Figure 4.7).

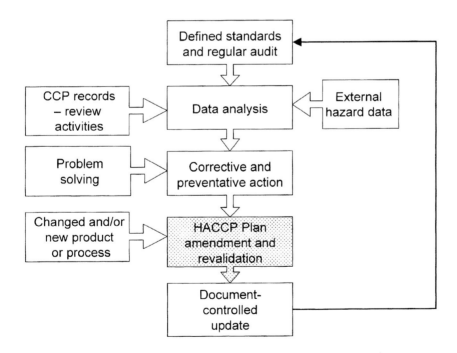

Figure 4.7 Key stage 4: Example of verification and maintenance of the HACCP plan (adapted from Mortimore & Wallace[60]).

Principle 7: Establish appropriate documentation concerning all procedures and records appropriate to HACCP principles and their application

In effect, principle 7 applies to all of the other six principles in terms of the requirement to maintain and retain records to demonstrate that the system is both established and working. Whilst there is no requirement to demonstrate that the maintenance of the system is sustained and under organic review, it is essential that the controlled exchange of product and process documentation in tandem with this is also documented.

Much of the global food safety legislation has not historically mandated HACCP for fear of the perceived problems inherent in documentation and maintenance of records. However, with the advent of new EU legislation in 2005 mandating the seven HACCP principles for all food businesses, and with the requirement in the USA to maintain product safety documentation under security-based legislation, the requirement to demonstrate the success of the HACCP plan through documentation and the maintenance of records grows ever more precise.

The types of records that should be retained should include:

- The HACCP plan, comprising process flows, HACCP control chart, hazard analysis, HACCP team detail and product description.
- Controlled exchange of documentation illustrating and detailing the amendments made to the HACCP and when they took place.
- CCP monitoring records.
- Traceability studies.
- Product recall procedures and testing.
- Staff training records.
- Internal auditing schedules and records of audit findings.
- Calibration records particularly for equipment controlling and monitoring CCPs.

> 'The effectiveness of a HACCP system in managing food safety is dependent on continuous maintenance.
>
> A HACCP plan will only achieve its purpose in managing food safety if it is kept up to date, i.e. through continuous maintenance.
>
> Operations change all the time due to factors such as new raw materials, new recipes and products, improved methods, updated equipment and structural changes in the kitchen or factory.'[61]

Whilst continuous maintenance is an established principle of compliance in the food manufacturing sector, I have yet to see an aviation catering HACCP plan be reviewed prior to or during menu rotation change. The introduction of a new product, even if the basic process remains the same, will invariably generate a requirement for an internal HACCP review. The sweeping generality used by the aviation catering sector to identify hazards has given way to product and process-specific evaluations, which is why many of the hazards, in my view, have the capacity to go undetected and uncontrolled.

Aviation catering HACCP versus manufacturing HACCP

I have maintained throughout this chapter and indeed throughout the book, that all current and attributable food safety and quality management systems in aviation catering need to embrace manufacturing protocols in order to assure that aviation catered food products meet all of the necessary standards of quality safety and legality.

We have looked at the fact that the HACCP cannot be effective unless it is underwritten by robust prerequisite programmes and GMPs, and that every single product and process in the operation needs to be considered in terms of its capacity to impact upon the safety of the finished product. I have been extremely critical of the absence of product and process assessments and evaluations in aviation catering and the adoption of a blanket, generic approach to food safety management, which is incapable of considering the actual hazard specifics of the product and the process without the incorporation of a modular plan.

My intention when I set out to write this book was to demonstrate the practical implementation of manufacturing HACCP in an aviation catering environment, but with so many other issues that needed to be considered first and which have formed the basis of the other chapters, I decided to focus my attention instead on the general inequalities inherent in the two approaches, and leave a complete discussion of manufacturing HACCP implementation for another time; indeed there is enough to fill another book. Some of the sector-specific implementation detail can be found in Chapter 12, 'Special Meals – Special Hazards' and Chapter 14, 'Food Safety In The Business Aviation Environment'.

So let us conclude this chapter by referring to several key areas where food safety management issues remain unconsidered in the aviation catering environment, when they have been proven over the course of the past 30 years to be critical to food safety in the food manufacturing sector.

The key alliance attributes that should determine the adoption of manufactured food standards in the aviation catered environment hinge on several key factors:

- The multiunit replication and mass production status.
- The capacity of the finished product to impact on the global food chain.

- The multicomponent nature of the product.
- The predominately high risk nature of the finished product and product components.
- The prepared meal status of the product.
- The requirement for verified shelf-life attributes.
- The requirement for consistent replication on a global scale.
- The inherent risks posed to aviation security by the products and the processes employed.
- The non-food service status of the product at the point of production.
- The dietary claims status of SPML provision.

In terms of the obvious operational hazards that are overlooked in catered environments vary hugely from unit to unit. However, the shortcomings displayed in the standard industry audit criteria and best practice guidance include the following:

- No operational, environmental or physical segregation of high risk and low risk activity.
- No consideration of transfer as a process step that needs to be documented and controlled, particularly in units where product flow is compromised by logistical considerations.
- No consideration of the hazards posed to food safety by food contact packaging materials including glass or ceramic when used as an end product receptacle and how that can be managed within the glass register prerequisite scheme.
- No consideration of a documented QMS as a vital tool in the implementation and maintenance of the HACCP system.
- No consideration of a product and process-specific end product technical specification documenting specific HACCP and process flow considerations.
- The adoption of a generic approach to HACCP formulation without the incorporation of a modular approach.
- The non-identification of process step hazards on generic airline catering process flows.
- No consideration of allergen controls as production and process hazards in the manufacture of special diet meals.
- No consideration of the hazards associated with the process step time delay post cook chill, before meal assembly. No critical limits for cooked product shelf-life set at this step.
- No acknowledgement of the risks posed to food safety management effectiveness by undefined and unscientific shelf-life.
- An amalgam of shared CCPs and shared attributable hazards in opposing environments and relating to different processes, e.g. the cooking step in

the production environment (the airline caterer) and in the reheating step in the service environment (on board the aircraft).
- The lack of distinction between process and product hazards in the flight catering unit and those on board the aircraft.

Figure 4.8 illustrates a typical airline catering operation process flow.

My experience of having worked in the sector as a manufacturer and supplier as well as a food safety consultant is that much of what is overlooked is born out of a fundamental misconception of what is both demanded and achieved by the adoption of manufacturing-style HACCP and quality

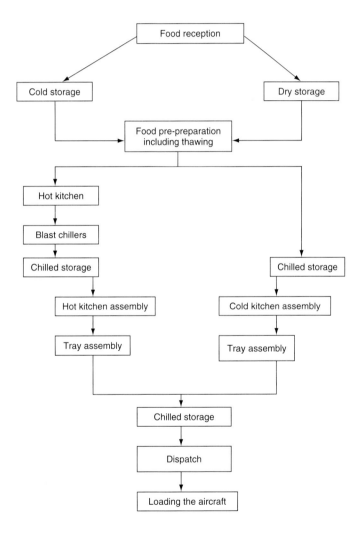

Figure 4.8 Typical airline catering process flow.

management systems. The traditional attitudes that pervade are that manufacturing processes and procedures cannot be integrated into catered environments. The simple fact is that aviation catering at point of production, not at point of service, has far more in common with food manufacture in terms of what is required in size, scale, product profile and quality assurance, than it ever has to do with hotel or restaurant catering.

The only preclusion to safer airline food through the adoption of manufacturing HACCP is not product profile or even fiscal constraints; it is the fundamental desire not to accept the harsh reality of what currently remains unconsidered by the industry in terms of product and process-specific hazards that have the capacity to impact on food safety assurance on a global scale.

5 Implementing manufacturing SOPs to achieve aviation food safety utopia

The whole essence of why I felt it was necessary to write this book is encapsulated within this chapter. If you read no other aspects of this book, I hope that you at least take heed of the disciplines laid down in this chapter.

The rudiment of much of what is wrong with the food safety management of aviation food provision is the fact that the systems are deeply entrenched in their catering roots. Whilst the reasons for this are perfectly understandable, in the 21st century it is essential that a systems management evolution finally takes place, in order to pave the way for assured food safety and integrity in these environments globally. I have always believed that so long as chefs drive the product attribute considerations and airline catering executives underwrite such methodology in advance of any food technology personnel, the road to food safety management utopia in the aviation industry will be an extremely long one.

Only a few days before writing this I was involved in a special meal presentation for a large North American carrier. I watched in dismay as dish after dish was brought under the scrutiny of a chef whose technical credentials in this area were rather limited. Having already taken great pains to explain the rudimentary technical issues that go to underwrite the safety attributes and nutritional composition of the meals, I then faced the battle against mixed component outsourcing, the utilisation of standard components on restricted diet tray sets and the queries over why every meal was devoid of added salt and why there was no fried chicken and chocolate on the children's menu! Trying to explain the wider context of manufactured meal production when specialist dietary claims were an issue, became an extremely arduous task.

The purpose of telling this anecdote is not to suggest that there is no useful place for chef input in the development of aviation food products, but merely to suggest that it must be done in tandem and in the context of the manufactured environment of the application. To suggest anything else is to assume that all catered products and components have no global replication application and no wider considerations in terms of their potential impact on the global food chain. To negate the technical proprieties and intrinsic factors that should accompany the development and production of all products bound for the aviation environment, is to place the industry in an impossible juxtaposition between safety considerations and perceived quality ones, whilst in fact the two should operate in tandem.

During the course of this chapter we will examine the manufacturing standard protocols required to be undertaken in the aviation food production environment in order to assure the safety and quality of the products worldwide. We will also examine how these systems can operate and amalgamate into a best practice scenario for the industry, directly attributable to the precise component and variable outsourcing nature of the products.

Total product safety is the aim. The emphasis must be on proactively seeking out management systems that ensure that getting it right first time is the status quo, not merely a lucky accident of fate which results in product deficiencies being underwritten by a reactive compensation culture.

Product development

This most critical of processes in any manufacturing environment has the capacity to set the agenda in terms of all other processes at this stage, if it is not undertaken in terms of the broadest possible picture of product attributes. Numerous issues must be given equal consideration at the product inception stage of the operation, in order that all safety and quality parameters can be fulfilled throughout the specific process flow and production processes.

In non-aviation manufacturing environments, the product development issues remain consistent and need not necessarily take account of the same set of safety and quality assurance parameters as aviation driven ones. Conversely some may display exactly the same sets of considerations. The essential thing to remember is that there will be both variable and non-variable considerations with regard to every single individual component that is made and it is essential at the product development stage to ensure that one has considered them all, each in turn in the broadest possible context of their potential application.

The following two sets of considerations are examples of where most product development work should begin. Identifying product-generic and product-specific considerations for each recipe under consideration will assist in predetermining the other critical safety, quality and production management issues appertaining to the product later in the process.

Product-generic issues for consideration

- Product name.
- Product type, i.e. entrée, salad, dessert, bakery, etc.
- Product code.
- Product weight.
- Batch size.
- Class of travel.
- Sector length.
- Return catering application.
- Cost.
- Load scale.
- Rotation.
- Manufactured in-house or bought in from supplier.
- Tray set component or bulk loaded.
- Global suppliers to be loaded from?

Product-specific issues for consideration

- Product packaging.
- Poduct labelling.
- Any restricted diet application?
- Nutritional application.
- End product quality parameters.
- End product safety attributes.
- Shelf-life attributes.
- Allergen declarations required?
- If a tray set component, relationship issues with other components.
- If bulk loaded, crew handling attributes to be considered.

Having established the issues for consideration, collating the information that applies to each of these will mean that decisions about the types of recipe dishes suitable for the application can begin to be made. This type of development process requires that all of the above issues are given equal consideration before any specific decisions about recipe development are finalised. This type of product development ergonomics is common in food manufacture, as all considerations are driven by the product parameters before any effective recipe formulation can begin. In food service environments the opposite is the case, with all new recipe dishes being created first in single unit format and then the issues of multiple replication and multiple application dealt with at a later date. Whilst this type of product development technique is perfectly appropriate in a restaurant-style environment, it will not effectively translate into multiproduct replication environments in terms of all quality, safety, transit and technical considerations being appropriately satisfied.

In order to effectively understand how this might work, let us take the considerations template piece by piece and apply it to a theoretical set of products.

Product 1 – generic issues

- Product name: Chicken Caesar Salad.
- Product type: appetiser/entrée salad – 'light' option on main menu.
- Chilled or frozen: chilled.
- Product code: (A.N.Other Airways) CCS.
- Product weight: total 350 g.
- Batch size: 350.
- Class of travel: business/first.
- Sector length: +6 hours.
- Return catering application: some flights.
- Cost: TBC.
- Load scale: business 50%, first 25%.
- Rotation: 1+4.
- Manufactured in-house or bought in: manufactured in-house at home base.
- Tray set component or bulk loaded: tray set in business class/bulk loaded in first class.
- Global supplier network to be loaded from: home base/USA and South America.

Product 1 – specific issues

- Product packaging: plated on china in business class, bulk in foils in first class.
- Product labelling: 'light' option in both cabins.
- Any restricted diet application: forms salad aspect of LFML in both cabins.
- Nutritional application: need to verify it meets LFML claims with dressing.
- End product quality parameters: lettuce must be crisp and evenly chopped throughout. No brown ends or tears in evidence. Parmesan shaved not grated, dressing supplied on the side. Croutons even cook, golden brown, crunchy, plated with salad in business, bulk packed in first. Chicken grilled and sliced showing grill marks. 1×200 g breasts per salad. Chicken plated in business, bulk packed in first.
- End product safety attributes: lettuce to meet ready-to-eat microbiological food standards. Chicken to be meat products licensed approved from verified source. Cooked and chilled rapidly in tandem with cook chill guidelines

Implementing manufacturing SOPs 97

(+75 °C–<5 °C within 90 minutes) safety parameters. Plated and stored at controlled temperatures, <8 °C plated, <5 °C storage. Raw egg not to be used in manufacture of dressing.
- Shelf-life attributes: P+3.
- Allergen/SPML declarations required: subject to content of dressing and rennet attributes of shaved parmesan.
- Relationship with other components: not to be served in tandem with chicken hot entrée selections.
- Bulk loading crew handling considerations: plated presentation photographed. Croutons, parmesan cheese and dressing to be added by crew.

Product 2 – generic issues

- Product name: Sicilian lamb.
- Product type: hot entrée.
- Chilled or frozen: frozen.
- Product code: (A.N.Other Airways) SL.
- Product weight: 250 g.
- Batch size: 2000.
- Class of travel: Economy.
- Sector length: any.
- Return catering application: all routes.
- Cost: TBC.
- Load scale: 50%+50% return application.
- Rotation: 2+4.
- Manufactured in-house or bought in: bought in.
- Tray set component or bulk loaded: tray set component.
- Global supplier network to be loaded from: all destinations.

Product 2 – specific issues

- Product packaging: disposable foil and foil lid.
- Product labelling: meat product licence number from supplier.
- Any restricted diet application: possible take-up on Moslem meal (MOML).
- Nutritional application: need to verify that nutritional content meets acceptable parameters in tandem with other components on tray set.
- End product quality parameters: prime fillet of lamb sourced from approved country of origin. Sauce to be made with red wine and tomatoes. Tinned tomatoes permissible. Seasoning to be basil and oregano – dried permitted. Sauce should cover meat. No fat or gristle visible.
- End product safety attributes: Lamb to be meat products licensed approved from verified source. Cooked and chilled rapidly in tandem with cook freeze

guidelines (+75 °C–<−18 °C in 120 minutes) safety parameters. Assembled with other components in controlled temperatures, <8 °C.
- Shelf-life attributes: best before 12 months @<−18 °C. Once defrosted use within 3 days if stored at <5 °C.
- Allergen/SPML declarations: contains alcohol.
- Relationship with other components: not to be used on medically restrictive diets. Other components to be alcohol-free and 'light' in classification. No more tomato-based components.
- Bulk, crew handling considerations: none.

Development issues

Having completed the exercise for this theoretical set of products, we now need to examine what product development issues this throws up that, during the normal course of catering-style product development, would not have been recognised as an issue until the product had found its way onto the menu.

Product 1 – generic issues

If we begin with the product-generic aspect, we can see that one of the intended applications for product 1 is as the 'light' offering on the main menu. If this application is to be fulfilled appropriately, further consideration will need to be given to the saturated fat content of both the Caesar salad dressing and the croutons. Traditional dressing will contain whole egg and cheese as well as anchovy oil. Traditional croutons may well be fried, not baked. Stipulations will need to be made in both the final product specification and the production method to verify the validity of these proposed nutritional attributes.

The decision to present this product in two different formats in two different cabins requires that the product HACCP documented on the end product specification takes account of all attributable risks inherent in the two differing process flows, i.e. those that account for the plated application and those that account for the bulk-packed scenario.

The high risk nature of the product and the length of sector will determine that the shelf-life attributes will need to be considered in the context of both. If the aircraft does not provide for refrigerated stowage, this will also need to be taken into account as part of a separate risk assessment before defining final product safety attributes. The added suggestion that this product may also be part of any round trip applications, whether as a direct turn around or as part of a pod loading, will need to be addressed in terms of the realistic nature of the required shelf-life.

Whilst the standard organoleptic considerations of outsourcing this product from outstations in the USA and South America may not seem insurmountable, the potential for some ports not to be in a position to meet the technical aspects of the product specification and process flows will need to be examined in fine detail before port capabilities to meet defined technical standards are assured and product replication requirements confirmed.

Product 1 – specific issues

Moving on to the product-specific issues and how they impact on the product development potential of product 1, we need first to focus on the product packaging requirements. Here we can see that the dual faceted application of the product will result in the same product being either plated on china or bulk packed in foils. Once again, whether this application will be undertaken at the same time and in the same production areas, or not, will determine the content of the process-specific HACCP that will be attributable on the end product specification. If the china plated salads become subject to breakages either during production, storage or transit, then the physical contamination issues will need to be documented effectively. If, on the other hand, the bulk-packed products are assembled in an alternative environment free from china or glass, then the process-specific HACCP for the bulk-packed product will not need to document this risk.

In terms of the 'light' and LFML meal claims being made, the nutritional data contained in the specification will need to bear out these attributes. In terms of the capacity to replicate authentically this product from other outstations, very specific production parameters will need to be determined in order to ensure that accurate reproduction is assured.

In terms of the quality parameters, effective production methods will need to be demonstrated to ensure that the grade ability of the chopped, washed lettuce meets both microbiological and quality standards. Assuring that no browning or tearing of the leaves occurs will be easier to achieve in the bulk-packed application but less so in the plated version when a three-day shelf-life is an issue. Particular attention will need to be paid to the raw material specification of the lettuce if it is processed in-house, with relative processing methods employed to assure both raw material quality and the required raw material shelf-life to achieve an end product that had P+3. In terms of the plated version it will be difficult to maintain the crunchy quality parameters expected of the croutons in the plated version, even with the dressing being left on the side.

In terms of the safety attributes of the product, the finished product specification will need to determine the microbiological standards in terms of the finished product application, taking into account the entire process flow.

Supplier audits will raise any relevant issues in terms of highlighting an outstation's potential to affect the safety measures required. Where this capacity is brought into question, product-specific process flows and HACCP may need to be altered in favour of outstation-specific standards, to assure the risks are dealt with and effectively documented. When centrally developed products are devolved to other ports, it is necessary to make sure that the specifications have the potential to be adhered to, and if not that they are altered in favour of the different process flows and HACCP required to be employed.

Any allergen or special product considerations will need to be documented on the final specification, but at the product development stage it is critical that they are considered in the context of the product in question. If a verified and consistent supply chain, in terms of raw materials, is an issue, then the assumption has to be that if an allergen or diet consideration must be made with the use of one type of raw material, then it has to be considered in the context of another. The ideal scenario is to assure the supplier base and the chain of raw material outsourcing to avoid out-of-specification raw materials. However, when global outsourcing is an issue that is not always possible. Better then to consider an allergen's inclusion and base the specification on it rather than not.

Product 2 – generic issues

Moving on to product 2, we can once again examine the issues that will impact on the product development credentials of a standard aviation meal product. In terms of the product-generic issues, we first need to examine the fact that this is a frozen product that is being brokered or outsourced from a producer other than the airline caterer. The requirement to trace this product through all of the relevant safety and quality parameters requires that the product is either developed in tandem with the supplier and the specification becomes an amalgam of supplier protocols and customer prerequisites, or the end product specification produced by the supplier has to be verified against the customer's quality and safety considerations, assuming knowledge of the intended application and in this case the requirement for global export to other outstations.

Product 2 – specific issues

In terms of the product-specific issues relating to this product, the requirement for it to be packaged in food-grade packaging suitable for regeneration in an aircraft oven will be a primary concern when outsourcing from caterers or manufacturers unfamiliar with such environments. In terms of this type of product development that allows for major components of tray set meals to be brokered elsewhere, it is essential that the supplier can fulfil full traceability requirements in terms of verifying the raw material supply chain attributes. Any meat products licensing attribution, in countries where this is a requirement of manufactured, prepared meals as well as raw materials, will result in a greater degree of raw material and end product assurance.

In terms of the special diet attribution of this meal, it is easy to make errors when the fine detail of the product specification is not documented and in front of you. Without detailed scrutiny it may seem as though product 2 could be capable of satisfying the MOML claim if the lamb was Halaal. It would only be as a result of detailed scrutiny of the end product specification that it would be obvious that alcohol was utilised in its manufacture and its meat products licence credentials rendered it incapable of meeting the dictates of ritualistic slaughter. It is essential therefore that in the development of airline meal solutions, all finished component specifications are scrutinised before a product's suitability for purpose can be assured.

Having looked at just a brief sample of products and the multitude of facets that would need to be considered when assessing their development potential and suitability for the purpose, it is clear that by adopting a manufacturer's view in advance of any decisions being made means that many of the potential technical issues are managed in advance. It is particularly important with outsourced products that the suppliers are given a clear technical brief in advance of presentation, in order to be sure that the products they are presenting can meet the parameters of the requirement exactly. For those products that are developed for production in-house the same rules apply; however, the technical information would need to be managed in tandem with the chef development aspects in order to verify that it has been fully considered by those involved.

Supplier outsourcing

Many of the manufacturing protocols appertaining to supplier outsourcing are covered separately in Chapter 7, 'The Airline Catering Supply Chain'. However, it is important in the context of this chapter to look again at how the influence of manufacturing standard systems can impact in this area to enhance total product safety and integrity.

Fundamentally in any aviation catering facility there will be several defined areas of supplier outsourcing potential. They can be classified in the following ways:

- Raw material suppliers – food.
- Finished product suppliers – food.
- Finished product suppliers – non-food.

- Product packaging suppliers – food.
- Catering equipment suppliers – non-food.
- Ancillary product suppliers – non-food but for use in food environments.

To take examples in each category we can define them broadly as follows.

Raw material suppliers (food)

- Fruit, vegetables and salad – prepared and unprepared.
- Meat and fish suppliers – raw and cooked – chilled or frozen.
- Ambient stores suppliers – tinned goods/packet mixes.
- Dairy products (chilled/frozen) – butter/cream/egg products/cheese/milk and milk products.

Finished product suppliers (food)

- Prepared meals – chilled/frozen.
- Prepared salads and fruit salads.
- Bakery products – ambient/frozen.
- Desserts – chilled/frozen.
- Branded goods – confectionery/ice cream.
- Baby food.
- Beverages.
- Ice.

Finished product suppliers (non-food)

- Newspapers and magazines.
- Amenity kits.
- First-aid supplies.

Product packaging suppliers (food)

- Cutlery packs.
- Meal foils.
- Toothpicks.
- Condiments and salad dressings.

Catering equipment suppliers (non-food)

- Trays and tray liners.
- Cups and glasses – rotable and disposable.
- Meal boxes and bags.
- China, glassware and linen.

Ancillary product suppliers

- Chemicals and cleaning equipment.
- Pest control services.

Having established that there are several food and non-food areas of potential product supply, it is for the manufacturer to consider how supplier protocols need to be established in each case to assure the consistency, integrity and safety of the supply chain base. In terms of the technical parameters that need to be in place to affect this, they too will fall into several broad categories of application:

- Supplier audit to predetermined standards dictated by the nature of the product supplied.
- Technical product specifications demanded, either supplier's own or to meet customer's standards if appropriate.
- Terms and conditions of business protocols encompassing delivery schedules, food safety and quality parameters, delivery protocols, penalty action for non-compliances.
- Hard copy of all suppliers' quality manuals and HACCP where appropriate held on site.

It is essential that each aspect of the supply chain satisfies defined supplier quality and safety parameters, in advance of the commencement of product outsourcing. If any part of the supply chain process involves the utilisation of food or non-food product brokers, then the technical standards required of them cannot differ from those demanded directly from the supplier themselves. It is for the broker to avail themselves of the information and for the airline caterer to demand it.

Chapter 7 focuses on the issues appertaining to traceability requirements in more detail, but much of the technical specification information will be relied upon to ensure that the supply chain detail is both structured and consistent and that the required traceability parameters can be satisfied extensively in all product supply areas.

It is important to recognise the limited compliance potential of some suppliers at certain destinations around the world. The aim then is to positively identify where suppliers are failing to meet defined technical standards and to assist them in the implementation of supplier systems management, to attain the required level of a safety assured supply chain. The demands will not be met however if the demands are not made, so it is crucial that potential suppliers are made fully aware of what is required of them in order to be deemed suitable to provide products to the industry. In any business the fiscal considerations are going to be instrumental in driving the quality standards of the supply chain, but they should never prove preclusive to supplier approval attainment in the context of predetermined product quality and safety parameters.

Every different category of product will be defined by relative quality and safety standards. It is essential that this exercise is undertaken by the catering procurement personnel in tandem with the technical personnel so that quality and safety and fiscal and logistical considerations all operate in tandem. Inevitably with the potential for some outsourcing considerations to be span the globe in terms of what is required and where, one has to be realistic in one's expectations. However, one also has to be realistic about the impact that a non-verified, inconsistent, poor quality supply base may have on the airline meal product itself. Quality standards, in the broadest sense of the term, must encompass all that constitutes or has some connection with product safety and integrity and not just those visual aspects that present themselves on the plate to the end consumer.

Raw material procurement

Having established the necessary credentials required of suppliers in order to be deemed fit to meet the defined technical criteria, our attention now turns to raw material specifications. This process is applicable in two formats depending on both the nature of the product and the likely technical credentials of the supplier. Raw material specifications *must* be held by the airline catering provider for every single raw material that is utilised in the production and formulation of airline food products. Without this centrally held data having been completed to appropriate technical standards, it is not possible to effectively produce end product specifications which also meet satisfactory defined technical standards. This situation would therefore leave one with a multitude of high volume, multiple application, components and finished goods that are produced to undefined food standards and are ostensibly untraceable.

The role that the raw material specification has to play in the manufacture of finished products is critical in order to assure quality and safety standards at every step in the supply chain. As I said before, the specifications can be achieved in one of two ways: if the product is a standard product supplied, the producer is likely to already hold a technical specification documenting concisely all the relevant product quality and safety parameters in one document. If, however, this is not something that the supplier holds, then it is for the procurer to supply a technical specification template to be completed by the supplier. This may well be necessary when dealing with small suppliers or those who are supplying goods in a non-standard production format at the request of the airline caterer.

Figures 5.1–5.4 illustrate the types of technical raw material specifications one would expect to receive.

COMPANY NAME
Address/Tel. No.

PRODUCT	PASTRY TART	GAA23
	COOKED PASTRY	
INGREDIENTS	FLOUR, OIL, WATER, SALT, E202	
PRODUCT STANDARDS	PER CASE OF 144	
	LENGTH	50mm
	WIDTH	55mm
	WEIGHT	500g
	CASE WEIGHT	8kg
ORGANOLEPTIC	COLOUR	
	FLAVOUR	
	BAKED FLAVOUR	
	BAKED COLOUR	
	PERFORMANCE	
COOKING INSTRUCTION	PLACE NO. ON TRAY, PREHEAT OVEN TO 180°C FOR 15 MINS, COOK FOR 15–20 MINS	
PACKING	NO. OF UNITS	8 PER CUTTER
	PALLET	18x8
	INTERNAL PACKAGING	560mm x 350mm
	OUTER CASE	560mm x 350mm

NUTRITIONAL INFORMATION

	RAW PER 100g	BAKED PER 100g
MOISTURE	30	7
PROTEIN	4	6
FAT	8	12
KJ	1500	2500

STORAGE	PRODUCT MUST BE KEPT IN A COOL DRY PLACE

Figure 5.1 Raw material spec sheet for pastry case.

Figure 5.5 illustrates a blank technical specification, which could be issued to suppliers to be completed by them in advance of supply.

The detail laid down in these types of technical specifications enables one to witness at a glance whether the product parameters meet all required quality and safety considerations. They also illustrate relevant aspects of the process flow and HACCP where they need to be considered and give information in a proactive fashion, which demonstrates that potential quality and safety concerns have been considered and redressed.

In order for the information to be verified appropriately it is useful to place copies of raw material specifications at the goods receipt area of the operation in order that the quality and safety attributes laid down in the specification are evident at the point of delivery, thereby allowing accept or reject protocols to be managed more efficiently. Laying down the parameters by which products meet acceptable limits prevents raw materials gaining access to the building whilst displaying sub-standard product attributes. This system also requires

MEAT SPEC FOR COMPANY

SUPPLIER

PRODUCT

Diced fore-quarter feather muscle meat selected from British Beef Waitrose Specification farm assured steers and heifers. Stock not fed animal meat and bone meal. Aged 30 months or less of UK origin.

PROCESS

Carcass fore-quarters are boned out between 2 and 4 days from slaughter at a maximum temperature of 4 °C. All bruises, stamp marks, discoloration, hair, blood clots, gristle, membrane are removed. All the separated muscles are trimmed of any visible excess fat and gristle.

PACKAGING

The product is diced into approximately 21 mm cubes on the cube king dice machine and packed into clear plastic pouches, vacuum packed, metal detected, passed through the heat tunnel, labelled, put into trays, weighed and sent to the chiller running between 0 °C and 2 °C **where they are stored until they are to be despatched.**

LABELLING

Inner pack label will show product description, batch number, pack date, use-by date, EEC number. White box label will show product description, pack date, use-by date, supplier, EEC number, tare weight, net weight.

DATE CODE

Kill date to pack date   2 to 3 days
Pack date to use-by date   +21 days
Minimum life into depot   11 days

STORAGE CONDITIONS PRIOR TO DESPATCH AND IN TRANSIT

Raw material chill   0 °C to 4 °C
Processing room   4 °C to 6 °C
Product chill   0 °C to 4 °C

PRODUCT RECEIPT AT DEPOT

Minimum -1 °C
Maximum $+4$ °C

METAL DETECTION

After vacuum packing and before boxing
Test piece size 7.50 mm ferrous, 8.0 mm non-ferrous, 8.0 s/steel
Method of detection audio alarm/belt stops

QUALITY ASURANCE

Check carried out	Frequency
Temperature checks	Each batch
Visual checks	Each batch
Label checks	Each batch
Weight checks	Each batch

MICROBIOLOGICAL STANDARDS

	TARGET	MAXIMUM
TVC	$<5.0 \times 10^5$	1.0×10^6
Enteros	$<1.0 \times 10^4$	5.0×10^4
E.coli 0157	Absent in 10 g sample	

Figure 5.2 Raw product meat spec.

SPEC FOR ICEBERG SALAD	SECOND TO NONE VEG COMPANY	
PICTURE OF THE SPEC SALAD	SPECIFICATION REFERENCE	
	S-TN 4444 LETTUCE	
	PRODUCT DESCRIPTION	
	WASHED ICEBERG SALAD	
	PRODUCT CODE	SALES UNIT
	RAP 2468	KILO/1000 G
	COMPOSITION	
	ICEBERG LETTUCE LEAVES	

GENERAL SPEC		
Leaf substitution may be necessary with customer approval if quality of supply cannot be guaranteed		
WASHING SANITISING PROCESS		
Product is submerged and tumbled through a two-staged washing process Clean water then chlorine wash Sodium hypochlorite solution is diluted using mains water Chlorine levels are recorded hourly		
MICROBIOLOGICAL SPECIFICATION		
MICRO SPECIFICATION	TARGET CFU/G	ACTION LIMITS
Total viable count	<10e/5	10/5 – <10/6
Total coliforms	<500	500 – <10/4
Escherichia coli	NOT DETECTED	<10
Listeria	NOT DETECTED	DETECTED
PACKAGING & PACK SIZE	**LABELLING**	**RECEIPT/STORAGE**
Pack size 1000 g Packed in airtight bags 500 mm wide × 500 mm long Bags are delivered in plastic crates	Named/weight Use-by/batch Code/store Temp	1–3 °C Shelf-life P 4

ISSUED TO (name of company)　　DATE　　　　　　　　　　SIGNED

Figure 5.3 Example raw material spec for washed lettuce.

that the staff operating the procedures at the back door will need to be familiar with the contents of the raw material specification and how they relate directly to assured product quality and safety standards.

In terms of the role that raw material specifications have to play in the development of end product specifications for products manufactured in-house, all of the essential quality and safety information should then be present ready for it to be devolved directly onto the recipe section of the finished product specification. Without effective raw material information at one's fingertips, the job of compiling the finished product technical information becomes impossible.

SPEC FOR IGRAPES		SECOND TO NONE VEG COMPANY	
PICTURE OF THE SPEC SALAD		SPECIFICATION REFERENCE	
		S-TN 4444 GRAPES	
		PRODUCT DESCRIPTION	
		WASHED GRAPES	
		PRODUCT CODE	SALES UNIT
		RAP 2468	KILO/1000 G
		COMPOSITION	
		RED OR WHITE GRAPES PICKED	

GENERAL SPEC		
Substitution may be necessary with customer approval if quality of supply cannot be guaranteed/product should be clean and fresh colour		
WASHING SANITISING PROCESS		
Product is submerged and tumbled through a two-staged washing process		
Clean water then chlorine wash		
Sodium hypochlorite solution is diluted using mains water		
Chlorine levels are recorded hourly		
MICROBIOLOGICAL SPECIFICATION		
MICRO SPECIFICATION	TARGET CFU/G	ACTION LIMITS
Total viable count	<1x10e4	>1x10 e%
Total coliforms	<10 e2	>10e3
Escherichia coli	<10	<100
Listeria	Not detected	Detected
PACKAGING & PACK SIZE	**LABELLING**	**RECEIPT/STORAGE**
Pack size 1000 g	Named/weight	3–5 °C
Packed in airtight tubs	Use-by/batch	
1 kg size	Code/store	Shelf-life P 4
Bags are delivered in plastic crates	Temp	

ISSUED TO (name of company)   DATE   SIGNED

Figure 5.4 Example raw material spec for washed grapes.

All raw material specifications should detail the following information:

- Product name.
- Product code.
- Date of specification compilation and issue.
- Product description: colour, aroma, appearance, texture.
- Photograph of product.
- Quality attributes.
- Safety attributes.
- Traceability information, i.e. country/countries of origin.
- Preparation detail and process flow.

Raw material specification

- **SUPPLIER DETAILS**

Name & Address					
Telephone		Fax		E-mail	
Technical Contact					

- **MANUFACTURER DETAILS** (if different from above)

Name & Address					
Telephone		Fax		E-mail	
Technical Contact					

- **EU LICENCE**

PRODUCTS LICENCE TYPE	EU licence number

- **PRODUCT DETAILS**

Product name		Supplier Product Code	
Description			
Country of origin			

- **PRODUCT ATTRIBUTES**

Appearance	
Texture	
Flavour	
Aroma	
Length		Depth	
Height		Weight	

Where possible please include a photograph of the product

Figure 5.5 Raw material/end product specification.

- **THIRD PARTY CERTIFICATION/ACCREDITATION**

Scheme	Level	Last audit date

- **INGREDIENTS:** include additives (with functionality and E number), processing aids and non-declarable ingredients

Ingredient	Percentage	Details

- **FOREIGN BODIES/EXTRANEOUS MATTER**

Material	Tolerance
Glass	
Wood	
Plastic/perspex	
Metals	

- **STORAGE**

Ambient		Chilled		Frozen	

Once opened

Ambient		Chilled		Frozen	

Figure 5.5 (Continued).

- **SHELF-LIFE**

Minimum at manufacture	
Minimum at delivery	
Shelf-life once opened	

Do you hold documentation to confirm the given shelf-life?	Yes		No		Details	

- **PACKAGING**

Unit per case			
Cases per layer		Layers per pallet	

	Yes	No	Details (% vacuum, gas mix)
Controlled atmosphere			
Vacuum packed			

Does all food contact packaging conform to current UK/EC regulations on plastics and other materials in contact with food?	Yes		No	

Food Contact Packaging

Length		Height		Depth		Weight	
Material							
Gauge				Percentage recycled			
Method of closure							

Outer Packaging

Length		Height		Depth		Weight	
Material							
Gauge				Percentage recycled			
Method of closure							

Labelling/Traceability

Ingredient label declaration	
Label claims (e.g. minimum meat content %, fat-free, etc.)	
Description of date coding	

Are all ingredients fully traceable to source?	Yes		No		Details	

Please attach a sample label (to follow)

Figure 5.5 (Continued).

112 Aviation Food Safety

- **PROCESS CONTROLS**

Please attach a full HACCP flow diagram identifying all CCPs

Critical Control Point (CCP)	Control Measure	Frequency of check	Tolerance

Figure 5.5 (Continued).

Do you carry out a system of supplier screening?	Yes	No	Details	
Are all critical instruments calibrated to a traceable national standard?	Yes	No	Details	
Is the manufacturing site covered by a full pest control contract?	Yes	No	Details	
Do you carry out regular glass and brittle plastic audits of your factory?	Yes	No	Details	

- **MICROBIOLOGICAL STANDARDS**

Test	Target (cfu/g)	Reject (cfu/g)	Test frequency
Total viable count			
Enterobacteriaceae			
Escherichia coli			
Salmonella spp.			
Listeria			
Bacillus cereus			
Clostridium perfringens			
Staphylococcus aureus			
Yeasts			
Moulds			
Aflatoxins			

- **CHEMICAL STANDARDS**

Test	Target	Reject	Test frequency
pH			
a^W			
Pesticide residues			
Trace metals			

- **NUTRITIONAL ANALYSIS**

Nutrient	Unit	Amount/100g	Nutrient	Unit	Amount/100g
Energy	kcal		Total fat	g	
Energy	KJ		- saturated	g	
Protein	g		- mono-unsaturated	g	
Total carbohydrate	g		- poly-unsaturated	g	

Figure 5.5 (Continued).

| - of which sugars | g | | Sodium | mg | |
| Fibre | g | | | | |

Is the nutritional information analysed or calculated?	Analysed		Calculated	☑
Please state source of calculated data				

- **FOOD INTOLERANCE/COMPOSITIONAL INFORMATION**

Is this product free from:	Yes	No
Peanuts (including any possible sources of cross-contamination)		
Nuts (including any possible sources of cross-contamination)		
Sesame seeds and derivatives		
Milk and milk derivatives		
Egg and egg derivatives		
Wheat and wheat derivatives		
Soya and soya derivatives		
Maize and maize derivatives		
Gluten		
Fruit and fruit derivatives		
Yeast and yeast derivatives		
Vegetables and vegetable derivatives		
Fish, crustaceans, molluscs and their derivatives		
Pork and pork derivatives		
Beef and beef derivatives		
Artificial additives		
Azo and coal tar dyes		
Glutamates		
Benzoates		
Sulphites		
BHA/BHT		
Aspartame		
Antibiotics		
Mechanically recovered meat (MRM)		
Comments		

Figure 5.5 (Continued).

- **GENETICALLY MODIFIED FOODS**

Does the product or any of its ingredients contain any genetically modified material?	Yes	No
Identify those ingredients which contain such material		

Is the product or any of its ingredients produced from, but not containing, any genetically modified material?	Yes	No
Identify those ingredients which contain such material		

Has the product or any of its ingredients been significantly changed as a consequence of the use of genetic material?	Yes	No
Identify any such ingredients		

Have genetically modified organisms been used as processing aids or additives or to produce processing aids of additives in connection with production of the food or any of its ingredients?	Yes	No
Identify those ingredients affected		

- **SPECIFICATION AUTHORISATION**

Specification completed by (print):	
Specification completed by (sign):	
Position:	

Figure 5.5 (Continued).

Date:	

For Castle Kitchens use only

Specification authorised by (print):	
Specification authorised by (sign):	
Position:	
Date:	

Figure 5.5 (Continued).

- Microbiological specification.
- Packaging detail and description – tertiary and primary.
- Pack size.
- Labelling detail.
- Critical limits upon receipt.
- Storage information.
- Durability coding.
- Signed and approved by technical representative.

In the absence of any of this information being readily available from the supplier, these data must be compiled in-house. Essentially if any of this information is not available from the supplier for whatever reason, then an alternative supply source needs to be found. It is sensible when contracting a supplier to include a blank raw material specification with the supplier audit questionnaire. In this way it is abundantly clear from the outset whether the supplier is going to be in a position to have the necessary technical resources in place to adhere to the requirement.

In addition to the technical data required it is essential to define the parameters of the other factors involved in the quality assurance of raw material procurement. Defined terms and conditions of trading will need to be established and signed off with each supplier. These will appertain to a variety of factors such as payment terms and failure to deliver penalties. Whilst many of these may seem to have little or nothing to do with product safety and integrity, in terms of assuring the long-term integrity and consistency of the supply chain it is essential that trading parameters are defined and performance standards met and documented. In this way it is easy to see at a glance which suppliers are performing in line with their obligations and which are not. This aspect of manufacturing protocols is particularly pertinent in the aviation environment when one considers the fiscal impact of a plane becoming delayed as a result of supplier non-compliances.

It is essential to understand that raw material specifications and assured supplier outsourcing form the basis upon which all manufacturing systems operate. To undermine one aspect is to undermine the whole process and underestimate the safety impact that poor quality raw materials may have upon the supply chain as a whole. If purchasing deals are being struck based on defined product attributes, then it is vital that the systems management protocols allow for those attributes to be verified effectively at point of delivery.

End product specifications

When I first became involved in manufacturing products for the airline as opposed to business aviation industry, I was fascinated by the constant references to product specifications. As a food manufacturer I assumed that they were indeed the finished product technical variety that manufacturers are so familiar with. To my surprise, the aviation industry's version of a finished product specification varies hugely from the manufacturing standard technical version. Figure 5.6 illustrates the point.

Aviation catering-style product specifications like those shown in Figure 5.6 have far more to do with documenting the perceived quality considerations than with taking account of the safety parameters and alternatively documenting quality and safety factors simultaneously. The only assumption one can make is that quality considerations are perceived as having nothing to do with safety considerations, and safety measures as having no impact on quality.

Whilst for logistical and operational reasons it is necessary to have a specification that amalgamates all the product presentation and packaging requirements reflected in the information displayed in Figure 5.6, its reliance on single unit documentation when the batch quantity is likely to be far in excess of that is somewhat irregular and can lead to production inconsistencies and non-conformities.

Manufacturing standard product specifications document all relevant safety, quality, process flow, HACCP, packaging, labelling, microbiological and nutritional aspects of the finished product in both a product and process-specific format. The batch size is documented and the recipe formulation is generated to reflect both a manufactured batch quantity and a finished product unit size and weight association. Figure 5.5 illustrates an example of an end product technical specification.

One can see from the detail in Figure 5.5 that being in possession of all relevant raw material technical information is essential in order to successfully

			MEAL NUMBER	
			CHICKEN TUSCAN	
PICTURE OF PRODUCT			ROTATION 4	ISSUE NO 2
			COMPANY	BIZ CLASS
				BATCH SIZE 1
			ROUTES USA	
CODE	INGREDIENTS	METHOD	NUMBERS	GRAMS
UUSS	LARGE FOIL	HAND	1	
UUAA	FOIL COVER	HAND	1	
	CHIX BREAST	HAND		100
	POTATOES NEW	HAND		50
	VEG. CARROTS	HAND		30
	VEG. BEANS	HAND		30
	SAUCE	HAND		
	ONION	HAND		10
	MUSHROOM	HAND		10
	TOMATO TINNED	HAND		30
SIGN		DATE		

Figure 5.6 Typical airline-style end product spec.

complete this kind of document. In this document we begin to see an amalgam of all relevant product information compiled into an individual set of information. The dictates of supplier outsourcing and raw material procurement begin to fall into place within the context of this single document. If the other prerequisite systems are in place, the compilation of this type of technical data is very simple; however, if any deviation from manufacturing systems compliance standards has occurred, completing this information accurately will be impossible.

Whilst this type of specification is very useful for single unit products, let us not forget the uniqueness of the aviation environment that dictates that the meal consists of not just one but indeed several different components. If one were to follow the technical detail to the letter, it would be essential that the end product specification contained all the technical detail for all the components in one finished specification. However, to begin with it is not essential unless the product is in the SPML category of provision and specific-restricted claims or allergen information are required. What is essential if the component format end product specification is not to be utilised, is that the individual end product specifications for all grouped products are centrally held grouped together. This includes what may be supplied as condiments or side dishes.

In the same way as the utilisation of raw material specifications engenders the requirement for goods receipt protocols to verify the parameters laid down, the introduction of this technical type of end product specification results in the requirement for production methods and process flows to reflect those documented also. This will result in the development of a new systems management culture, which places perceived quality and genuine quality considerations in the same box as total product safety aspects. The production process methodology and respective systems management paperwork can be seen later in the chapter but it is important at this stage to gain an understanding of how all manufacturing systems should integrate throughout the whole process flow and critical paths.

The essential differential in terms of greater product safety assurance, between this type of technically-based end product specification and those of the traditional airline catering variety, is the individual consideration given and documentation applied to the specific production method and process flow attributable to each and every product. It is this aspect also that makes the multicomponent specifications difficult to tackle without a sound experience of the single component specifications first. By making the HACCP product and process-specific, instead of environment generic, all attributable safety aspects can be considered and documented appropriately. It is inevitable that in any airline catering unit, the volume of products procured, manufactured and assembled will be huge and genuine quality considerations will often dictate a deviation from documented, generic HACCP systems and process flow. The utilisation of an end product specification allows for such deviations to be considered, safety parameters identified and established and any necessary action documented.

All end product specifications should contain the following information:

- Product name.
- Product code.
- Issue date.
- Product description: colour, aroma, taste, etc.
- Raw material supply information including country of origin.
- Recipe breakdown in Quantitative Ingredient Declaration (QUID) format.
- Sub-recipe breakdown.
- Compound ingredient breakdown.
- Nutritional composition based on defined batch quantity.
- Allergen declarations.
- Product-specific process flow.

- Product-specific HACCP.
- Packaging detail – tertiary and primary.
- Life attributes and durability coding scientifically verified.
- Labelling information.
- Quality parameters.
- Microbiological parameters.
- Reheat instructions where applicable.
- Storage information.

Whilst inevitably the major culture shift from catering standards specifications to manufacturing ones cannot happen overnight and can only be achieved with a firm understanding of the broader picture, this should not preclude effective sea changes in systems management. The entire regenerated modus operandi will need to be established in draft format first and then applied individually to each and every process throughout the supply chain followed by the production and despatch processes until the new systems management roll-out is complete. It is critical that the overhaul of current systems is conducted brutally, and the manufacturing ethics are not compromised by a mere watering down of existing systems when in fact a complete dissolution is what is required. The main focus has to be that if all the systems structuring can happen in tandem, each aspect should, in theory, roll into another more effortlessly than first imagined.

Each area of operation needs to be established against a defined set of parameters and compared in catering versus manufacturing format so the obvious work to be done is evident from the outset. In terms of the end product specifications required of suppliers who are supplying finished goods to the airline caterer, the technical detail needs to be comparable with that used on in-house products. Essentially, in the same way as for raw material suppliers, a blank specification needs to be issued, to be completed by the producer/supplier if a standard specification generated in-house is not available. Often the end product specifications are far too brief to suit the purposes of the requirement, at which point a generic in-house version will need to be issued and signed off. In any case it is helpful to have some kind of specification uniformity, so no essentially required information is left out of the loop.

Goods receipt

The next process in the technical chain of events that needs to be considered and managed in terms of its impact on total product safety, is the goods receipt procedure. In terms of the airline catering supply chain, at this stage let us not forget the list of potential products that this aspect may incorporate. All received goods, whether food or non-food, will need to be managed through a documented back door procedure, which will take account of all attributable risks posed by goods coming into the catering unit.

The size and scale of the unit and its flow and critical paths will determine whether segregated receipt of food and non-food goods can occur. It is essential to segregate clean food product and raw materials from the incoming path of dirty food waste from decatered aircraft. Having established which of these protocols can be adhered to, it is essential to take account of the receipt of goods flow in the documented HACCP. Finished goods and non-food goods that form aspects of the completed tray set, like condiments and toothpicks, need to be stored and received into appropriate areas as near to or adjacent to their required production area as possible. If the flow does not allow this then a documented process of finished products/goods transfer from storage areas, through production areas and onto assembly or further storage areas, needs to be established in order to take account of any attributable risks posed.

Having developed a protocol that ensures all products brought into the production facility have a technical specification attributable to them, it is essential that the goods receipt paperwork takes account of the documented safety and quality parameters laid down in the specification in order that these attributes can be verified effectively at the point of goods receipt, and any non-conformities identified. These types of quality assurance checks can be documented on a single set of generic goods, with the relevant checks being made where appropriate. Figures 5.7a and 5.7b illustrate a working example of the type of goods receipt paperwork that should be in place in order to satisfy these criteria.

The following information should appear on all goods receipt paperwork:

- Supplier name.
- Product name.
- Product description.
- Date ordered.
- Date to be delivered.
- Description and classification of goods, i.e. food/non-food.
- Delivery and storage parameters, i.e. chilled, frozen, ambient.
- Delivery temperature parameters.
- Batch code.
- Durability code, i.e. best before/use by.
- Visual and physical checks.
- Vehicle hygiene inspection and temperature checks.
- Packaging conformity.
- Weight checks.
- Labelling compliance verification.
- Delivery window adhered to.
- Accept/reject.

	PURCHASE ORDER			

PRODUCT+ CODE _____

SUPPLIER: _____   **Order no.** _____

Phone: _____   **NON-FOOD ITEM** _____
Fax:
email: _____   **FOOD ITEM** _____

RAISED BY: _____   **DATE PLACED** _____

Item	Description	Date due	Quantity	Price
1				
2				
3				
4				
5				
6				
7				
8				
9				
10				
11				
12				
13				
14				
15				
16				
17				
18				
19				
20				

Note:
Goods not meeting the agreed specification will be rejected
Time is of the essence
All products must be GM free
Chilled goods must be delivered at temperatures < 5 °C
Frozen goods must be delivered at temperatures < −18 °C

Figure 5.7a Purchase order linked to goods receipt requisition.

Having defined the parameters for all of the above in the product specification, it will be necessary to cross-reference these at the point of goods receipt. This will be a simple task for back door personnel if copies of the product specifications are centrally held at the back door, filed in supplier indexes. Aspects of the HACCP plan attributable to this exercise will need to be thoroughly

GOODS INWARD NOTE

SUPPLIER				FOOD/NON-FOOD			
				DATE RECEIVED			
				ORDER NO + DATE			

Item	Batch No.	Average Weight EEC NO.	C	Use-by Date B/before	Condition of packaging		Condition of goods	
					Accepted	Rejected	Accepted	Rejected
1								
2								
3								
4								
5								
6								
7								
8								
9								
10								
11								
12								
13								
14								
15								
16								
17								
18								
19								
20								

Time delivered: _____ Vehicle temperature °C _____ Chiller -
Date delivered: _____ Freezer -
Delivery note/invoice: _____ Condition of vehicle: _____ Acceptable   Unacceptable why
Accepted by: _____ Condition of driver: _____
Free from damage: _____
Free from pests: _____
Signature: _____ Labelling correct: _____

Figure 5.7b Example of goods receipt paperwork linked to a purchase order.

trained for and understood by personnel undertaking this task, so that they are empowered to identify and effectively document non-compliances as soon as possible and report them to the respective supervisory and/or technical personnel. Examples of these types of non-compliance can be issues such as product delivered outside of the documented primary and tertiary packaging specification, e.g. in wood or cardboard when the specification stipulates returnable plastic, or goods delivered outside of defined quality parameters such as 'washed and spun', when the products arrive 'wet and soggy'. With defined technical detail devolved to the back door personnel, the risks of out-of-specification goods finding their way into the supply chain are reduced dramatically and non-compliance grounds for product rejection are easily defined and understood.

There will be several factors documented on the specification which, despite the fact that they form aspects of the quality and safety parameters, will not be identifiable by virtue of visual checks alone. It is necessary therefore to ensure that the back door regime follows other protocols for product non-compliance identification. Product checks initiated at the back door should also include the following:

Visual checks

- Packaging integrity – intact and to spec.
- Labelling integrity – intact and to spec including durability coding.
- Product integrity – meets defined quality parameters.

Physical checks

- Packaging integrity – no breaks, tears, dents or leaks.
- Product integrity – to spec with no signs of physical or mechanical damage or chemical contamination.
- Hygiene – vehicle and packaging if in returnable crates.

Temperature checks

- Product.
- Vehicle.

Whilst the microbiological integrity of finished products and raw materials coming into the building will not be discernable without lab testing, supplier conformity in all other areas will give an excellent indication of how effectively their systems are being managed and controlled. It is essential that supplier performance is documented and assessed, so that a supplier's ability to meet defined criteria is constantly under review. The emphasis is on building systems that demonstrate a tangible measurement of a supplier's capacity to get it right first time.

It is difficult to imagine how the full scope of product verification checks can occur if all product parameters are not defined in advance on product specifications and then the attributable receipt systems do not allow for the documentation of non-compliances. For the most part my airline catering colleagues will tell me that if it is cheap and it turns up at the back door they will accept it! A damning indictment of the restaurant-style ethics the industry has adopted, which have no place in consideration of the modern product requirement.

In terms of the impact that systems which do not allow for these types of product, packaging and raw material verification procedures, may have on the long term integrity of the supply chain, it is essential to view the goods receipt aspect of the supply chain in terms of the perceived impact on food safety and indeed security that is posed by quality and safety deficient raw material supplies. It is also essential that any deviation from policy-defined terms and conditions, such as delivery window slots, on-site health and safety adherence, driving speeds and delivery drivers' hygiene and behavioural standards, are recorded as non-compliances along with product deficiency ones. In this way the supplier performance criteria can be used to harvest a supplier quality assessment based on all round performance values and not just those directly and obviously attributable to perceived product quality and safety standards.

In the overall scheme of things the manufacturing sector places a huge focus on defining supplier attributes in terms of their operational capacity to undertake every aspect of the supply function. This is based on the requirement for consistent supply chain service standards as well as product quality ones. With a verified supply chain being a must in the manufacturing sector, there is no opportunity to outsource from elsewhere should the specified and documented supply chain fail. It is essential therefore to consider all supplier attributes as part of the bigger picture before final product outsourcing decisions are made.

By adopting this kind of supply chain quality assurance techniques and defining the parameters by way of goods receipt documentation, a continuation and assurance of supply chain integrity is undertaken which should reflect the documented safety and quality standards laid down in the product specifications. Without in-depth product assurance and verification systems in operation at the back door, all of the detailed work that has been undertaken in the compilation of the specification and supplier audit criteria will be in vain and supplier performance will go unpoliced.

Once non-compliances have been identified at the point of delivery, it is essential that the system of documentation is capable of taking account of all types of potential product non-conformity. These need to be documented on a product reject sheet with the course of corrective action clearly documented also. Training of all back door personnel in the application and management of these procedures will be required, and the level at which a specification deviation is permitted before automatic rejection is the only available course of action, will also need to be defined product by product. Figure 5.8 illustrates an example of product rejection paperwork that could be utilised.

The essential application of this type of documentation is to make sure that the systems at point of goods receipt do not allow product non-conformities to go unnoticed and unrecorded. In this way the goods receipt prerequisite systems are capable of underwriting the HACCP at this step in the process, with both critical limits and corrective action a natural and integral part of the systems management, rather than an uninitiated adjunct at the point that product or performance non-compliances are first identified.

REJECT SHEET NON-CONFORMITIES DELIVERIES					
PRODUCT SUPPLIER	TYPE OF NON-CONFORM	CORRECTIVE ACTION	ACCEPT		SIGN
			YES	NO	

Figure 5.8 Example of product rejection sheet used at point of goods receipt.

Production protocols

The key to comprehending how the systems described throughout this chapter critically function is to focus on how each process discussed connects with the next and how the established protocols in each area relate effectively to satisfy a perfectly integrated end result. There is no other aspect of the supply chain where this is more critical than in the protocols that govern the production process.

Firstly let us recap. Product assurance processes up to this point have taken the following steps:

- Responsible and technically considered product development.
- Responsible and technically considered supplier outsourcing.
- Responsible and technically considered raw material procurement.
- Technically considered end product and raw material specifications.
- Goods receipt protocols dominated by quality and safety parameters defined in and devolved from technical product specifications.

The five steps discussed above have predetermined the following factors in order that the next process step can happen effectively and safely:

- Product and recipe development renders the product fit for purpose in the intended production and service environment.
- Either the finished product or raw material supplier is proven capable of achieving consistency, safety, quality and service standards defined as required, in order to supply.
- All raw materials utilised in the production of products are predefined in terms of their origin, safety and quality attributes, packaging credentials and microbiological criteria.
- All attributable quality, safety, nutritional, microbiological, packaging, labelling, durability, storage and labelling criteria are established and documented in a standardised format product specification.
- Only quality and safety assured finished product and raw materials received into the operation for use in the manufacture of finished products or as components of finished products.

In theory then all processes so far have led to a situation where the quality and safety attributes of the raw materials or finished product used as part of the end product are defined, documented and quality assured. The key at this point of the process is not to lose sight of the continued requirement to verify the previous system's success and document its potential failures, and not assume that all is as it should be. The requirement for a continued systems management culture throughout the ongoing production, assembly and despatch processes, where all traceability aspects of the documentation are transposed onto the paperwork associated with each process in turn, is the best possible way of assuring that all of the systems amalgamate effectively into one cohesive process.

Having established the detail of the product and production process flow in an end product technical specification, the first step is to devolve the recipe, HACCP and process flow into a bulk production record that reflects the batch quantity to be produced. Figures 5.9–5.11 illustrate the type of templates required to form part of the documentation process depending on the types of products to be manufactured.

Here we can see that every aspect of the process-specific HACCP contained within the product specification is devolved onto the production paper so even those with no specific HACCP knowledge can control and document the CCPs throughout the production process as each documented step instructs them when and how to do so. Whilst certain aspects of this paperwork will be generic to all products, production records will still need to be devolved product by product from the specification. What is also clear from this paperwork is that the safety and quality parameters of the process are documented jointly so that safety and quality aspects are recognised as having equal status at this crucial stage of production. In some instances there may be several ways in which one product is manufactured. Whilst the recipe may remain the same, it is essential that the process flow and quality and safety attributes are documented differently in each case to reflect the genuine hazards and required quality measures associated with each method of production.

The raw material batch coding information for the products that amalgamate to make up the finished product, should be attributed to the paperwork at the point of decant so that all raw material usage is documented for traceability purposes. In this way the supply chain details are followed through the entire process, ensuring that all batches are fully traceable. Where mixed batch raw materials are used, that must also be documented effectively but should be avoided if at all possible.

The decanting process requires that an assembly of the recipe breakdown is undertaken in advance of the transfer of raw materials into the production area. At the decant stage all of the following steps have to occur:

- Raw materials are removed from their primary packaging.
- Raw materials are weighed out against a defined batch quantity and recipe for the batch.
- Raw materials are decanted into food grade containers.
- All raw material durability and production coding information is devolved onto the production paperwork which will accompany the production process.
- Batch recipe raw materials are covered and transferred into the production area.

BULK PRODUCTION RECORD

DISH................... VEGETABLE LASAGNE (Sauce)   PAGE 1 OF 3

%	PRODUCT	USE BY	BATCH CODE	WEIGHT	RAW TEMP.	BATCH	COOKED TEMPERATURE	TIME
	Spinach			kg			CCP1	
	Diced onions (5 mm)			kg				
	Olive oil			ml				
	Garlic purée			g				
	Cornflour			kg				
	White wine			ltr				
	Whipping cream			ltr				
	Salt			g				
	Pasta sheets			kg				

BATCH SIZE

BATCH CODE

METHOD

1. Prepare cheese, lasagne sheets and cherry tomatoes ready for assembly (page 3 of 3)
2. Batch code trays
3. Heat oil in pan
4. In LOW RISK add garlic and sweat off (approximately 1 minute)
5. Add onions and sweat until soft
6. Add white wine, stir and reduce by 50%
7. Remove from heat and add cream
8. Mix and gradually bring to boil (approximately 2 minutes)
9. Preheat brat pan in readiness for vegetables (see page 2 of 3)
10. Once sauce is boiling, place hand mixer in pan
11. Switch on and add spinach leaves
12. Mix and simmer until thoroughly blended
13. Dilute cornflour
14. Bring sauce to boil and add diluted cornflour to achieve desired thickness **PROBE 1 CCP1**
15. Add salt and stir
16. Transfer 1/3rd of mix (8 trays' worth) to high risk area for assembly (see page 3 of 3) **CCP1A**

DATE MADE

PRODUCE BY

CHECKED BY

DATE

Current issues: Page 1- no 3 Page 2- no. 3 Page 3- no.3

USE CAPITAL LETTERS ONLY

Figure 5.9a Bulk production template for lasagne production – sauce.

BULK PRODUCTION RECORD

DISH.................... VEGETABLE LASAGNE (Vegetable base) PAGE 2 OF 3

%	PRODUCT	USE BY	BATCH CODE	WEIGHT	RAW TEMP.	BATCH	COOKED TEMPERATURE	TIME
	Mixed peppers diced (25 mm)			kg			CCP1	
	Courgettes sliced (15 mm)			kg				
	Button mushrooms			kg				
	Olive oil			ltr				
	Red onions diced (25 mm)			kg				
	Red wine			ltr				
	Tomato purée			kg				

METHOD

1. Add oil to brat pan and preheat to 200 °C
2. Preheat pass through oven to 220 °C for assembly stage (see page 3 of 3)
3. In LOW RISK add red onions and mushrooms to brat pan
4. Stir regularly whilst sweating off for 5 minutes
5. Add courgettes, mixed peppers and stir in
6. Sweat off for a further 10 minutes or until all the ingredients have softened
7. Add red wine, tomato purée and mix in thoroughly.
8. Turn brat pan down and hold at 80 °C–82 °C **PROBE 1 CCP1**
9. continue on page 3 of 3.
10.
11.
12.
13.
14.
15.

DATE MADE

PRODUCE BY

CHECKED BY

DATE

USE CAPITAL LETTERS ONLY

Figure 5.9b Bulk production template for lasagne production – vegetable base.

Implementing manufacturing SOPs

BULK PRODUCTION RECORD

DISH.................. VEGETABLE LASAGNE (Assembly)   PAGE 3 OF 3

%	PRODUCT	USE BY	BATCH CODE	WEIGHT	RAW TEMP.	BATCH	COOKED TEMPERATURE	TIME	BLAST TEMPERATURE	TIME
	Red Leicester cheese – grated			kg			CCP1		CCP2	
	Mozzarella – grated			kg						
	Cherry tomatoes – halved			each						
	Parsley – dried			g					BLAST FREEZE	
							BATCH SIZE			
							BATCH CODE			
							TUXEDO BATCH CODE			

METHOD

1. **MOST IMPORTANT** – during assembly, ensure plastic sheets separating raw lasagne sheets are removed and disposed of
2. In HIGH RISK Add 1 × 20 oz ladle of vegetable base to the first 8 trays
3. Add 1 sheet of lasagne to each tray
4. Add second 20 oz ladle of vegetable base to each tray
5. Add second sheet of lasagne to each tray
6. Add 1 × 10 oz ladle of sauce to each tray
7. Spread mixture of cheeses proportionately onto each tray
8. Add 8 half cherry tomatoes to each tray
9. Sprinkle each tray with proportionate amount of parsley, each tray 3 kg
10. Place completed trays in preheated pass through oven until cheese is melted and slightly browned
11. Transfer to blast chiller
12. Blast chill from 75°c to 5°c in < 90 minutes. **PROBE 2 CCP2**
13. Repeat above operation with next batch of 8 trays
14.
15.

DATE MADE

PRODUCE BY

CHECKED BY

DATE

Temperatures			
Holding	Hot		
		SAUCE	
		MIX	
		SAUCE	
		MIX	
		SAUCE	
		MIX	

USE CAPITAL LETTERS ONLY

Figure 5.9c Bulk production template for lasagne production – assembly.

BULK PRODUCTION RECORD

DISH................... FISH CAKES (Fish) (250 cakes)   PAGE 1 OF 4

%	PRODUCT	USE BY	BATCH CODE	WEIGHT	RAW TEMP.	BATCH	COOKED TEMPERATURE	TIME
	Smoked haddock			kg			CCP1	
	Milk			ltr				

BATCH SIZE

BATCH CODE

METHOD

1. Defrost haddock for 24 hours in fish chiller
2. In LOW RISK tray up at approximately 6 fillets per gastronome tray
3. Set pass through oven to oven and preheat
4. Add milk proportionately to each tray
5. Place trays in oven for 15 minutes
6. Probe 5 samples to ensure they are at 80°C and hold temperature for at least 2 minutes **PROBE1CCP1**
7. Whilst fish is cooking, start making the sauce (see page 2 of 4)
8. Remove fish trays from oven in HIGH RISK
9. Transfer liquid to the sauce
10. Transfer the fish to the boiler
11.
12.
13.
14.
15.

DATE MADE

PRODUCE BY

CHECKED BY

DATE

Current issues: Page 1-no 2   Page 2-no. 2   Page 3-no.2   Page 4-no. 3

USE CAPITAL LETTERS ONLY

Figure 5.10a Bulk production template for fish cake production – fish.

BULK PRODUCTION RECORD

DISH................... FISH CAKES (Sauce) (250 cakes)   PAGE 2 OF 4

%	PRODUCT	USE BY	BATCH CODE	WEIGHT	RAW TEMP.	BATCH	COOKED TEMPERATURE	TIME
	White wine			ltr			CCP1	
	Garlic purée			kg				
	Salt			g				
	Horseradish			ltr				
	Cornflour			kg				

METHOD

1. In LOW RISK place white wine into pan and add garlic
2. Gently bring to boil and reduce by 75%
3. Add liquid from fish and combine
4. Bring to boil and reduce by 25%
5. Whilst reducing, gradually add salt whilst stirring
6. Dilute cornflour
7. Add to sauce to achieve a thick texture
8. Add horseradish and stir in, check temp. and record using **PROBE 1 CCP1**
9. Decant sauce into boiler with mash
10.
11.
12.
13.
14.
15.

DATE MADE

PRODUCE BY

CHECKED BY

DATE

USE CAPITAL LETTERS ONLY

Figure 5.10b Bulk production template for fish cake production – sauce.

BULK PRODUCTION RECORD

DISH.................. FISH CAKES (Mashed potato) (250 cakes)   PAGE 3 OF 4

%	PRODUCT	USE BY	BATCH CODE	WEIGHT	RAW TEMP.	BATCH	COOKED TEMPERATURE	TIME	BLAST TEMPERATURE	TIME
	Peeled potatoes			kg			CCP1		CCP2	
	Butter			g						
	Spring onion – chopped			kg						
	Salt			g						
	Parsley			g						

METHOD

1. Put potatoes in boiling kettle from LOW RISK and cover with water
2. Boil potatoes at 104°C
3. When cooked, drain potatoes
4. Set up mixer and select programme 2
5. When mixer 'beeps' after 10 minutes, add butter, spring onions, salt, parsley
6. Continue to mix until smooth
7. Add fish/sauce and mix
8. **Probe 1** to ensure temperature is >80°C and hold temperature for at least 2 minutes **CCP1**
9. Transfer to blast chiller on HIGH RISK
10. Blast chill from 75°C to 5°C in <90 minutes **PROBE2 CCP2**
11.
12.
13.
14.
15.

DATE MADE

PRODUCE BY

CHECKED BY

DATE

USE CAPITAL LETTERS ONLY

Figure 5.10c Bulk production template for fish cake production – mashed potato.

BULK PRODUCTION RECORD

DISH.................. FISH CAKES (Assembly) (250 cakes)

PAGE 4 OF 4

%	PRODUCT	USE BY	BATCH CODE	WEIGHT	RAW TEMP.	BATCH	COOKED TEMPERATURE	TIME	BLAST FREEZE TEMPERATURE	TIME
	Liquid egg			ltr			CCP1		CCP2	
	Milk			ltr						
	Cornflour			kg						
	Polenta			kg						
	Vegetable oil			ltr						
							BATCH SIZE			
							BATCH CODE			

METHOD

1. Place the fish/mash mixture into a tub and transfer to the cold preparation area in LOW RISK
2. Set up 1 tray of flour, 1 tray of mixed liquid egg and milk and 1 tray of polenta
3. Remove fish/mash mixture from tub in 180 g–200 g portions
4. Shape each portion in a 75 mm mould
5. Pane the cakes (i.e. dust in flour, dip in egg/milk mix and coat in polenta)
6. Reshape the cakes in 80 mm mould, ensuring they are symmetrical, fully coated and free from cracks
7. Add oil to brat pan, ensuring the depth is approx. half the fish cake thickness
8. Preheat oil to 200°C
9. Fry one side of cake until golden brown, then turn and fry the other side until golden brown
10. Probe and record a sample of 10 temperatures **PROBE1 CCP1**
11. Transfer to gastronome trays in HIGH RISK and record batch code on this sheet
12. Blast freeze – 18°C in <90 minutes **PROBE2 CCP2**
13. Check cakes visual quality against spec
14. Check cakes weights against spec, place in foil container four to foil box + label
15. _____
16. _____

DATE MADE

PRODUCE BY

CHECKED BY

DATE

Temperatures

USE CAPITAL LETTERS ONLY

Figure 5.10d Bulk production template for fish cake production – assembly.

BULK PRODUCTION RECORD

DISH................... STILTON AND PECAN CHEESY BASKET

% 1x	PRODUCT	USE BY	BATCH CODE	WEIGHT	RAW TEMP.	BATCH	FINISHED TEMPERATURE	TIME	BLAST TEMPERATURE	TIME
	Mango chutney			g			CCP1		CCP2	
	Maple syrup			g			MIX			
	Pecan nuts			g						
	Water			mm			Nuts			
	Stilton cheese			g						
							BATCH SIZE			
							BATCH CODE			

METHOD

1 Decant transfer **CCP1A**
2 Pass crumbled stilton, mango chutney and pecan nuts through dip tank into high risk area
3 Mix stilton and mango chutney thoroughly
4 Ensure the temperature is held at <5°C + Document **PROBE 2 CCPA2 2A**
5 If mix is above <5°C blast chill to below 3°C + Document
6 Cover mix + batch code/use-by label, place in WIP fridge, label
7
8 Mix maple syrup/water together and pour over pecan nuts. Roast off until golden/blast chill
9
10 Fill required amount making sure there is an even distribution of ingredients in each canapé
11 Ensure the temperature is held at <5°C + Document **PROBE 2 CCPA2 2A**
12 If mix is above <5°C blast chill to below 3°C + Document
13 Place canapé tray in bag and seal ready for boxing
14
15

DATE MADE

PRODUCE BY

USE-BY DATE

CHECKED BY

USE CAPITAL LETTERS ONLY

Figure 5.11 Bulk production template for canapé production.

The transfer process between storage and decant and decant and production is likely to be a hazard that is rarely identified while operating under catering-style HACCP. It is certainly not considered in many best practice guides to aviation food supply that I have seen. However the geography and size of many airline catering units should dictate that the physical contamination hazards associated with the transfer process, if not effectively considered and controlled can ultimately lead to all sorts of problems with end product contamination.

Once in the production area the segregation of high and low risk activities needs to be established in order to ensure that the 'dirty' and 'clean' activities are totally separate. Whilst low and high risk separation is a commonplace manufacturing SOP and whilst many manufacturers that service the aviation sector operate this type of process segregation, it is extremely uncommon to find a low/high risk divide in a traditional airline catering unit. This is yet another example of how mass catering protocols rather than manufacturing standard SOPs have permeated the sector despite the obvious requirement for extended life attribution on products such as prepared meals.

The essence of this type of product and process separation is based on the theory that low and high risk activities are identified and documented in the process flow attributable to the production of all products. High risk finished products are handled in an entirely separate area to low risk raw food handling and cooking areas. While in order to accommodate the successful and efficient throughput of products, low and high risk areas may well be adjacent to each other, they are separated by both physical and operational barriers which include the direct separation and segregation of the following:

- Production personnel.
- Production equipment, utensils and storage receptacles.
- Production areas.
- Workwear.
- Cleaning equipment.
- Air flows.
- Drainage flows.

The process is designed to avoid the recontamination of high risk foods postcooking when shelf-life is a product requirement. The design of the unit will need to accommodate this divide and with the older style premises synonymous with aviation catering this is unlikely to be in evidence. It is possible, however, to take account of these principles in any unit and apply them to a degree, even within the confines of a restricted environment. The critical aspects are the geography of where cooking, chilling and assembly operations are undertaken and how the flow of that can be organised to accommodate the low/high risk divide. Whilst the physical separation issues may be more difficult to accommodate in a unit that has not been specifically designed for the purpose, the good practice aspects such as separate and obviously denoted personnel, equipment and workwear can indeed be established. It is essential, however, that there is a strong systems management culture established in order to underwrite this type of premises design failure so that the process can still operate effectively.

In production processes that require no cooking the same type of segregation must occur, and these types of chilled, high risk products must be manufactured in temperature controlled areas. Once more the decant and transfer aspects of these products also has to be effectively considered in terms of the attributable hazards associated with them, whilst the production paperwork, as seen in Figure 5.11, should reflect the defined criteria by which products are assembled, held and temperature and quality checked throughout the process.

What is critical and unique in the aviation environment is the requirement for the end product to be represented by not just one but several components. To this end it is critical that production schedules are managed efficiently so that component batches are made within defined time-frames and prepared meal and chilled component production and assembly schedules correspond. In this way the finished tray set will be represented by a series of products reflecting similar life spans and production cycles. For example, where the prepared meal comprises several components in its own right, e.g. meat, potatoes, vegetables and gravy, the production schedule should accommodate the production of all four within the same time-frame so that at the point of assembly the end product is not represented by mixed life cycle components.

Where aspects of the tray set comprise a frozen product, the defrost schedules need to operate in the same way as the production ones. This type of system can be complicated, which is why the premises ergonomics need to be considered at the product development stage, to ensure that the technical requirement to assemble tray sets with corresponding life parameters over a range of products is possible.

Overall, the process flow, premises design and operational transfer of products requirements are the factors that are going to govern how well the production protocols stack up against manufacturing SOPs. The critical aspects are a management culture that demands it and an operational team that works hard to accommodate any premises design shortcomings to achieve the satisfactory establishment of the technical requirements.

Assembly protocols

The aspects of product assembly protocol that need to be considered in these environments vary significantly between individual product assembly, such as prepared meals and salad components and the assembly of what constitutes the end product, which may be anything from a tray set to a deli bag to a snack box.

Defining clearly both aspects of assembly assists us in defining the hazards associated with both the safety and the quality issues that need to be addressed by the establishment of appropriate systems management. Invariably the component assembly protocols will need to be documented in the same way as the bulk production records, with raw materials making up aspects of the composition in some cases and in other scenarios cooked components doing the same. The assembly paperwork will need to document the following:

- Batch code of raw materials to be utilised.
- Use by time of raw materials to be utilised.
- Name of component.
- Name of end product.
- Assembly area.
- Assembly start time.
- Assembly finish time.
- Size of batch.
- Random temperature checks throughout assembly process.
- Finished temperature.
- Despatch temperature to chilled or frozen storage.

Another term for this type of assembly detail in the industry is 'portioning' and this is used to denote the operational differences between tray set-type assembly and component product assembly. In manufacturing terms there is no difference and the requirements for documenting both processes would be the same. Figures 5.12 and 5.13 illustrate the type of assembly information required for both a ready-to-eat salad product and a ready meal style entrée.

In order for the portioning aspect of product assembly to operate effectively and to manufacturing standards, it is necessary to ensure that the production schedules and assembly schedules operate in tandem, so that food is being manufactured and assembled straight away to a defined schedule and not manufactured and then left for extended periods of time before being assembled.

This seems an obvious statement but in my experience these kinds of protocol differentials are the most common distinctions between catering and manufactured standard products. It is easy to understand how the above situation can happen in a catering environment, when many of the products

BULK PRODUCTION RECORD

DISH...................NICOISE SALAD

% ×1	PRODUCT	USE BY	BATCH CODE	WEIGHT	RAW TEMP.	BATCH	COOKED TEMPERATURE	TIME	BLAST TEMPERATURE	TIME
	Green beans			50 g			CCP1		CCP2	
	Olives black			3 × no.						
	Cherry tomato			3 × no.						

BATCH SIZE

BATCH CODE

METHOD

1. Decant all items in low risk cold room
2. Blanch beans off in boiling hot water, transfer over to high risk
3. Transfer to blast chiller CCP5 and chill to <5°C within 30 mins
4. Lay out bowls and place beans in bowl followed by toms, olives
5. Lid and transfer to assembly
6.
7.
8.
9.
10.
11.
12.
13.
14.
15.

DATE MADE

PRODUCE BY

CHECKED BY

DATE

Temperatures	
Holding	Chilled

USE CAPITAL LETTERS ONLY

Figure 5.12 Bulk production template for salad production.

BULK PRODUCTION RECORD

PRODUCT.................Chix veg pot   main meal   Category Rotation 1.

%	PRODUCT	USE BY	BATCH CODE	WEIGHT	RAW TEMP.	BATCH	FOOD ITEM	START TIME	COOK TEMP.	FINISH TIME
	Chix breast			100 × no.		×1	Chix			
	Potatoes diced			3 kg		×1	Pots			
	Carrots			1 kg		×1	Veg			
	Beans			1 kg		×1				
	Asparagus			1 kg		×1	Sauce			
	Tom purée			2 × 800 gm		×1				
	Tom chopped			1×		×1				
	Onion 6 mm			1 kg		×1				
	Mixed herbs			50 g		×1				

BATCH SIZE: 100

BATCH CODE:

PRODUCTION AREA:

METHOD

1. Verify clean-down activity as documented on schedule for production area and equipment to ensure allergen regimes have been applied   CCP1
2. Decant and weigh out recipe
   Document confirmed usage CCP2
3. Cover and transfer into production area   CCP3
4. Blanch vegetables in boiling water   CCP4
5. Steam potatoes in oven   CCP4
6. Sweat off onions in blast pan 1, add tomatoes, herbs, tomato purée and add water   CCP4
7. Steam chicken breasts in oven   CCP4
   Oven probe with probe 2 > 75°C and document
8. Transfer all cooled products into high risk and decant into high risk containers and document
9. Transfer into blast chiller CCP5 and chill to < 5°C within 90 minutes
10. Transfer into high risk pan
11. Lay out 'code light green' foils and assemble, lid and label
    Transfer into chilled storage

SPECIAL NOTES

Ensure all high risk are wearing code black PPE throughout all

DATE MADE:

PRODUCE BY:

CHECKED BY:

USE BY:

Temperatures		ASSEMBLY		
Holding Random 5	<5°C LIMIT	10 MINS	20 MINS	FINISHED TEMP.

USE CAPITAL LETTERS ONLY

Figure 5.13 Bulk production template for chicken meal production.

being processed are being utilised as components in a variety of different end products. Vegetables are a great example as they may form a component part of a variety of different finished meals. It is essential, however, that the same batches of processed raw materials are being assembled within the same time-frame. If end product specifications are in place, demonstrating product and process-specific HACCP and process flow, then the opportunity for this kind of catering standard assembly is unlikely to occur.

Here we see then how each process and protocol in turn has to be established in tandem to achieve the effective end result. It is essential also that the assembly systems are set up in unison with the production throughput to ensure that the product traceability is not compromised despite the multi-component nature of the finished product.

In terms of the protocols that need to be established to underwrite the effectiveness of the end tray set assembly, the following information will need to be documented on the tray set assembly paperwork for every component including ambient goods and condiments:

- Product name/code.
- Product use by/best before.
- Product batch code.
- Flight number/s.
- Number in tray set run.

This information should then be traceable to every component assembly detail, production record and raw material goods receipt. Here at last we can see how an amalgamated systems approach renders the component-by-component traceability application a relatively easy task. Figure 5.14 illustrates the type of assembly documentation required to document the finished tray set components.

The term 'tray set' can be applied in the broader sense to any airline food product that comprises an amalgam of components packaged together to represent the meal. Other contemporary product lines that may fall into this arena could be snack or deli bag offers as well as snack and meal boxes.

In terms then of the premium cabin components that may be bulk loaded and selectively served from a menu delivery system on board, the traceability and assembly documentation can be reflected in single unit not multicomponent format. Whilst the same information needs to be documented, the assembly paperwork merely comprises a list of component detail. Figure 5.15 illustrates the type of paperwork that could be utilised to this effect. This type of paperwork can also be used to document other single unit components to be loaded, such as beverages, savoury snacks and ice creams.

DATE	FLT NO		ASSEMBLY TRAY SET-UP ROTATION 1		
MEAL	FOOD ITEM	TEMP	BATCH CODE	No TRAY SETS	USE BY DATE	SIGN
B/FAST	BAKED FRUITS					
ECO	MUFFIN	AMB				
	YOGHURT	AMB				
	WHOLE MILK JIGGER	AMB				
	MEAL 1 B/FAST					
M/MEAL	NICOISE SALAD					
ECO	DRESSING	AMB				
	FRUIT BAR	AMB				
	WHOLE MILK	AMB				
	RAISINS	AMB				
	NAAN BREAD	AMB				
	MEAL 1 CHIX CURRY					
B/FAST	FRUIT SALAD					
ECO	SOYA MILK	AMB				
	PROMOVEL	AMB				
	FRUIT BAR	AMB				
	MUFFIN	AMB				
	MEAL 1 VEGAN					
M/MEAL	SAMOSA					
ECO	SOYA MILK	AMB				
	RAISINS	AMB				
	NAAN BREAD	AMB				
	BAR	AMB				
	MEAL 1 VEG CURRY					

Figure 5.14 Tray set assembly documentation.

DATE			FLT NO	BULK ASSEMBLY ROTATION 4		
MEAL	FOOD ITEM	NO	TEMP	BATCH CODE	USE BY DATE	SIGN
B/FAST	POACHED EGG	x6				
BIZ	BEANS	x6				
	BACON	x6				
	SAUSAGE	x6				
	ROLLS	x6				
	BUTTER	x6				
	JUICES	x6				
	B/F BREAD	x6				
M/MEAL	CHIX TUSCAN	x6				
BIZ	POTATOES	x6				
	VEGETABLES	x6				
	SAUCE	x6				
	BREADS	x6				
	BUTTER	x6				
	SALAD	x6				
	DRESSING	x6				
	WATER	x6				
	DESSERT	x6				

Figure 5.15 Premium cabin bulk loading paperwork.

I am always staggered by the industry's resistance to either document the traceability detail of the end tray set content or the component detail in an effective manner, to ensure that all components can be traced, tray by tray, flight by flight. As I have said before, the difficulties in achieving this level of traceability are not necessarily inherent, if manufacturing standard procedures are adopted. The unique difficulties faced result from the multicomponent nature of the end product, which needs merely to be viewed in the same way as a manufacturer producing an end product comprising multiple raw materials which are actually all finished products, e.g. a chicken mayonnaise sandwich. In this case the manufacturer would document the production detail of the chicken, the mayonnaise, the bread and the spread into an assembly document that represents the finished sandwich. In the case of a tray set, the component list may well be longer but the method of traceability and the principle remain the same.

Labelling and shelf-life attribution

In the same way as the durability assembly protocols fell into two distinct camps throughout the process, so too do the durability coding issues. Attributing a life to manufactured components in isolation is one issue; the added difficulties inherent in attributing a corresponding life to the completed tray set is another.

In my experience, most life attribution of airline catered products when the products are made in-house is based on the ancient mass catering directive of the 72-hour rule. It is as it sounds a non-verified, non-scientific blanket approach based more on operational convenience than sound, impartial scientific data. As in any process of life attribution the specific detail of the potential life cycles of the products in question will need to be given full consideration before an adequate testing regime can be devised and implemented.

Shelf-life testing in a manufacturing environment concerns verifying every aspect of the microbiological product safety attributes, following production under parallel environmental conditions, with consideration given to the likely life cycles of the products in question. This process is always a precursor to the product being deemed acceptable for launch.

Unfortunately the culture in aviation catering is to apply any aspect or combination of aspects of the 72-hour rule until a food poisoning complaint or random microsample identifies a problem. Unless a manufacturer's approach is adopted to life attributes at the product research and development stage, with product suitability for purpose having as much to do with microbiological issues as perceived quality ones, then accurate durability coding of airline catered products will pose a major problem.

It is not acceptable to use verification microtesting of products to endorse one's assumptions about non-verified shelf-life, as inevitably verification microtesting on finished products usually occurs within the first few hours of life, directly postproduction and potentially a long time before despatch to the aircraft. In this way testing is only carried out once in isolation and not, as in shelf-life testing, during several stages of life over an extended period and under variable conditions designed to replicate the likely life cycle of the products.

In consideration of the above, the variable applications of each and every product will need to be considered before a decision can be made over durability requirements and an adequate testing regime suggested. The considerations that need to be made may include any one of a number of the following unique life pattern possibilities affecting airline food products:

- Possible breaks or potential for extended duration outside the chill chain.
- Return or back catering requirement.
- Capacity for reheat products not to achieve optimum temperatures during in-flight reheating.
- Existing or predetermined life attributes of collective product raw materials.
- In-flight storage or handling abuses.
- Complete tray set life attribute requirement.

Once verified, the product life parameters can be documented on the end product specification and the product's profile and suitability for each application for which it is to be considered, can be assessed.

At this point it can clearly be seen how the airline caterer's ability to achieve an accurate and acceptable level of manufacturing standard durability testing, is made easier by the integration of manufacturing systems that are likely to deliver the required level of product integrity.

In terms of the durability coding of the end tray set, various considerations will also have to be made. These will focus on how many components are required to meet the 'use by' on the tray set or in the compilation container and how many possess no chilled life considerations. Assuming that at least two or more of the components are governed by chilled life parameters, then the durability date marking of the tray will be dictated by the component bearing the shortest life.

I am aware that in many airline applications, specific life codes are not directly applied to individual components; instead a rather elementary system of colour-coded dots is applied, with each colour representing a day of the week when the product was made. In terms of traceability to a specific batch, this system is completely ineffectual unless no more than one batch of any given product is going to be made in any given 24-hour period. Knowing the typical types of volumes involved, the 'single batch in a day' scenario is unlikely and therefore specific batch traceability is not possible.

Along the same lines as the coloured dots, the durability coding may take the format of a date applied to a bulk pack or container containing a large number of identically dated components. This system is fine so long as all of the components are utilised together and left-overs are not dispensed into other containers containing the same products but bearing a different durability code and batch code.

Having established the parameters for durability-type labelling, our focus needs now to turn to the wider labelling issues that ultimately impact on the airline product. Once more we are faced with a dual aspect approach: one which governs component labelling requirements and the other which governs the end product tray set or compilation products. Arguably, if the ultimate service environment of the food is classified as restaurant-style or food service, then neither the tray set nor the individual components need to be individually labelled unless special dietary claims are being made (see Chapter 13). However, all product attribute information needs to be held on a central specification which represents both individual components and, where necessary, the complete tray set or compilation meal product. In this way the spec acts as the product label and can be called upon to verify the product attribute information effectively.

Despatch protocols

At the point of despatch, we reach the final process step that separates airline catered products from their ultimate service environment. At this point the processes need to be devised in order to effectively verify the success of all previous process steps. The verification processes will be an amalgam of both quality and food safety attributes and will need to bear out all of the information represented on the specification. Certain logistical checks will need to be made also in terms of ensuring that the load scales have been met and the inventory for the flight has been completely satisfied.

Verification checks will all be recorded on accompanying documentation, which will require the following to be checked and countersigned by the dispatcher and/or loading supervisor:

- Product name.
- Batch code and durability code within limits.
- Packaging to spec and intact.
- Labelling to spec and intact.
- Visual aspects of product meet specification.
- Quality parameters documented in spec are satisfied.
- All components are in place.
- Temperature parameters are satisfied.
- Load scales met.
- Any deviations from spec recorded and reported.
- Catering load security protocols verified.

Without all of these vital checks being carried out and documented at the point of despatch, the line between purveyor (airline caterer) and vendor (airline) becomes blurred and any subsequent problems that arise cannot be traced to any particular point in the supply chain. If deviations from spec are noted and recorded at point of despatch, it makes the source of the process failure far easier to pinpoint. To the same extent, if all despatch documentation is in place and verified as to spec, then any subsequent problems will be traced to issues that may have arisen during transit or in-flight preparation and service.

It is essential that the airline caterer pays particular attention to the despatch protocols that govern the point of release from their premises to the aircraft. The shift of liability at this point from the purveyor to the vendor has to be effectively documented and for logistical and operational reasons any deviation from appropriate specification or load shortfalls must be reported. Temperature and quality checks are also critical at this point, to ensure that at the point of handover all is as it should be.

Verification microbiology and product recall

The benefit of having all product attributes documented on an end product specification is that the microbiological critical limits are documented also. Having defined the parameters for each product, in a product and process-specific fashion on the spec, it is then for both the caterer and vendor to take the necessary steps to verify that the safety parameters are being consistently met.

Microbiological food standards should meet those prescribed by the food industry as akin to those acceptable in a manufactured environment. Invariably these standards will differ internationally, but where best practice becomes the guide then the bar must be set as high as possible, not as low as acceptable. The basis of any verification testing regime has to be formulated on the assumption that all aviation food products have an extended life capacity throughout the supply chain, and not based on the assumption that product is of the food service variety and as such is engaged in a cook and serve application.

In the manufacturing sector the emphasis is always on the desire to develop systems that result in a 'right first time' approach as opposed to relying on verification microbiology to identify a process lapse. With a full and given knowledge of the life cycles of airline food products, it is more helpful and less costly to verify the shelf-life application of the products in advance. Clear indicators of the product's capacity to achieve extended life under life cycle conditions will then assist in determining the likely schedule of verification microbiology that needs to be undertaken.

To adopt a blanket verification microbiological approach is always a mistake and each schedule of testing needs to inaugurate a system of testing of products during different stages of life, not just at the point of production as is currently most common. It is for the airlines themselves to verify also the in-flight safety and integrity of their products by adopting a sampling regime for products taken from on board, in the same way as retailers do. The problem with the sampling of most airline food products is that if they have not been produced and specified in the same way as manufactured food products then the ingredient quality and traceability will vary, as will the production and process methods. To this end, the requirement to verify the end product more often than traditionally manufactured goods is all the more vital as it will have been derived from an inconsistent raw material supply chain and process flows.

As with all verification microbiology procedures, whether they are for raw material or finished products, it is crucial that a testing schedule is adopted and documented on the product specification, and adhered to. Let us not forget that the main purpose of this type of testing is to verify scientifically the success of the production and process flows, as well as the integrity of the products in terms of their ability to meet ready-to-eat food standards. Results that fall below defined safety parameters cannot be ignored and must be investigated to ensure that all systems are in place and functioning efficiently. At the very least, the ability of the caterer and the airline to trace the likely cause of the problem has to be displayed and if it is not, this has to be taken as an indicator that all is not as it should be. Any delay in identifying the root cause of a microbiological deficiency can result in the problem persisting, ultimately culminating in a food poisoning outbreak.

It is critical that international carriers who are serviced by a variety of outstations establish their microbiological reporting expectations at the point of catering contracts being negotiated. Consistent deviations from critical limits should automatically result in a suspension of contracts pending investigation by all parties concerned.

I have to admit that my personal experience of dealing with the airlines in this regard has been that there is a wholly disinterested approach to verification microbiological regimes, particularly where an extensive schedule may have an impact on product cost. Even when I have known problems to be identified, tolerance levels of certain micro-organisms have been variable to say the least. It is essential that at the point of audit microbiological verification schedules are defined and documented and then reviewed after an ongoing assessment of results.

In manufacturing environments, the capacity to instigate a product recall is a little easier than in the aviation sector. By the time a problem has been identified, the products can have been loaded on a variety of flights destined to travel to the four corners of the globe. With the non-product and component-specific batch-coded nature of the products, they may prove impossible to trace unless the production, assembly and despatch protocols have been followed under manufacturing guidelines.

It is interesting to note that in the IFCA/IFSA World Food Safety Guidelines[62] there is a manufacturing standard procedure for product recall and product hold. However, the rest of the prescribed systems that would need to be established in order to ensure that a successful recall was possible, are not described. Where certain contributory aspects are mentioned as a process, such as in the date coding section, they remain totally ineffectual. It is not possible to effectively recall without batch-coded production processes and traceability by component, and without a consistent and verified raw material and finished product supply chain.

The emphasis has to be on manufacturing SOPs that allow airline caterers the best opportunity to 'get it right first time' or at the very least ensure that in the event of a problem arising, the products involved are fully traceable, flight by flight, component by component, tray set by tray set.

In-flight documentation

One of the most fascinating aspects of food safety systems management is that all processes move consecutively throughout every stage until, at the point of despatch, the transfer of responsibility from one supplier to the next and potentially from one supplier to the end purveyor, becomes apparent. At this point the entire system of quality and safety management begins again.

So too, then, should this be the natural turn of events in the aviation sector. At the point at which goods are placed in transit between the catering provider and the aircraft, postdespatch, the procedures on board from goods receipt to service should also follow the same documented protocols. Whilst the variations between producer and purveyor will be focused on the production versus service aspects, the requirement to document all on-board procedures is as essential. Prerequisite systems need also to be established in flight, in the same way as every food service and production environment, and these should include everything from adequate chilled storage facilities for high risk foods to appropriate and separate hand-wash areas for crew. In the absence of such rudimentary prerequisite issues being satisfied, it is unlikely that food safety systems management on board has a chance of succeeding.

So let us look at each of the in-flight systems that need to be established.

Goods receipt

All catering supplies arriving at the aircraft require the following safety and product integrity verification and documentation:

- Temperature checks.
- Visual checks.
- Product to spec.
- Packaging intact.
- Load scale meets requisition.
- Durability coding apparent and in date.
- Vehicle hygiene and temperature.
- Labelling intact and to spec.
- Foreign bodies/mechanical damage/chemical contamination.
- Security sealed.

Product storage

All catering stores and supplies arriving at the aircraft require the following safety information to be verified and documented:

- Product name.
- Durability code.
- Chilled or ambient.

- Storage unit – mechanically chilled/dry ice/frozen dry ice/ambient/oven racks.
- Time of storage.
- Temperature at storage.

Product service

All catering stores and supplies require the following safety information to be verified and documented at the point of service:

- Product name.
- Product temperature – core temperature if reheated, service temperature if chilled.
- Time of start of service (local).
- Finish time of end of service (local).
- All ingredients and products to spec and visually acceptable.
- Spec and visual non-compliances documented.
- All temperature parameters achievable.

I am always astounded by the reaction I receive when I suggest that the above in-flight systems need to be established on every flight in order that the total integrity of the extended supply chain can be verified. The reaction is always that the crew cannot possibly be expected to carry out such checks in the time-frames provided and in any case are not suitably trained to do so. Without these protocols being established and undertaken by the crew in the same way that they are required to be undertaken in food service establishments on the ground, there is no way of identifying the potential causes of a food safety crisis. In the same way, if the crew are not documenting service temperatures then any mechanical failure of the equipment on board is likely to go undetected and unreported. Chapter 9 documents that the specific role the crew have to play in food safety on board, by their impact on the success of so many aspects of food safety systems management, is a crucial one.

So we come to the end of the chapter and all that it has entailed to take the aviation catering sector into the product safety driven world of food manufacturing SOPs. The critical aspects are that everything that is done is interconnected and the drive has to be total product safety, quality and integrity. Effective and efficient replication of products is also a critical issue, as is component traceability. Unless all systems are rolled out in tandem it will be extremely difficult to make them work. Any systems overhaul has to be all encompassing and whilst every individual catering operation will require some degree of systems modification, the general rules will apply in all cases.

I hope that throughout this chapter I have made an effective contribution to those seeking to establish a greater degree of product safety compliance, which ultimately results in a greater level of producer confidence in the systems that underwrite the products and that go to make up a greater proportion of what constitutes airline catering concepts.

6 Liability issues – protecting the airline brand

The focus of this chapter is to look at the issues that drive airline liability and to investigate how the unique structuring of air law with regards to airline liability influences the establishment of reactive approaches to aviation food safety.

We have already looked in Chapter 2 at how the consumer view drives aviation safety initiatives and in that chapter we set the tone for the sharing of liability issues between major airlines and major food brands. What remains largely undetected in the psyche of the flying public, however, is the limitation to airlines' liability provided in the conventions that govern international air law and which have the capacity to supersede all other nationally directed legal conventions.

We will look at the basics contained within these conventions in this chapter as well as at how the safe provision of food on board becomes a liability issue under the terms of these conventions. We look in detail in Chapter 7 at the liability sharing arrangements between catering providers and their airline customers, and at how food safety liability is managed through the catering contracts established between airlines and their catering partners. Meanwhile, in this chapter we focus more closely on the true essence of airline brand liability and what impact safety deficiencies in the food service arena may have on the airline itself.

While I am not a lawyer, I have had a fair degree of experience in working with claimants and their legal representatives in cases where food or food provision on board aircraft has resulted in personal injury to the passenger. These cases have involved not only food poisoning but in one of the most recent that I was associated with and which implicated a major international carrier, the physical injuries sustained by the plaintiff were life threatening and were caused by the passenger consuming a foreign body contained within the meal itself. This particular case is discussed in detail later, when we examine how the airline defence collapsed following an assessment of the HACCP plan which clearly showed that many of the attributable hazards associated with physical contaminates affecting food safety had neither been considered nor controlled by the catering provider or the airline itself.

Aviation liability

Establishment of the Warsaw Convention/Montreal Convention

Shortly after World War I, the lack of international uniformity in the field of private air law became a matter of major concern. There was a chaotic set of rules and regulations enforceable on a national, not international, basis. It was recognised by the international community that in the absence of some degree of international uniformity, it would be difficult to develop the entire business of civil aviation.

The first International Air Law Conference was convened in Paris in 1925, where a draft was prepared for consideration before a diplomatic conference in Warsaw, 'Unification of Certain Rules Relating to International Air Carriage'. The Warsaw Convention was firmly established in 1929 and provided for the principle of air carrier's liability for damage caused to passengers, baggage and goods and also for damage caused by delay.

The conference was attended by representatives from 43 states, who all examined a number of problems and resolved to constitute a committee of legal experts, known as Comité International Technique d'Experts Juridiques Aériens (CITEJA). The purpose of CITEJA was to undertake a detailed study of the problem relating to Private Air Law and to prepare a draft convention for consideration by the subsequent international diplomatic conference. The experts involved in preparation of the draft, dealt with the problem of liability of the 'air carrier' (airline) engaged in the international carriage of passengers and goods. A draft text was submitted for consideration at the Second International Conference in Private Air Law, which met in Warsaw in 1929. The diplomatic conference approved on 12 October 1929 the Convention for the Unification of Certain Rules Relating to International Carriage by Air, which came to be known subsequently as the Warsaw Convention[63].

The Convention came into force on 13 February 1933, after it had been ratified by 33 states, and as of 30 June 1998 it had been ratified by a total of 144. The primary objective of the Convention was to establish uniform rules governing the rights and responsibilities of both the air carrier and passengers as well as consignors and consignees of goods in countries that were party to the Convention. It also set limits to the liability incurred by air carriers for passenger injury, death or loss or damage to goods or baggage carried. It is interesting to note that at this first stage of ratification, the USA was not represented at the conference and continued to raise issues over liability limits even after amendments were facilitated in 1955 by The Hague Protocol[64].

It was the criticism of the Warsaw Convention 1929 that led to various amendments being made to it from time to time by way of protocols and bilateral agreements, until eventually it was superseded by the Montreal Convention 1999. These amendments included the following.

The Hague Protocol 1955

This protocol proved only a patch applied to amend the Warsaw Convention. It redefined international carriage and specifically excluded mail and postal packages. It imposed obligations on the carrier to issue tickets and baggage checks. The most significant change related to the revision of the upward limit of liability recoverable for the death or injury of a passenger. Further, it also made provision for the limit of liability to extend to the servants and/or agents of the carrier.

The Guadalajara Convention 1961[65]

This convention was an adjunct as opposed to an amendment to the Warsaw Convention, necessary with the expansion of charter air travel after World War II. It drew up specific rules appertaining to the charter air carriers not considered by Warsaw and came into force on 1 May 1964 after ratification by five states. Under this convention, travel agents and tour operators are held liable, especially those operating charter flights.

The Montreal Agreement 1966[66]

This was a bilateral agreement between the USA and international air carriers operating from, to or via the USA but only in so far as US citizens are concerned. It ushered in a new concept of air carrier's liability in the field of international air transport law, by changing the concept of fault liability to risk liability. It is not a protocol attached to the Warsaw Convention but a private agreement between the air carriers and the US Civil Aeronautics Board. The claimant had no longer to prove that the carrier was at fault, and the maximum liability of the carrier was fixed at US$75 000. The wording on passenger tickets was also changed from reference being made to 'wounding or bodily injury,' to 'personal injury'.

The Guatemala City Protocol 1971[67]

The Montreal Agreement 1966 manifested a clear discrimination in favour of US citizens and against other passengers using air carriers internationally. Proposals were made which cut across all aspects of the Warsaw Convention, not just on the issue of liability limitations. The essential elements of these proposals were incorporated into the Guatemala City Protocol 1971 and it was signed by 21 nations, including the USA, on 8 March 1971. As yet it has not come into force due to the fact that only 11 of the 21 states ratified it.

The Four Montreal Protocols 1975[68]

These additional four protocols substituted the Special Drawing Rights (SDRs), as defined by the International Monetary Fund (IMF), for the Poincare franc to be used as the monetary unit in which liability limitations are expressed. The additional protocol No 4 provided that provisions of the Montreal Protocol 3 should prevail over the Guatemala Protocol and established the risk as opposed to fault liability; the air carrier could no longer therefore invoke the defence of due diligence afforded them under Article 20 of the Warsaw Convention.

The Inter Carrier Agreement 1977[69]

On 30 October 1995, IATA adopted at its 51st Annual General Meeting in Kuala Lumpur an Inter Carrier Agreement (ICA) which provided for a single, universally applicable scheme documenting specified limits of liability and for the recovery of actual proven damages in accordance with the law of the domicile of the passenger. IATA also suggested that the due diligence defence allowed for under Article 20(1) of the Warsaw Convention should be waived.

These amendments, subsequent to the ratification of the original 1929 agreement, were conducted and facilitated by the various bilateral agreements and protocols listed above and were designed to primarily revise the liability limitations of the carrier. These limits had been fixed under the Warsaw Convention with no provision for a mechanism to revise the limits without amending the convention in its entirety; however, due to inflation, various amendments simply had to be made.

During the early stages of the development of civil aviation, the main advantage to the contracting parties was to minimise their risk by specifying their limits of liability. The codification of Private International Air Law was advantageous because countries could take advantage by incorporating the provisions of the Warsaw Convention into their international laws. The cumulative effect of the Convention was the development of a code of commercial air law for the civil aviation industry. However, as we have already seen, the USA, a major player on the international aviation stage, had always been unhappy with the limited liability clauses and was the driving force behind many of the subsequent amendments.

Liability under the Warsaw Convention

The principles of liability under the Warsaw regime could be summarised as follows[70]:

- Substantive principles
  - risk liability
  - fault liability
  - exculpating factors
  - mitigating factors.
- Procedural principles
  - notice of liability limitations
  - other terms of contract
  - delivery of documents including air ticket and airway bill.
- Descendant claimants
  - defendant air carrier agents or servants of the carrier.

We have already seen that the original objectives of the Warsaw Convention were to achieve international uniformity in air carriers' liability and documentation of air transportation. The convention introduced a uniform system of strict but limited liability for international carriage of passengers, baggage and cargo. Uniformity was desirable to facilitate transactions across borders, languages and cultures and in order to avoid inevitable problems with conflict of national law. Liability was made strict to avoid the problems of proving fault and to compensate for the imposition of limits of claims. It was argued that it was necessary to limit claims at the time, for fear that in the early days of air transportation a single disaster could bankrupt an airline and insurance premiums would prove cost prohibitive both to the carrier and the passengers to whom the insurance costs would have to be devolved through an increase in the ticket price.

The primary problems to beset the Warsaw Convention can be summarised by the following.

Limits to liability

Severely limits the ability of a passenger or the surviving descendants to recover damages fully resulting from injury or death on board an international flight.

Domestic versus international air disasters in the USA

Huge disparity between damages awarded in international and national crashes in the USA as there is no liability capping of damages under US law for domestic air crashes. Conversely, international disasters implicating US carriers are subject to the limited liability provisions of the Warsaw Convention.

Jurisdiction

The Warsaw Convention not only curtails the amount and type of damages recoverable but also the forum for filing the claims against the air carrier. The jurisdiction under which an action for damages must be brought by the claimant is:

- where the carrier has its principal place of business, or
- where it has an establishment, or
- where the contract of air carriages was made, or
- before the court having jurisdiction at the place of the destination.

Difficulty of amendment

The Convention is a treaty and can only be amended by its signatories in accordance with the procedures provided by the Convention itself. Even the airlines cannot amend it.

Difficulty in drafting and interpretation

Ambiguity in requirements for drafting of air travel documents. Huge diversity in interpretation.

Wilful misconduct

Varying interpretations of the term wilful misconduct as it applies to air carriers' liability under Article 25 of the Warsaw Convention.

Delay

Under the provisions of Article 19 of the Warsaw Convention, the carrier is liable for damages caused by the delay in the carriage of passengers and baggage. The liability of the carrier is the same for delay as it is for death and injury.

Advances in aviation technology, safety and insurance have made many of these arguments void in the 21st century and whilst subsequent bilateral agreements and protocols have served to attempt to bring the Convention up to date, particularly in terms of liability capping, a complete replacement for the Warsaw Convention to define a new regime of private International Air Law seemed like the best solution. The demand for a total overhaul of the system has been huge from passengers, governments and even the carriers themselves. The attempts made in the past have resulted in a chaotic amalgam and superimposition of international treaties and bilateral agreements that have all served to undermine and jeopardise the viability, certainty and uniformity of the global legal framework initially sought to be satisfied by the Warsaw Convention at its inception in 1929.

In May 1999, under the direction of ICAO, a diplomatic conference was convened with the sole purpose of replacing the Warsaw Convention of 1929. Every since its inception and acceptance as the treaty governing international air carrier liability in 1933, it has been a major source of consternation, criticism and debate. In view of the lack of progress with the ratification process of the Montreal Protocols 3 and 4, and at the risk of an outright breakdown of the Warsaw System, a fundamental reappraisal of the situation was necessary, by ICAO. A study group was formed to consider the adoption of a single legal instrument as the preferred solution and how such an instrument could be used to consolidate all the positive and useful elements of the Warsaw system, as well as modernise the regime comprehensively encompassing liability for passengers, baggage and cargo.

Liability under the Montreal Convention

The Montreal Convention 1999[71] is designed to replace the Warsaw Convention and all of its related instruments as well as to eliminate the requirement for this patchwork of regulation and private voluntary agreements. The most significant features of the new Convention include the following:

- The removal of all arbitrary limits in the receipt of compensation for death or injury caused to passengers.
- The imposition of strict liability on carriers for the first 100 000 special drawing rights (SDRs) equivalent to US$138 000 of proven damages in the event of passenger death or injury.
- An expansion of the basis for jurisdiction for claims relating to passenger death or injury to permit states in the passenger's place of domicile or permanent residence subject to certain conditions.
- It presumes all key benefits achieved for the air cargo industry by the Montreal protocol No. 4 enforceable on 14 June 1998.
- It provides an in-built scheme and mechanism of review of liability limits as well as calculates the conversion of monetary units.
- The Convention is applicable to commercial international air carriage between two states, out of which at least one should be a contracting state.

The Montreal Convention represents therefore a vast improvement on the liability regime established under Warsaw and its related instruments. It clarifies the duties and obligations of carriers including those engaged in code sharing agreements, with respect to cargo, passenger and baggage documentation. Whilst it has been seen to be a great improvement and a modernisation of the outdated Warsaw system, many commentators have claimed that it still fails to redress many key issues of international air law liability that have been witnessed throughout the history of international air carrier liability litigation.

The Montreal Convention has been viewed by many as not being able to fully address the following broad categorisation of issues, as follows.

Burden of proof

The question of who will have the burden of proving the air carrier's negligence, necessary to claim full compensation without limitation, has not been formalised under the new convention.

Burden of costs

The question of who shall bear the burden of costs due to the increased liability of the air carrier has yet to be seen and is not determined by the Convention.

The early settlement of claims

The Warsaw regime had been under constant and genuine criticism for delays in the settlement of claims, and the same provision has gone unresolved under Montreal 1999.

Consumer protection

Whilst recognising 'the importance of ensuring protection of the interests of consumers in international carriage by air and the need for equitable compensation based on the principle of restitution', Montreal Convention 1999[71], the new Convention does not provide for any representation for consumers via ICAO or IATA deliberations. The consumers therefore do not have a body or association to represent their interests in the same way that the aviation industry has.

So it would seem that despite over three-quarters of a century of debate and attempted amendment, the long awaited, total revision of the Warsaw Convention has not been completely successful in addressing all the issues of air carriers' liability thrown up by real life cases. The ratification of the new Convention is in its early stages. Only time and legal history will provide us with the answers as to how successful it will be in resolving the aviation liability issues that have permeated the industry for so long.

Food safety as an aviation liability issue

Having established the protocols by which the liability of air carriers is defined, it is now time to put all of the legal ramifications of International Air Law into the context of what it means in terms of the provision of food on board aircraft.

Under the provisions of the Warsaw Convention the liabilities associated with the provision of catering and catering supplies are referred to in Article 25 and have formed the precedent upon which many liability cases related to food poisoning or food product-related personal injury claims have been fought.

The principles upon which the concept of liability of the air carrier were founded under Warsaw are based upon the fault theory, which differs fundamentally from the risk theory associated with most national common law. Articles 17, 18 and 19 sought to create a uniform regime and a presumption against the carrier in the case of damage affecting passengers, baggage and goods and in the case of a delay. Article 17[72] states:

> 'The carrier is liable for damage sustained in the event of the death or wounding of a passenger, if the accident which caused the damage so sustained took place on board the aircraft or in the course of any of the operations of embarking or disembarking.'

Article 20(1) lays out the concept of due diligence that forms the basis of the fault as opposed to risk theory. The Paris invention of 1925 had approved the fact that if the carrier could establish that 'it had taken all reasonable and normal measures' then the carrier would not be held liable. So as we can see, under Warsaw the carrier had to prove due diligence to be relieved from liability. This principle has formed one of the base measures upon which aviation law has been amended under the Montreal Agreement 1999, which reverses the fault theory liability and due diligence defence to that of risk, but only with regard to claims arising out of death, wounding or other bodily injury of a passenger not relating to damage to baggage. Thus Montreal altered the liability regime with regard to passengers only.

Article 21 of the Warsaw Convention relates to the issue of contributory negligence and allows the carrier to be partially or wholly exonerated from liability if it can prove that the damage was caused or contributed to by the negligence of the claimant. Under the provisions of this Article the onus was on the carrier to prove that the damage was caused or contributed to by the injured person. The carrier must prove two fundamental elements for Article 21 to apply:

- Negligence of the passenger.
- The causal role of negligence in the resulting injury or damage that occurred.

Article 25 of the Warsaw Convention provides that the carrier is not entitled to avail itself of the other provisions that exclude or limit its liability, if the damage or injury is caused by intentional misconduct or fault by the carrier or any servant or agent of the carrier acting within the scope of its employment. This concept of wilful neglect is the one most often cited by claimants in cases associated with death or injuries to passengers through the consumption of unsafe food, because if it can be proven that the carrier or representative of the carrier, i.e. the airline caterer, has been negligent, it not only overturns the provisions of Article 20, due diligence, but it also overturns the convention precedent to limit the amount of compensation that can be paid to the injured party.

The historical outcome of many cases citing wilful neglect led to the adoption of the words 'intent to cause damage' in the Rome Convention of 1952, Article 12. The Convention provides that if the person who suffers damage proves that it was caused by a deliberate act or omission of the operator, done with intent to cause damage, the liability of the operator should be unlimited.

It is fascinating to witness how these precedents could be applied in cases of personal injury resulting from the consumption of unsafe food and drink on board, and how under Montreal, with the due diligence defence under air law now void, the liability issues surrounding the carrier in this regard are potentially immense. Traditionally under Warsaw, not only did the fault of the carrier have to be proven by the claimant, the ensuing compensation was capped. The nature of aviation catering provision is immensely detailed and complex, as we have seen, and historically the capacity of many claimants to win cases against the carrier in this area have been hampered by the inability of the claimants' legal representatives to disprove due diligence and to prove wilful neglect. The contractual nature of the relationships between air carriers and their catering representatives are entrenched in the concepts surrounding liability limitation under Warsaw and rely on a sharing of both liability and wilful neglect. It has therefore been in the interests of both contracting parties to ensure that in the event of a claim, the due diligence can be upheld.

The implications under Montreal of the disbanding of due diligence as a defence under international air law, leaves the air carrier and their catering partners having to disprove liability without the benefit of this fault precedent. Whilst the opportunity still remains to utilise this defence under some food safety legislation, historically international air law pervades in cases of personal injury aboard aircraft.

With the ratification of the Montreal Convention and the subsequent loss of the due diligence defence, the legal liability issues surrounding the provision of aviation catering are enormous. At this time there has never been a better reason to ensure that the provision of aviation catering is as safe and secure as possible, and with the inherent problems that we have witnessed throughout the book associated with outdated food safety management systems, potentially challenging times lie ahead. The 21st century trends towards a compensation culture mean that unless airlines and their catering providers sharpen up their defences a litigable minefield looms.

Case study

In September 2002 I was contacted by an American lawyer representing a gentleman who was wishing to pursue a personal injury claim against a major international airline. The gentleman in question had taken a flight from London to San Francisco and it was claimed that as a result of his consumption of the in-flight meal he suffered serious internal injuries.

It transpired that he had in fact swallowed a small plastic toothpick whilst consuming the meal, which consequently punctured his digestive tract and caused him to suffer serious internal bleeding, resulting in his near death.

It was difficult to imagine how he could have inadvertently consumed the toothpick until it became clear that he had undergone complex dental surgery several times during his life and as a result much of his mouth was desensitised.

The case was fought on the grounds of Article 17[72] of the Warsaw Convention as it applied then, which states:

> 'the carrier is liable for damage sustained in the event of the death or wounding of a passenger, or any other bodily injury suffered by the passenger, if the accident which caused the damage so sustained took place on board the aircraft or in the course of any of the operations of embarking or disembarking.'

Under the provision of Article 20(1) it was for us as the claimant's representatives to attempt to disable the air carrier's 'all necessary precautions defence', as it still stood. The nature of food safety legislation in the UK, whilst it allows for the same due diligence defence to be cited, was superseded in this instance by the provisions of international air law and as such the Warsaw Convention. However, the primary objective was the same. It was for us to provide enough evidence to support our claim that all necessary precautions had not been taken by either the air carrier or their catering representatives to assure product safety.

The contaminant was a physical one whose nature of risk would have been managed by prerequisite programmes had the environment been manufacturing. There were three possible sources: it may have been a used toothpick generated from refuse material collected by the airline caterer and processed at the catering unit; it may have been an unused toothpick which formed part of the cutlery pack on the tray set up itself; or it may have been deliberately and maliciously placed in the meal as part of an act of wilful neglect by the airlines' catering representatives.

The main focus of attention had to be to prove that the airline had not considered all of the attributable hazards associated with the risks posed to food safety by the handling and transit of the toothpick through the unit. A simple assessment of the HACCP laid bare several factors, which were key to the disabling of the airline's defence:

- HACCP was a bland generic statement which did not take account of the specific product and process hazards inherent in the production of this particular meal.
- No consideration had been given anywhere to the specific hazards associated with the receipt and throughput of food packaging or food contact material.
- The company held no technical specification on the toothpick itself.
- 'Transfer' as a process step associated with the potential for physical contamination of product was not documented anywhere.
- The product specification did not document any of the associated food safety hazards devolved from the main plan as work instructions to the operatives charged with the responsibility of monitoring and controlling the product and process-specific hazards.
- The intrinsic product characteristics impacting on the particular food safety attributes of the product had not been considered.
- There was no documented procedure for the transit of the cutlery packs containing the toothpicks into the working area.

As a result of all of the above factors being highlighted, the airline's defence of 'all necessary precautions' under Article 20(1) was thrown out by the pretrial judge and any attempt by the airline to claim contributory negligence of the plaintiff under Article 21 would presumably have been disproved by the details surrounding his dental condition.

It is interesting to note that despite a substantial financial settlement being reached, which presumably would have been met under the shared liability precedents existing between the airline and its catering partner, a recent look at the industry HACCP still reveals no reference to the very food safety considerations which, left unconsidered, resulted in the near death of a passenger and a substantive financial loss to the airline whose position was viewed to be indefensible.

It seems remarkable that even faced with a case as significant as the one I have just described, the actions of the industry to deal with the issues have remained wholly uninspiring. With the removal of Article 20(1) from the new international air law convention under the Montreal Convention 1999, the airlines and their catering providers will have to ensure that their systems are more robust than ever. The only conceivable way to facilitate this is to refrain from looking at food safety systems management from a caterers' perspective and to start to consider the whole process of aviation catering provision as a manufacturing process where all necessary risk management is done in the context of the specifics of the products and the processes.

Brand liability in the aircraft environment

In Chapter 2, 'Consumer Perceptions', we looked at the potential increases in food brand marriages in the aviation catering environment. Today, businesses and consumers are placing an ever increasing importance on brands. Brands give a product a unique identity, and successful brand affiliation can pave the way for increased consumer satisfaction and confidence.

In terms of the impact on food safety liability of brand marriages in the aviation catering environment, there are several key factors for food brand leaders to consider. The multicomponent nature of the in-flight food service product gives a huge capacity for a multitude of mainstream food brands to be exposed at once. It is essential then that branded product owners are assured of the integrity of all of the other components associated with their brands in-flight.

The capacity for a major international food brand to be implicated in an in-flight food safety scare may be immense. Purveyors of high-risk foods are particularly susceptible. A branded ice cream, for example, may be subject to all manner of temperature abuses during its life cycle on board. The normal food service parameters are intrinsically compromised by the inherent restrictions in the cabin environments. With the ever increasing trend towards buy on board service styles, it is even more essential that the safety and therefore liability issues surrounding major food brands are considered in the overall context of what it may mean to be implicated in a food safety crisis on board.

With the advent of buy on board has also come the requirement to present food and drink products for sale that have historically formed an integral aspect of the on-board service offer. The required level of product safety and legality information on the packaging of food product offered for sale on board will be the same as in any retail environment.

The concept of 'brands in hands' has never been more obvious than in the modern day on board food service offer. It is essential that food brand leaders understand all of the associated liability issues that pervade the aviation environment and that are unique to the legal precedents set by international air law. The prospect of a food brand leader not being able to offer up due diligence as a defence in a claim of food poisoning seems incomprehensible, but its potential application in the aviation environment is almost assured.

The obvious benefits of food brand marriages with the aviation sector must be viewed in the wider implications of what an association with aviation catering may mean in terms of liability. Far from brand marriage being the focus, it is for brand managers to consider brand damage first and foremost and how it may affect the sales potential of the product on the ground.

In 2002 my company was approached by a major airline to research and develop on their behalf a range of special diet products for first and business class cabins. The project was undertaken in collaboration with their catering partner and the current provider of their special meal products. As the conclusion to the project drew near, I respectfully enquired of the airline's representative as to how the attributable costs associated with the nutritional analysis verification testing were to be divided. He asked me what I meant and when I explained that we had to verify the menu development theoretical nutritional data scientifically to ensure that we had due diligence, he responded by asking me how I would feel about doing nothing.

The point of this story is to demonstrate the type of brand jeopardy the airlines are potentially prepared to put themselves in, perceivably to save costs. My company's brand was, and still is, nothing compared to theirs, but the attributable risks inherent in placing our products in the aviation environment were immense, had I allowed myself to be compromised.

The whole issue of airline brand liability is intrinsically linked to the precedents laid down in international air law, as we have seen. In terms of what this means to the production and supply of food products, the liability issues are unique to aviation like no other food service or retail environment. The precedents allow for all manner of charges outside of normal food safety and security legislation, and it is essential that anyone involved in the quality and safety assurance of food products in the aviation environment has a clear concept of aviation food safety liability issues and how these have the ability to affect their businesses.

As the climate for litigation grows year on year, the focus of attention will continue to be protecting the airline brand and all associated brand marriage opportunities to ensure that the financial benefits are not compromised by the risks inherent in aviation food manufacture and supply.

7   The airline catering supply chain

I believe that in order to achieve the broadest possible perspective on what aviation food safety really means, and in order to ensure a full understanding of its critical impact and unique connection with the wider issues of global health, bio terrorism, vector, water and food-borne disease and the economic devastation attributable to any perceived breakdown in aviation safety mechanisms, it is essential to focus in detail on every aspect of the aviation food chain.

In this chapter we unravel the different dynamics and modus operandi associated with the extremely complex nature of aviation food product outsourcing. It would be impossible to account for the finite detail in each case; however, in order for the aviation catering novice to gain an insight into how the most frequently utilised supply chain mechanisms potentially impact on food safety and integrity in this environment, we will focus in turn on some of the well established supply chain protocols.

Airline catering and airline caterers

The catering supply chain has historically been littered with misconceptions as to the precise nature of its functionality and the manner in which it meets the constantly changing demands of the industry. Whilst 30 years ago a greater percentage of all food and drink supplied to aircraft operations was subject to total processing and manufacture through the catering unit itself, much of the modern day operation sees a more logistical focus being applied to airline kitchens. Much of this has had to do with the burgeoning, eclectic nature of the requirement, with airlines attempting to create and deliver a satellite-style restaurant in the sky, particularly in premium cabins and on many long haul routes. At the other end of the spectrum the low cost and charter operators have opted for non-integral food service opportunities, to which the catering industry has had to respond appropriately. This has resulted in a perpetually evolving, inventive process of product outsourcing, when in-house manufacture has became either commercially unviable or logistically unfeasible.

To understand the full spectrum of supplier responsibilities embodied by the airline caterer function, it is helpful to break them down broadly into food and non-food groups (Table 7.1).

Table 7.1 Supply function spectrum of aviation caterers

Product	Food or non-food	Requirement
Hot meals TSU	Food	All flts
Cold meals TSU	Food	All flts
Cold boxed meals	Food	Some flts
Beverages	Food	All flts
Ambient snacks	Food	All flts
Alcohol	Food	All flts
Crew meals	Food	All flts
SPMLs	Food	Some flts
Ice	Food	All flts
Chilled snacks	Food	Some flts
Disposables	Non-food	All flts
Rotable equipment	Non-food	Some flts
Newspapers	Non-food	Some flts
Duty free goods	Food/non-food	Some flts

It is clear to see that the term 'catered' in the context of aviation supply is not confined to the food and beverage arenas. Whilst Table 7.1 is something of a generalisation of the overall requirement and by no means an exhaustive representation, it pretty much covers the broader categories of potential supply. It is true to say that on many domestic carriers the requirement for specific crew food or special diet meals may not be met by the airline caterer, but nonetheless if the catering supplier is servicing more than one airline, all of the above will have the potential to come into play at some point.

Defining an airline caterer

So what exactly constitutes an airline caterer? Well, in order to fully understand the term, one first has to look at other possible connotations of the same function. They include the following, all of which could be considered to have a connection with aircraft food supply but are not in the strictest sense of the term airline caterers:

- Food brokers.
- Catering logistics partners.
- Manufacturers of food products for the aviation environment.
- Buy on board food service providers.

The major distinction between those who merely contribute to the aviation catering supply chain and those who are truly classified as airline caterers, is the combined application of all of the above functions attributed directly, via contractual liability to the carrier themselves. The other types of catering supply relationships would not fulfil the same direct, contractual and indemnity obligation to the carrier. More often than not the carrier would sub-contract the other types of supplier relationships through the airline caterer, thereby predetermining the airline caterer as the central framework upon which the overall supply chain infrastructure is crafted.

I am concerned about making a careful distinction between catering relationships and the carrier themselves at this point, so as to establish the parameters by which the hazards associated with the product provision are attributed. Product safety liabilities ultimately lie with the entire supply chain but the assurance of those liabilities is essentially ring fenced, and therefore determined, by the particular dynamics of the carrier/airline caterer/airline catering product supplier relationships.

In the past, many large airlines operated their own catering facilities as either wholly or partly owned subsidiaries of the airline company. This was considered to be the best way of assuring the quality and safety of the supply chain, as the carrier airline had total control. In the WHO *Guide To Hygiene and Sanitation In Aviation*[73], James Bailey writes:

> *'Aircraft meals are supplied by kitchens that are:*
>
> *(1) under the direct control of the airline; or*
> *(2) staffed and controlled by a catering concessionaire but permanently supervised by the airline; or*
> *(3) owned by a catering concessionaire and only partially supervised by the airline.*
>
> *Category (1) is the most desirable arrangement but is not always feasible. Category (3) is not to be recommended except in instances where the food uplift is small and no other system is practicable.'*

In the same vein he goes on to say:

> *'Quite frequently an airline finds itself in the position of having no choice in the selection of premises from which food can be uplifted – for instance, when a particular caterer holds a monopoly. This is a very unsatisfactory arrangement and one that can easily result in poor hygienic standards. When an alternative caterer is available the competition provides an incentive to improve.'*

These days there are very few airline-owned and operated catering facilities; many are worldwide subsidiaries of four or five of the major airline catering conglomerates. So are Bailey's fears justified in the 21st century when the dreaded option (3) is now a startling supply chain reality, or have supply chain safety attributes evolved in tandem with the new breed of outsourcing opportunities?

There is no doubt in my mind that supply chain issues are led more by cost and menu opportunities than by safety considerations. At no time in all the years that I have dealt with the aviation sector as a supplier, have product safety attributes been a major consideration over fiscal ones and the airlines embalmed notion of 'quality'. Quality issues have long been dominated by restaurant-style comparisons and chef's perceptions of what quality means, rather than by the wider context which focuses on how quality really amalgamates with total product safety and quality assurance issues.

Factors governing product development

To truly appreciate the essence of the different dynamics that exist between the airline caterers' view of product quality and safety attributes, we need to look at illustrated examples of not just product differentials but of functional differentials. In theory, the airline caterer/airline catering product supplier relationship should be determined by the same issues as the retailer/retail product supplier relationship, particularly if the retail supplier is manufacturing retailer branded products. However, as we can see below from the examples given, the situation is quite different.

Table 7.2 illustrates the varying product development credentials required by the airline caterers and the food manufacturers to achieve the same end result. Ultimately the route taken to achieve a roast beef dinner ready meal is determined by the same factors:

- quality
- cost
- consistency
- shelf-life
- reheat capabilities
- safety.

However, the routes taken by each sector to achieve them, are vastly different.

What can be witnessed from the above isolated example is that the factors required to truly govern quality, safety and the requirement for global outsourcing and replication are not witnessed in the catering example, only in the manufacturing one.

The typical emphasis is devolved from the notion that each product is developed in isolation by a chef, in the same format that one develops a menu item in a restaurant. The quality parameters are dictated by the size and shape, as well as the physical attributes of the product, and quality ideals are all linked to how it may look and taste in isolation, without wider consideration being given to the ensuing escalation of batch quantity, life parameters and service environment.

Table 7.2 Product development differentials between airline catering and food manufacturing

Product	Supplier	Product attributes

Roast beef dinner	Airline caterer	Recipe determined by R&D chef
		Recipe approved by airline catering rep.
		Costs approved by airline catering rep.
		Finished product spec determined by photo
		Weight recipe ×1 of product
		Quality parameters = organoleptic
		Meet spec = (photo and weight)
		Safety parameters = shelf-life (non-scientific)
		  HACCP generic to all similar products
		  Process flow generic to all similar processes

Roast beef dinner	Retail food manufacturer	Recipe determined by R&D chef
		Full technical specification approved by food technologist
		Costs approved by retail buyer
		Packaging approved by food technologist
		Finished product spec determined by:
		  Raw material specification
		  Raw material traceability
		  Nutritional composition
		  Recipe, sub-recipe and compound ingredients composition
		  Shelf-life attributes (scientific data)
		  Labelling compliance
		  Heat trials
		  Organoleptic trials
		  Finished weight including cooking loss attribution
		  Recipe × batch of product
		  Process and product specific HACCP
		  Product specific process flow

The safety issues should be intrinsically linked to raw material outsourcing traceability and consistency, and the wider safety issues cannot be assured without the evidence of process-specific HACCP and documented process flows. The same process attributes are applicable to all aviation-bound products be they hot or cold, ambient or frozen, in the same way that they are to manufactured ones. In the absence of these types of technical specification requirements, the true safety and quality parameters cannot be determined and the potential success of multibatch replication is severely compromised.

The key to manufacturing standard quality and safety issues being resolved, hinges upon the determination of a consistent and verified supply chain. Typically the catering-style attitudes and rationale associated with this concept are entrenched in the industry's historical roots. The restaurant-style product development ethics do not demand the quantified and qualified aspects of raw material outsourcing, and in many instances the non-specified brokering of raw materials is the culture. This remains the status quo and a means by which caterers achieve contractually specified price point directives for extended periods of time.

The 'integrated cost' versus 'pay as you go' cultures, which separate retail sales from mainstream aviation food provision, has much to do with this; however, with the wider product integrity issues at stake, raw material outsourcing and traceability are critical food safety issues for aviation caterers to redress on both a national and international scale.

Operational catering issues and safety

We will look at product traceability issues in isolation at the end of the chapter, so in the meantime let us move on to other issues impacting on the airline caterer in terms of product safety and integrity issues.

Table 7.1 shows that the operational food-related aspects undertaken by the airline caterer are impacted by many non-catering functions. The supply of many non-food products and services and the logistical functions of aircraft de-catering and equipment washing all have to be considered in the overall product safety attributes of such a multifaceted operation. They include the following:

- Handling and washing equipment, rotable and disposable.
- Controlled disposal of food waste.
- Supply of non-food and drink products.

The above potentially 'dirty' activities within a food manufacturing environment provide ample opportunity for product and process contamination if they are not managed appropriately within that environment. Documented risks attributable to the through processing of dirty equipment can be defined and documented in a variety of ways depending on the catering premises geography and throughput proportionality. The handling of dirty equipment poses both physical and environmental contamination risks to food products, and the prolific use of chemical solutions required to both clean and sanitise the plethora of rotable equipment characteristic of airline operations, introduces the third breed of contamination risk. Consideration must also be given to the ventilation ducting in these environments to ensure that positive pressure ducting is in place to prevent air-borne contaminants from impacting on safety parameters. Filters must be cleaned and replaced regularly and commensurate with throughput, whilst air change requirements will need to be balanced to take account of food manufacturing protocols and health and safety directives. These should also be proportional to throughput and linked to the level of personnel required in each area at any given time calculated on maximum saturation levels.

The controlled disposal of food waste is an essential aspect of many airline catering operations and is often regulated by agriculture departments in tandem with Port Health Authorities around the world. The protection of the international food chain is what primarily drives the nature of the directives, to ensure that contaminated food waste from an aircraft does not reach the food chain either by cross-contamination or by food waste being processed into animal feed. It seems amazing that major aspects of the initial application of such a highly risk-sensitive area of food waste disposal are undertaken by the very operational facilities also charged with the obligation of manufacturing safe food products within the same environments. It is inconceivable to imagine this kind of disposal of food waste from unknown sources being undertaken by a food manufacturer in any other environment.

For standard airline catering operations, it is essential to categorise and segregate food waste by-products generated by their own production environments, from international waste by-products generated by incoming flights, to assure the correct controlled waste disposal activities. International catering waste will normally be subject to a different set of waste disposal categorisations then normal food waste generated on site. Upholding the safety and integrity of the food chain is essential in this regard to avoid animal disease epidemics such as foot and mouth.

The prospect of a manufacturer of retailer branded food products taking in and compacting internationally derived food waste from any combination and variety of sources, is totally inconceivable, yet in the aviation environment it is the status quo. Once again we see an example of an industry system established over 40 years ago never having been challenged in the contemporary context of food safety risk assessment and management. Ultimately, postcompacting, food waste will be transferred by licensed contractors to landfill or incineration, but the design and management of waste throughput has to be conducive to GMPs which would demand total operational segregation. This is rarely evident, particularly in older-style catering units and fundamentally is not encouraged in the design of newer facilities, as the operational requirement for the swift and efficient transfer of cleaned equipment from waste compacting and washing areas to proximate food production areas remains the overriding consideration.

The airline caterer's involvement in the supply of non-food products can involve the handling and throughput of a variety of products from newspapers and magazines to amenity kits, toothpicks, duty-free goods and alcohol. This activity is in itself open to a variety of risk factors if process methods do not allow for total product segregation. In tandem with the glass risk attribution associated with non-food products, there is the added difficulty of devolving glass, plastics and ceramic control policies into an environment where the food products for supply are packed in precisely these materials. A large percentage of preplated, prepacked dishes will be housed on china plates or in glass bowls and this renders the physical contaminant control issues of such products and in such environments particularly challenging.

In the manufacturing sector food-safe packaging rarely involves the utilisation of glass, china or ceramics and this is driven by the two key aspects of the application: the requirement to exclude such products from the food production and storage environments, and the attributable breakage risks of transporting food goods in such materials without tertiary housing. Whilst it is not impossible to monitor and control the risk factors in controlled, purpose-specific environments, the large volumes of products associated with this application in the aviation sector, and the absence of tertiary packaging on the finished products, make this a huge issue. The effectiveness of a glass register, in terms of the maintenance aspects of its application, is rendered ineffectual by the prevalence of such materials. Once more we see a living example of where catering ethics, driven by restaurant ideals, have the capacity to impact on product safety in these environments.

Overall, the modern airline catering supply chain concept is less about catering and more about the successful brokering of components and services which by the nature of their outsourcing will have an impact on the safety attributes of the catered end product. In predetermining the risks, the systems have to be flexible enough to either account for the total proliferation of potential product outsourcing possibilities or be constantly evolving in tandem with emerging product solutions.

In conclusion then, the generic safety concerns encountered by airline caterers have to be raw material and finished product traceability, controlled disposal of food waste, raw material and finished product transfer issues, control of non-food product throughput and the amalgam of catering/restaurant-style development issues with total product safety attribution. These are the issues in the broadest possible sense, without considering the product or process-specific issues or any of a prerequisite nature. Given little or no consideration in the wider context of the provision requirement, these issues have the capacity to impact significantly on total product safety.

Food brokers

There is no doubt that, given the huge requirement in any airline catering operation to outsource raw materials as well as finished goods, the requirement for certain products to be brokered through other suppliers is inevitable. The buying function in any airline caterer will be focused in several broadly different areas:

- Raw materials for product manufacture in-house.
- Finished products for assembly in-house.
- Finished goods for direct sale in-flight.
- Non-food goods.

It is conceivable that products in every category have the capacity to be brokered; indeed many smaller airlines choose not to have buying teams in situ and outsource all products through food brokers. So what distinguishes a food broker from any other provider of aviation food products? Ostensibly brokers do not have any direct connection with the transit, storage or despatch of food products; indeed, they do not actually see many of the products they 'sell'. Their role is to outsource suppliers and to set up the parameters by which products will be ordered and despatched to the airline caterers who are to use them. Brokers need no food premises, merely an operating address from which to administer the buying and selling of food products. Most food brokers will have benefited from a direct connection with the airline catering industry at some point, either as a buyer or catering manager, in order for them to understand the product requirement parameters and the types of businesses who will be willing and able to supply the desired products.

Often brokers will assist airline caterers who, for whatever reason, have experienced a supply chain failure and need to receive a consignment of goods urgently. The brokering function is entrenched in airline catering supply chain dynamics and has been for decades. Whilst there is no disputing the industry reliance on brokering activities, the contemporary food safety management issues appertaining to the assurance of total product safety and traceability render the focus on certain aspects of brokering activities of particular interest.

My experience in dealing with food brokers has led me to the opinion that although they accept the safety and legal compliance responsibility for the products they are outsourcing and supplying, they take no documented steps to verify the appropriateness of the supplier themselves. The critical parameters for orders placed are cost and the ability to fulfil the order within the time-frame determined. What this means for an industry that makes such prolific use of this system of supplier outsourcing is quite alarming. It is essential that all supply chain activities are verified by every link in the chain and that the availability of both raw material and finished product technical specifications is devolved throughout.

Supplier audit information on brokered food products needs to be undertaken by the brokers themselves and retained for their catering clients to refer to in the event of a problem arising. I believe that is also essential for those both auditing and enforcing standards in airline catering facilities to fully appreciate the airline broker dynamics and not assume that all supplied goods are outsourced or manufactured directly.

When modern food safety and quality assurance systems focus so profoundly upon reducing the links in the supply chain, the brokered food concept is looking remarkably out of sync with the burgeoning safety requirements. The need to maintain a consistent, sustainable and verifiable supply chain, and to achieve appropriate safety and quality parameters, is well documented in food manufacturing systems management and the concept of brokering does not sit well with this. It seems particularly inappropriate for brokered food products to be cost attributable to the brokers themselves, rather than directly with the suppliers, thus making the supply chain picture appear distorted. In an audit environment one would examine goods receipt paperwork, which would in essence be attributable to suppliers who actually manufacture, store and distribute nothing and who operate from non-food premises that may not be detectable by the relevant inspection authorities. This is not the same situation and should not be confused with food service companies who may not manufacture anything but who do store and distribute from approved premises.

The likelihood of an airline catering food broker having a documented quality manual and brokering specific HACCP plan is unlikely, as is the likelihood that they will centrally hold product and process-specific specifications on the products they supply. In the context of the nature of the activity this is not surprising; although many brokers will specialise in particular types of product requirements such as bakery products, dairy products or economy class hot meals, much of their activities are customer requirement driven.

Verifying safety systems

So, having established the inappropriateness of the food brokering systems, it is for us to determine how they could be constructed in order to work effectively and satisfy assured product integrity and traceability requirements. In order to verify the safety systems of any food brokering activity, the following steps would need to be taken by the caterer and the broker respectively.

Step 1 – Product requirement evaluation

- Product name.
- Product category – chilled/frozen/ambient.

- Quantity.
- Frequency.
- Delivery schedule.
- Product quality attributes.
- Product safety attributes including shelf-life.
- Product packaging attributes.
- 'Known' or 'unknown' stores classification.

Step 2 – Product supplier evaluation

- Supplier name.
- Supplier address.
- Key personnel flow diagram.
- Supplier classification
    - caterer
    - manufacturer
    - distributor.
- Types of products supplied.
- Licensed or unlicensed premises.
- Third party accreditations held.
- Audit approval met.
- Own or third party distribution.

Step 3 – Product safety evaluation

- Product classification – low/high risk.
- Shelf-life verified scientifically.
- HACCP documentation reviewed.
- Raw material technical specifications.
- End product technical specifications.
- Supply chain verified.
- Country of origin determined.
- Food safe packaging.
- Premises inspected.
- Micro verification available.
- Staff trained.
- Prerequisite issues satisfied.
- Vehicles inspected.
- Labelling compliance fulfilled.

It is clear from the above that the steps food brokers need to take to assure product safety and traceability are no different from those required by those they are supplying. For the airline caterer not to insist on the food broker verifying the product safety attributes by providing all of the technical information, is to assume that the supply chain is technically not robust. In many cases, the supply chain is further complicated by food brokers who outsource products indirectly. This can occur either via other distributors or caterers and therefore not directly with the manufacturers themselves. This has the direct result of elongating the supply chain even further and making the product traceability issues even more complex.

It is critical that the global supply potential aspects of food brokering activities are not lost sight of and that the complexities of the food brokers' obligations are fully understood both by those from whom they procure and by those to whom they supply. In the context of supply chain issues also, it is important to recognise the potential safety implications of combining brokered and non-brokered components on the same tray set, or brokered and non-brokered raw materials as part of the same finished product, particularly if the three documented steps outlined above have not been undertaken and verified.

Logistics and catering operations

Whilst it may seem at this point in the chapter that all airline caterers are actually more catering logistics partners to the airline rather than catering providers, due in part to the vast amount of finished products outsourcing, there are actually dedicated catering logistics operations who fulfil that function exclusively. This relatively new concept is based on the principle that the logistical and food production activities traditionally encompassed under the same umbrella in the airline catering facility, are completely separated. In reality this means that the airline caterer fulfils all of the operational aspects of the catering function but outsources all of the food products and components from food manufacturers. The functions are separated in the following way.

Operational functions

- De-catering and wash-up.
- Controlled disposal of waste.
- Equipment handling and supply.
- Non-food procurement and soft services provision.
- Tray set assembly.
- Product QA.
- Product delivery to aircraft.
- Catering purchasing and billing.

Catering functions

- Supplier identification.
- Supplier outsourcing.
- Supplier auditing.
- Product presentation.
- Product outsourcing.
- Product receipt and storage.
- Tray set assembly.
- Product monitoring.
- Alcohol and beverage supply.
- Contractual catering liability and indemnity.

The crucial divide between operational and catering functions is made by the fact that every single food product is manufactured elsewhere, outsourced from food manufacturers and brought into the unit for tray set and meal assembly only, whilst the operational functions are completed in-house. The wider aspects of meal component manufacture and raw material outsourcing are not an issue for these types of airline caterers and they tend to rely on the fact that the airlines will assume responsibility for dictating the menu specifications for each supplier to fulfil. Whilst the contractual liability aspect of this arrangement still intrinsically links the logistics partner to the catering function in the same way as mainstream airline caterers, the overall focus is on logistical reliability instead of catering quality, which is predetermined by the nature of the outsourcing dynamics.

So let us look at the implications on food safety management issues of this type of operation. In theory, looking at the list of catering functions undertaken, it seems like airline catering safety utopia! By virtue of the operational and physical segregation of both dirty and clean functions in different units, i.e. the food manufacturing functions carried out in food dedicated environments and the de-catering, wash-up and refuse disposal functions undertaken in the dedicated logistics environment, inherent safety issues are dealt with at source.

In order for this type of airline catering operation to fulfil its conceptual promise in terms of what it aspires to deliver in the safety arenas, several key objectives have to be achieved in a very specific fashion. Any deviation from the following prerequisite issues renders the concept an elaborate way of achieving nothing more than the same food safety management shortfalls demonstrated by mainstream airline caterers.

Critical issues to be considered are:

- Product outsourcing only from audit-approved manufacturers who meet retail branded product standards.
- Audit compliance to be a prerequisite to trading agreements being established.

- Full technical finished product specifications.
- Full raw material traceability.
- Individually not collectively date-coded and batch-coded components.
- Retail standard labelling and full nutritional breakdown.
- Supplier HACCP to be product and process-specific.

By being in a situation where all products are outsourced from elsewhere, this type of logistics-only airline caterer is afforded a unique opportunity to ensure that the appropriate product quality and safety parameters are met. Deflecting the pressure to produce and quality assure from within to suppliers outside can be an extremely effective way of achieving an end product that is quality assured to manufacturing, not catering, standards.

Whilst a swift glance at the prerequisite issues results in the feeling that nothing should prove a problem if sourcing from outside manufacturers and not in-house from a catered environment, unfortunately experience has taught me that other factors often tend to detract from this picture of perceived utopia.

Invariably manufacturing product within technical parameters has an impact on cost and with the added complication of the logistics partner having to levy operational fees for their role in the operation, this type of airline catered food can prove extremely expensive. The other issue is the packaging predetermined by airline catering styles and types of service. Many airlines require the food products to be packed into equipment that is glass, china or ceramic. In any manufacturing environment the impact of having to pack and handle products in the very types of materials excluded from their operations by safety policy is a non-starter, unless they are in the rather unique position of total product and process segregation.

Once again we see the impact that cost parameters can have on food safety as many of the large food manufacturers struggle to meet airline specifications and price points simultaneously. The resulting situation is one where product is not sourced from manufacturers at all and instead is farmed out to the very airline catering kitchens that the concept saw fit to distance itself from in the first place. In many cases, despite the fact that the suppliers fail to meet the logistics partner's very own audit standards, they are still utilised as suppliers because they can meet the appropriate price points and satisfy the perceived quality standards when quality is based on organoleptic considerations alone.

Without the traceability initiatives being fulfilled to manufactured standards, the onus on the logistics-type caterer is to assemble tray sets and meal carts from a whole variety of components without having the traceability information to underwrite their aspects of the operation. It is essential that these types of airline caterers insist on audited systems from their suppliers that can guarantee batch and product traceability to each and every component. Having made that assertion to the supplier, it is then for the logistics partner to establish assembly paperwork to document which components come from which suppliers, demonstating which batch codes are brought together to comprise the meal.

So it is clear that whilst this concept promises much in terms of assuring the safety issues inherent in traditional systems of airline catering, unless the boundaries of the concept remain intact, it will prove ineffectual in tackling the critical supply chain safety problems.

To buy in or make in-house – the manufacture of airline catering products

It does not really matter which supply chain mechanisms airlines employ to outsource their food products as long as the supply chain integrity is assured. The assumption that the required quality and safety parameters are more effectively met by logistics-style catering providers as opposed to traditional airline caterers is an obvious one to make, but if the concept is not underwritten by responsible supplier outsourcing and managed in-house through robust, documented storage, assembly and delivery protocols, then the same safety issues are at stake.

Whether the decision is to buy in or to outsource from brokers or suppliers elsewhere, the safety considerations are the same. There is no reason why, given the right technical back-up, mainstream airline caterers cannot achieve the same quality and safety standards as food manufacturers. The crucial issue is to recognise what is required and then embrace the requirements as a positive aid to quality assurance and product integrity. The major focus on airline-style product specifications, denoting the product in a one-unit format in terms of its size and presentation without any technical references being made to batch quantity, process flow or assembly detail, is the major differential that separates products manufactured in-house from those bought in. It is essential in order to assure accurate and effective replication that technical product detail is formalised in a finished product specification.

In terms of moving away from catering ethics, specifications need to represent the whole product picture in terms of quality parameters and organoleptic considerations as well as safety ones. Once this has been formalised as part of an entire menu rotation, the amalgam of HACCP and process flow will become evident and will set the agenda in terms of what process techniques may or may not need to be employed to accommodate its successful throughput.

Whether catering products are bought-in or made in-house, the development of the documentation needs to reflect the safety considerations of each and every component that goes to make up the meal. Assuming that the meal equals the entire contents of the tray set or menu items listed, the assumption has to be that all items offered will be consumed by the recipient. On that basis the distinction between outsourced and in-house manufactured products has to be made and the technical information for all components brought together into one amalgamated technical specification. It is a difficult concept to grasp when one considers that most products made to manufacturing standards are single-unit end products. The specifications have to be built in terms of the following criteria:

- Ingredients of components.
- Compound ingredients of components.
- Sub-recipes of end products.
- All components as a sub-recipe of the entire meal if the meal is a tray set or special diet meal.
- If meal is not a tray set or special diet meal then only the first three criteria apply.

For the manufacturers of these catered products it is essential to apply the same production protocols to components bound for aircraft as it is for all other manufactured items. It is essential that whether it is demanded by the airline and the airline caterer or not, maximum assurance technical standards, not minimum requirement specifications, are provided.

Catered standards have no place in such a volume driven environment. It is crucial not to confuse the food service aspect of the operation on board with the manufacture of the products that comprise the menu on the ground. Catering standards are those that apply in flight; manufacturing standards are what are required in-house at the point of both product development and production, to underwrite the successful and safe application of what is ultimately served on board.

All in all the supply chain aspects of the catering operation can be defined in much less broad categorisations than those discussed in this chapter. However, whilst the broad picture throws up some supply-specific considerations, the smaller picture will no doubt throw up some product-specific ones. Gaining an understanding of the types of product outsourcing dynamics potentially employed by the industry assists anyone attempting to predetermine the safety parameters of products used in this application. In consideration of the facts appertaining to safety in the supply chain arena, it is critical that whatever methods are employed, full product traceability and defined technical product data are a prerequisite to supply. In this way the wider issues of quality assurance and global replication will prove far less of a logistical mountain to climb.

In terms of airline caterers making decisions as to which products they will manufacture and which they will outsource, the considerations need to be defined by their capacity to compete with the technical abilities of large manufacturers and successfully attribute shelf-life or dietary claims where necessary with all the required back-up. It is essential also when raw material and finished product traceability is such a key factor, that where applicable the relevant meat, fish and dairy products licensing requirements are in place. Under EU food safety directives, the requirement for all food businesses, whether caterers or manufacturers, to meet full traceability requirements, will become law in 2006. All the more reason then for airline supply chain parameters to be predetermined by a supplier's ability to comply to food safety dictates, rather than to meet airline price point directives.

The basis of the justification for anything less than full manufacturing technical standards in the aviation food provision environment is that the food service nature of the product renders its production environment a catered one. Unless the requirement is for immediate service, proportionality of scale and shelf-life attribution dictate nothing less than manufacturing standard protocols.

I have sat through many airline catering product presentations in my career and I am always overwhelmed by the frustration inherent in watching product parameters and quality attributes defined and judged on the basis of single-unit presentations and non-scientific data. The crux of the matter lies in the industry's tolerance of product attributes being driven by fiscal considerations masquerading as quality ones, and chefs with no technical knowledge determining, what are in essence, the technical considerations of the products.

Whilst in a manufacturing environment research and development chefs create dishes, they do so in tandem with technical personnel who dictate the safety parameters and operational personnel who dictate the throughput and process flow. Product presentations are done in the context of multibatch trials to ensure consistency and to demonstrate and verify the production methodology. Unless this approach is applied to new product development in the aviation catering environment, then the quality assurance and safety parameters will be continually underachieving and consistent technical verification will remain a pipe dream.

Traceability – the critical issues

Throughout this chapter I have focused on the essential issues of product traceability through the application of technical specifications and documented systems management. The difficulty the industry faces in this regard is the component nature of the meal. This requires systems to be devised that allow for traceability, component to component and tray set to tray set. The documented management techniques required to achieve this are described in Chapter 5, but the principles need to be understood in the context of supply chain resource management.

The essence of traceability assurance is entrenched in the requirement for each product to have an attributable end-product specification. The requirements for raw material outsourcing will be documented in this and will include supplier detail as well as country of origin attributes. The raw material supply chain must be assured and consistent within the parameters laid down in the product specification, and verified by goods receipt documentation and production paperwork which verify such supplier detail throughout every stage of the process. Standardised raw material outsourcing is essential, and if outsourcing the same raw material from various suppliers then the production paper needs to demonstrate that mixed-batch raw materials may have been used.

Following the individual production documentation through to the tray setting stage requires a culmination of production data onto an end product assembly sheet which attributes component data onto an end product or batched tray set. The end result is a situation where each and every component on each and every tray can be traced exactly.

Production paper needs to allow for the following data to be documented in a standardised format in order that the quality, safety and technical detail of the product specification are evident to those engaged in its manufacture:

- Batch size.
- Recipe per batch.
- End product weight.
- Assembly detail in % format.
- Documented process flow.
- Quality parameters to be amalgamated into the production method.
- Operative and location sign-off.
- Time and temperature application – chilling/hot/cold holding.
- Batch codes and use by/best before data of raw materials.
- Finished product temperature.
- Finished product batch code.
- Finished product 'use by' or 'best before'.

Figures 7.1 and 7.2 illustrate the manner in which this can be devolved onto standardised production data sheets that allow for raw material traceability to be documented throughout.

Figure 7.3 illustrates the type of assembly documentation that is required to represent the cumulative production data of each component onto one tray set that constitutes the meal.

BULK COLD PRODUCTION RECORD

DISH....................NICOISE SALAD

% ×1	PRODUCT	USE BY	BATCH CODE	WEIGHT	RAW TEMP.	BATCH	COOKED TEMPERATURE	TIME	BLAST TEMPERATURE	TIME
	Green beans			50g			CCP1		CCP2	
	Olives black			3 × no.						
	Cherry tomato			3 × no.						

BATCH SIZE

BATCH CODE

METHOD

1. Decant all items in low risk cold room
2. Blanch beans off in boiling hot water, transfer over to high risk
3. Transfer to blast chiller CCP5 and chill to <5 °C within 30 mins
4. Lay out bowls and place beans in bowl followed by toms, olives
5. Lid and transfer to assembly
6.
7.
8.
9.
10.
11.
12.
13.
14.
15.

DATE MADE

PRODUCE BY

CHECKED BY

DATE

Temperatures			
Holding	Chilled		

USE CAPITAL LETTERS ONLY

Figure 7.1 Standardised production record for salad.

BULK PRODUCTION RECORD

PRODUCT................Chix veg pot   main meal   Category Rotation 1.

%	PRODUCT	USE BY	BATCH CODE	WEIGHT	RAW TEMP.	BATCH	FOOD ITEM	START TIME	COOK TEMP.	FINISH TIME
	Chix breast			100×no		×1	Chix			
	Potatoes diced			3 kg		×1	Pots			
	Carrots			1 kg		×1	Veg			
	Beans			1 kg		×1				
	Asparagus			1 kg		×1	Sauce			
	Tom purée			2×800g		×1				
	Tom chopped			1×		×1				
	Onion 6 mm			1 kg		×1				
	Mixed herbs			50 g		×1				

BATCH SIZE: 100

BATCH CODE:

PRODUCTION AREA:

METHOD

1. Verify clean-down activity as documented on schedule for production area and equipment to ensure allergen regimes have been applied   CCP1
2. Decant and weigh out recipe
   Document confirmed usage CCP2
3. Cover and transfer into production area   CCP3
4. Blanch vegetables in boiling water CCP4
5. Steam potatoes in oven CCP4
6. Sweat off onions in blast pan 1, add tomatoes, herbs, tomato purée and add water CCP4
7. Steam chicken breasts in CCP4
   Oven probe with probe 2 > 75°C and document
8. Transfer all cooled products into High Risk and decant into high risk containers and document
9. Transfer into blast chiller CCP5 and chill to <5°C within 90 minutes
10. Transfer into high risk pan
11. Lay our code light green foils and assemble, lid and label
    Transfer into chilled storage

SPECIAL NOTES
Ensure all High Risk are wearing code BLACK PPE throughout all

DATE MADE:

PRODUCE BY:

CHECKED BY:

USE BY:

Temperatures		ASSEMBLY		
Holding Random 5	<5°C LIMIT	10 MINS	20 MINS	FINISHED TEMP.

USE CAPITAL LETTERS ONLY

Figure 7.2 Standardised record for main meal.

DATE		FLT NO		ASSEMBLY TRAY SET UP ROTATION 1		
MEAL		FOOD ITEM	TEMP	BATCH CODE	No TRAY SETS	USE BY DATE	SIGN
B/FAST		BAKED FRUITS					
ECO		MUFFIN	AMB				
		YOGHURT	AMB				
		WHOLE MILK JIGGER	AMB				
		MEAL 1 B/FAST					
M/MEAL		NICOISE SALAD					
ECO		DRESSING	AMB				
		FRUIT BAR	AMB				
		WHOLE MILK	AMB				
		RAISINS	AMB				
		NAAN BREAD	AMB				
		MEAL 1 CHIX CURRY					
B/FAST		FRUIT SALAD					
ECO		SOYA MILK	AMB				
		PROMOVEL	AMB				
		FRUIT BAR	AMB				
		MUFFIN	AMB				
		MEAL 1 VEGAN					
M/MEAL		SAMOSA					
ECO		SOYA MILK	AMB				
		RAISINS	AMB				
		NAAN BREAD	AMB				
		BAR	AMB				
		MEAL 1 VEG CURRY					

Figure 7.3 Tray set assembly documentation.

For items manufactured or brokered from outside and not subject to this type of in-house production documentation, assuming that the technical specification for the finished product manufactured elsewhere is in place, the assembly paperwork will represent the bought-in component as a batch code and a 'use by' which will tally with the paperwork at the point of goods receipt. If the product is a bought-in item, which then has a further application attributable to it in-house, e.g. frozen part baked bread rolls or a dessert that is to be plated and garnished in-house, then the appropriate production paperwork will have to be formulated to represent the further application or assembly work and correlate with the rest of the component production paper to be represented on the end tray set assembly sheet. It will also be essential that the supplier of the bought-in product assumes responsibility for manufacturing standard production documentation also. This will be evident at point of supplier audit or from looking at the end product specification.

It sounds a difficult task to complete in terms of the work involved. However, as with all successful systems management, if the rest of the protocols are correctly established then it will work in tandem with them. If there is non-compliance in terms of what is done at point of specification or at goods receipt, then indeed the systems will be difficult to manage.

It is essential that these types of traceability diligent systems are managed effectively in order to assure the sustainable integrity of the airline catering supply chain which ultimately has an application to food safety management worldwide.

8 Fitness to fly

One of my main objectives in writing this book was to somehow formalise the issue of flight-deck fitness to fly. I aim to bring some thoughts together, based on my own work and industry data, in order to approach the issues from a rather different perspective than that proffered by the aviation industry currently.

Among all of the typical subject matter for a book such as this, some of the real aviation food safety and security issues that remain unexplored and unregulated confront the flight-deck and their food safety fitness to fly.

Industry statistics from many airlines and indeed studies run by the International Civil Aviation Organization (ICAO) and pilot representative organisations such as the British Air Line Pilots Association (BALPA), from time to time illustrate the threat to aviation safety posed by pilot incapacitation. Statistically many recorded cases cite the root cause as food poisoning.

The prospect that those flying an aircraft, with safety at the forefront of their minds, could actually be frequently prevented from effectively carrying out that function by a food-borne illness, seems incomprehensible. Do flight-deck crew, by the nature of the job, render themselves naturally predisposed to food safety crisis more often than those who work on the ground? If so, why? Do airlines, having collated years of pilot incapacitation data, act on the information collectively and develop SOPs to deal with such prolific occurrences? What protocols are adopted by flight kitchens in tandem with the airlines they service to ensure that the manufacture of flight-deck fare is carried out to the highest possible standards of safety and security?

In this chapter we seek answers to these questions and attempt to place the answers squarely into the appropriate context. The issues have been sidelined for too long and it is only by making the debate a topical and relevant one that it is possible to set a better agenda and climate for change.

Pilot incapacitation and its link to in-flight food safety

Since the first studies into in-flight pilot incapacitation were commissioned in the 1960s[74], the indisputable connection between incidences of food-borne illness and this most dangerous of all in-flight operational circumstances was forged. The data are underwritten more recently by the UK CAA figures for pilot incapacitation between 1990 and 1999, which illustrate clearly that just under 54% of incapacitation incidents were linked to gastrointestinal symptoms. In direct contrast to those, cardiac-related incapacitations are shown at less than 3%.

Most of the reported cases cited systematic, self-limiting illness represented by acute gastroenteritis and its accompanying symptoms, vomiting and diarrhoea. Without exception all reported cases were linked directly to the consumption and ingestion of contaminated food and/or water.

The operational safety impact of any in-flight pilot incapacitation is immense and should never be underestimated in terms of its aviation safety significance. Not only is the single pilot operation of a multi-pilot aircraft dangerous, but emergency diversion and the ensuing operational disruption to both passengers and crew can be extensive and expensive.

It is critical therefore to establish a connection between the term 'in-flight pilot incapacitation' and what it means to the safety of the passengers and crew on board an affected aircraft. If there is a link to contaminated food being causative, then why and how best to tackle the problem is something that has to be redressed for the long-term benefit and safety assurances of the industry.

A more realistic and regularly documented approach to aviation incident reporting data can be seen in the monthly Mandatory Occurrence Reports (MORs) collated by the UK-based Civil Aviation Authority (CAA). The events categories are a type of broad coding system that was introduced into the MOR database at the beginning of 2000 and include every kind of incident from landing gear problems to systems malfunctions and unruly passengers. The monthly reports are designed to provide a snapshot of the types of problems encountered within the UK aviation industry. A brief look at a random month, March 2003, documents four incidents of pilot incapacitation, with all citing food poisoning as the suspected cause. In October 2002 there are also several other reports of gastrointestinal illness among the flight-deck crew, with one reading as follows:

> *'During flight P1 consumed a sandwich that was supplied with the crew food. Just prior to top of descent, P1 started to feel sick with a constant sandwich after taste. Mild food poisoning was suspected although the sandwich was in date with undamaged packaging. The nausea eventually abated and P1 was able to continue his duties. The reporter comments that previous reports have been made to the contracted catering company concerning out of date food and damaged packaging and suggests that environmental health personnel should inspect their facilities.'*

The most frustrating aspect of the types of aviation safety incidence reporting mechanisms such as the MOR system in the UK, is that many of the follow-up protocols are left in the hands of the airlines themselves, particularly if the regulatory authorities duly consider that:

'the (incident) occurrence should be adequately controlled and monitored by the reporter's action, coupled with existing procedures, control systems etc.'

Once more we can see an example of how the safety agendas in aviation are not viewed as an amalgamated picture. If food poisoning has continued to be implicated in the pilot incapacitation statistics that span three decades, why are the causative factors not being dealt with?

The answers to this question are various but stem from the industry's unwillingness to formalise the methods of data reporting in the numerous studies that have been carried out into pilot incapacitation. The studies that have been commissioned have opted to use different cross-sections and sources of incidence reporting in order to ensure that the results delivered suited the intended safety purpose or crusade which the data were to be used to support. It is my view that it has remained in the industry's interest to suggest that the major causes of incapacitation amongst pilots are longer term health defects such as cardiovascular malfunction, epilepsy, diabetes and renal problems, as in this way they can be seen to be dealing with the probability of incidence by pilot medical reporting certification. Indeed, there have been some suggestion among pilot's representative groups that pilot incapacitation data have been used to terminate pilots early when they reach certain age groups or medical classification.

Such data can be witnessed in a study of career termination due to loss of licensure insurance in members of the US Air Line Pilots Association (ALPA) from 1955 to 1966[75], which found 891 cases of career termination broken down into the following categories:

- 229 due to accidents.
- 662 due to disease.

The rate of death and disability due to accidents was 2.07 per 1000 pilots per year, whilst the rate for disease was 8.05 per 1000 pilots per year. Although the overall rate for cardiovascular disease was only 2.91 per 1000 pilots per year, the age-specific rate ranged from zero for pilots under 30 years of age to 27.33 for pilots between 55 and 58 years of age. Using the incidence of incapacitation rates for career termination due to disease by age, and the age distribution of active ALPA pilots, the authors of the study estimated the probability of serious in-flight incapacitation by age. Their estimates ranged from 1 per 58 000 pilots for the 30–34 age group, to 1 per 3000 pilots for the 55–59 age group.

The incapacitation studies used, that indicate statistically higher levels of incidence due to long-term health defects, are based on data formulated from in-flight deaths and aircraft accidents. In 1969, Buley summarised three sets of pilot incapacitation data[74]. First he reported on a collaborative study initiated by ICAO and performed by the International Federation of Air Line Pilots Associations (IFALPA) and the International Air Transport Association (IATA). Buley examined in-flight deaths of airline pilots between 1961 and 1968 and made the following findings:

- 17 reported cases of airline pilot deaths.
- 100% caused by heart disease.
- 5 out of the 17 resulted in aircraft accidents.
- 4 of the 5 accidents were fatal, resulting in 148 fatalities.
- A further 5 cases resulted in near accidents.
- Of the 17 events, 7 occurred on the ramp, 5 en route, 4 during approach and 1 during landing roll out.

Buley next reviewed 42 cases of non-fatal in-flight incapacitation in pilots of IATA member airlines between 1960 and 1966 and found that in 24 out of the 42 cases causal organic disease was diagnosed, and the most common categories of incapacitation were:

- Epileptiform manifestations (6)
- Coronary occlusions (4)
- Renal/ureteric colic (4).

As part of the same set of research, Buley reviewed the results of a questionnaire administered to pilots of IFALPA member associations in 1967, in which the following findings were concluded and summarised:

- 27% of 5000 respondents reported 2000 incidents of significant in-flight incapacitation.
- 4% of cases reported that safety of flight was affected.
- 50% reported incapacitations occurred in the en-route phase of flight.

In 1975, Raboutet and Raboutet[76] reviewed 17 incidents of sudden incapacitation in French professional civil pilots between 1948 and 1972. They stated that for an incapacitation incident to occur, the incapacitation must:

- affect the pilot at the controls
- be sudden
- be total
- take place during a critical stage of flight.

This classification and definition was significant in the interpretation of data that followed, as it now allowed for pilot incapacitation incidences classification to be broken down into two distinct categories: impairment and incapacitation.

A definition was assigned to each, and certain medical classifications attributable to each helped form the basis of the breakdown of data studied to form the Federal Aviation Administration (FAA) study, *In-Flight Medical Incapacitation and Impairment of US Airline Pilots 1993–1998*, conducted in 2004[77].

In 1971 Lane[78] updated the 1967 IFALPA questionnaire data analysed by Buley[74] with IATA data from 1962 to 1968 that was provided to ICAO. Lane added 51 additional non-accident cases to Buley's original 17 cases, for a total of 68 cases. Lane then calculated that the probability of an incapacitating event resulting in an accident would be 5/68.

In 1991 James and Green replicated Lane's 1967 IFALPA survey with very similar results[79]. Of 1251 respondents, 29% reported at least one incident of in-flight incapacitation severe enough to require another crew member to assume their duties.

The most common causes of incapacitation were:

- Gastrointestinal – 58.4%.
- Blocked ear – 13.9%.
- Faintness or general weakness – 8.5%.

The most common phases of flight where incapacitations occurred were:

- En route – 42.1%.
- Climb – 18.4%.
- Descent – 17.3%.
- On the ramp – 11.4%.

With the modern day amendments to pilot incapacitation data classification, the new statistics are divided by impairment versus incapacitation.

Pilot impairment is classified as when the pilot affected can still perform limited in-flight duties, such as reading checklists or performing radio communications, even though their performance may have been degraded.

Pilot incapacitation is classified as when the pilot concerned can no longer perform any in-flight duties. It is interesting to see that the associated medical conditions linked to impairment versus incapacitation in the most recent studies, cite food poisoning and gastrointestinal complaints as likely protagonists in both types of classification.

So it would seem that even the statistics can be interpreted to alter, and some may say obscure, the view of pilot incapacitation and attributable, causative food-borne illness. It is for the industry to acknowledge that however the classifications are altered, since the early days of Buley and the first data-based surveys, food poisoning and associated gastrointestinal complaints have been, and continue to be, instrumental in a huge proportion of reported incidents of pilot incapacitation.

Are pilots predisposed to food-borne illness?

Having unearthed and then examined the strong statistical link between pilot incapacitation or impairment and food poisoning, it is for us to begin now to understand the reasons why this may be the case and then look at whether what pilots eat and drink on duty really does compromise their fitness to fly.

To begin with we must at least acknowledge that there are generally accepted aviation safety rules associated with pilot incidence of food poisoning and the aviation safety SOP requirement for pilots to eat separate meals during in-flight was established for this very reason. Whilst accepting the rationale behind this ruling, I find it difficult to comprehend that anyone really believes that such a blanket, piecemeal ruling has gone anywhere near far enough in dealing with the real issues affecting cockpit crew and their food safety fitness to fly.

There has been much industry and media speculation in recent years as to the reality of both the long and short-term health effects on crew members. Cabin health issues have been at the forefront of many aviation safety debates. Everything from pilot deep vein thrombosis (DVT) and cabin air quality to blood-borne pathogens and cosmic radiation, have found their way onto the platform for debate when discussions concerning cabin crew and cockpit crew health issues have risen to the fore. Interestingly enough, I can find no industry research that draws the same kind of personal health effects comparisons between pilots' incidence of gastrointestinal illness and incidence of gastrointestinal illness in workers on the ground.

Bearing in mind that even the most recent pilot incapacitation data are still citing food poisoning as the most common cause of pilot impairment whilst at work, and that despite the various prescribed methods of analysing the data being subject to conjecture and debate, the bottom line is that the situation of pilot illness, impairment and incapacitation due to gastrointestinal problems has not improved in nearly four decades since studies began. I cannot think of another cabin crew/technical crew health issue that would escape detailed scrutiny, media attention or aviation safety regulation.

So are pilots naturally predisposed to food-related illness more than other groups of workers, and if so why? Whilst not wishing to state the obvious, it is perhaps inevitable that pilots engaged in long haul flying activity are more likely to encounter difficulties in eating and drinking safely while away. Invariably their potential to visit a series of high risk destinations during the pattern of their working lives leaves them at greater exposure to variable and questionable standards of food and water quality.

It is impossible to determine whether pilots are more at risk from the foods they consume in-flight or the foods they consume at designated crew hotels. What is certain, however, is that the aviation industry is anxious to proffer the view that all attributable technical crew illness through consumption of poor quality food or drink, has more to do with the questionable safety and integrity of what is voluntarily consumed by crew members whilst down route, than it has to do with anything provided by them and consumed by crew members in-flight.

Despite the fact that the tech crew are officially on duty while down route, many airlines will make no stipulation about what tech crew eat and where they eat, even in high risk destinations around the world. Whilst the quality and integrity of the food and water available in the designated crew hotels is often subject to scrutiny by the airlines, this is not always the case and in terms of ensuring that the crew do eat in audit-approved facilities, the airlines would never mandate to that effect, despite the proven connection between pilot impairment in-flight and food-borne illnesses.

So the conclusions to be drawn lead us to believe that by virtue of the travelled nature of their job, pilots are naturally at a greater predisposition to food-borne illness and gastrointestinal complaints. This may well be due to the poor quality of the aircraft crew food or may have a similar connection with the poor food choices made by the crew themselves when down route. Whilst there is some level of procedural surveillance and control over the food consumed by both pilots in-flight, there is little if anything to mandate the meal choices made by either one or both pilots on the ground regardless of the port of call and associated and acknowledged risks posed to health and therefore safety.

Having examined the bare bones of how it could be possible that food poisoning could affect the pilot demographic so virulently, let us turn now to the level of industry acceptance of these data and the steps taken to counteract its impact on aviation safety.

We have already looked at the existing industry ruling that allows for a situation whereby pilots must consume different meals in-flight. This is a wonderful example of the industry offering a less than ideal solution to what is a big problem. It is obvious to anyone who knows anything about the likely causes of food poisoning (including the airlines themselves!) that it will matter not a jot that the pilots have consumed different meals if there is found to be an inherent hygiene problem at the catering unit from whence both meals were ultimately sourced. The nature of airline catering logistics provides for a situation where the same personnel can pack different meals in the same unit. In this circumstance there is no real protection for the tech crew from an incidence of food poisoning.

This precise point was acknowledged at the time of the introduction of the recommendation that pilots eat different meals and is borne out by Bailey in his 1977 *Guide to Hygiene and Sanitation in Aviation*[80]:

> 'When flight-deck personnel eat during the flight, it is absolutely essential that the Captain should be given a completely different meal from that served to the co-pilot,

prepared from food obtained from different sources. The same principle must apply if they eat in ground catering premises a few hours before take off. This is an essential safety precaution to reduce the possibility of their both eating food contaminated by a pathogen that causes a disease with a short incubation period to which they might succumb during the next flight.'

Ultimately, a far better safeguard would mean that all crew meals were generated from a separate unit with separate facilities for their packaging and handling. The different meal for pilots rule, often on long sectors where more than one meal service occurs, does not extend to the consumption of any other foods in-flight other than the main hot meal. Crew sandwiches, fruit and snacks are many times ill-considered in terms of their potential safety impact on the cockpit crew fitness to fly. Drinks too, juices in particular, are left to crew discretion.

In terms of food security, there are issues also in terms of the safety impact on food and drink designated for tech crew. We look at these issues in more detail in Chapter 14, in terms of the security implications inherent in the manner in which crew food is produced, packed, labelled, stowed and served.

Overall, having established the very real trend in pilot incapacitation and/or impairment due to food poisoning, we have now concluded that there has been no upgrading of cohesive industry initiatives to deal with the problem in over four decades of compelling data. The overriding perspective I put forward throughout this book is that aviation safety as it appertains to food does not in any circumstance become an amalgamated aspect of the larger aviation safety debate. The connections between mainstream aviation safety issues and aviation food safety are never made, even when tech crew fitness to fly becomes an issue. The picture needs to be considered not just in sound bites of data, but as an overall perspective of interrelated safety issues, many of which implicate the quality and integrity of aviation food provision at their core.

The benefits of food safety training for tech crew

Having established that there is little the aviation industry is willing or able to do in respect of safeguarding the quality and integrity of what tech crew eat and drink down route, it would seem appropriate to turn one's attention to the positive benefits of inducting pilots in the rudimentary rules of food safety fitness to fly and how they may impact on their ability to carry out their duties effectively.

Pilots, by the nature of their job, expect to undertake a great deal of competency training to ensure that the crucial decisions they make are the right ones in terms of the safety function that they have to perform. It would seem sensible then to empower them with the training and tools to ensure that they have the ability to recognise the potential impact that certain foods in certain destinations around the world may have on their ability to carry out their jobs effectively.

The curriculum for such training need not be long or complicated. My experience in training pilots in food safety is that the more they know, the more they want to know. Many issues that, to a food safety professional, may seem rudimentary, have been a total revelation to the pilot community. Very often foods that they consider to be healthy, such as freshly squeezed juices and salads, can pose the greatest dangers to them while they are away from home and it is these issues that need to be part of an integral training initiative aimed specifically at their issues.

It is always useful to gather together groups of relatively lone workers, such as pilots, to share their experiences in this regard, whilst ensuring that the subject matter remains firmly within their particular agenda. Any hint of a training initiative that may be viewed as a more appropriate tool for food handlers, or their cabin crew colleagues, will find the pilot fraternity heading for the hills!

Such training initiatives would prove a valuable tool in enhancing the understanding that needs to occur, so that pilots can be encouraged logically to make the right choices in terms of the food and drink they consume while on duty. Mandates are fine but normally meet with resistance. Far better then to encourage through empowerment and a relative exampling of the issues, a culture of awareness and responsibility to duty, that has proved a historically successful basis for compliance in other areas of the industry.

Whilst there is no amount of training or empowerment techniques that can dissuade the determined pilot from defying the food safety odds in the street markets of the third world, a genuine awareness and context citation of the issues in terms of fitness to fly need to be made. Rather than viewing food safety fitness to fly as a personal liberty limitation, pilots may well discover it instead to be the best possible tool to ensure protection from the incapacitation curse they have come to dread.

It is not to say that all cases of food poisoning experienced by pilots and culminating in in-flight incapacitation are self-inflicted and spawned from an intuitive desire to flout food safety convention and fly in the face of protocol. Indeed, many of the reported cases on which in-flight incapacitation data have been formulated cite the food provided to the crew and served in-flight by the airlines themselves. Invariably, because more than one incidence of food poisoning has to occur to constitute an outbreak, and by the nature of crew meal provision the other members of the crew are unlikely to have eaten the same meal, the true picture of food poisoning data associated with crew food consumed in-flight becomes skewered.

Any awareness training programme for tech crew must also empower them to look for hazards which may be present in the foods they are served in-flight, and encourage them to take an active interest in the quality and safety standards of the catering on board generally. The pilot in charge assumes overall responsibility for the safety, security and integrity of everything on board after the doors have closed, so why not a training initiative to qualify and quantify the safety issues inherent in the catering supply chain, that may have the capacity to impact on the overall safety status of the flight?

I have listened to many crew members over the years recount tales of how the PIC will often insist on eating something from the passenger menu as opposed to their designated crew meal, without really appreciating the potential compromise to safety incurred by such a request. If the protocols are not formalised and galvanised into a training initiative, in the same manner as all other safety-impacting issues, then how much respect are the tech crew expected to have for the procedural aspects of the catering supply chain as it appertains to them?

Developing aviation industry SOPs to assure food safety fitness to fly

To conclude this chapter, I thought it helpful to formalise the threads of the content into some sort of wish list for the industry to ponder. It seems to me that the issue of tech crew fitness to fly is such a crucial one in terms of aviation safety and its direct link to food safety, that I do perhaps have some chance of striking a note and provoking some kind of debate. Exactly how successful crew food initiatives really are in this area requires some kind of formal examination, along with an investigation of how precise the industry recommendations need to be to ensure that cohesion and adherence to the directives become fully visible.

The following constitutes a set of issues for further consideration to assist in the greater proactive assurance of pilots' food safety fitness to fly:

- An industry-commissioned survey into ten-year incidences of food-related illness in pilots whilst on duty, whether related to incapacitation and impairment reporting or not. (Cross-section of data should include those still holding commands as well as a percentage that are retired.)
- A comparison between the numbers of flight crew affected by food-related disease and the number of non-crew workers affected by food-related disease where food is provided at work on the ground.
- An immediate review of the pilot meal provision directives. This should include all foods and beverages consumed by tech crew in-flight and also all

foods and beverages consumed during a determined time-frame period prior to take off on the ground.
- A review of the audit procedures and audit parameters for tech crew accommodation in all ports around the world.
- A proactive approach and incentive to tech crew to eat at audit-approved premises only whilst down route.
- A risk-based review of the manner in which crew meal provision is determined and the parameters by which it is procured, prepared, packaged, stowed and served, formalised into a cohesive industry document.
- An industry commitment to developing and initiating relevant food safety-related fitness to fly programmes endorsed by a food safety agency, which should be pilot-specific and not food industry generic.
- Pilots' representative groups to be proactive about ensuring appropriate standards of safe, wholesome food provision and initiating independently led assessments to scientifically verify the quality, integrity and nutritional completeness of the foods provided by the airlines to their members.
- A specific industry reporting system to deal with food complaints with a formalised procedure for follow-up and resolution.

The issues surrounding tech crew and their food safety fitness to fly will continue to disturb me, not just as a member of the flying public but as someone ideally placed to comprehend the magnitude of the food safety and quality shortcomings of much of what is provided to the aviation sector.

It is crucial that the context of the issues is reviewed satisfactorily and that 40 years of industry data are used to the best possible effect. An overhaul of crew food provision has long been overdue and with the advent of a new security focus and expansive culture, since 9/11, there has never been a better time to deal with the problems cohesively and with a view to assuring the highest possible levels of crew food safety, quality and security on a global scale.

9 Cabin crew – the missing link

This chapter gives me the opportunity to formalise all my opinions about the critical role that cabin crew have to play in aviation food safety.

The industry disagrees with me and always has, which gives me even greater reason and impetus to illustrate the issues that need to be raised in this regard, and demonstrate the undeniable cabin crew link to effective food safety management protocols.

My initial relationship and subsequent burgeoning connection with cabin crew was where my first obsessions with aviation food safety began. Early in my catering career, I had professional involvement with those charged with the dubious responsibility in the business aviation arena of procuring, cooking, and serving food on board corporate aircraft to the movers and shakers of the world.

As a catering provider of such fare, I watched and learned with fascination about all of the techniques that these individuals were forced to employ to get the job done. The focus was that as long as the catering arrived at the aircraft on time and was vaguely palatable, there was no need to question its safety or integrity or indeed the potentially questionable nature of its origin.

Chapter 14 on business aviation food safety reveals all the possible sources of catering supply in the business aviation environment and makes striking comparisons with the commercial aviation sector. However, for the purposes of underwriting my assertion that the cabin crew ultimately have a massive impact on the effectiveness of aviation food safety, be it business or commercial aviation, standard provision or buy on board, the following two stories illustrate my point beautifully.

In July 1998 I was called upon to cater a corporate aircraft at London Heathrow. The long-range business jet was scheduled to carry one of the world's leading figures in film back to Los Angeles, following surgery in London to remove a diseased kidney. I had spoken to the flight attendant the day before to ascertain the precise catering requirements of both the passengers and the crew for the flight home. In addition to the expected, medicinally-based menu selections like soup and casserole, came the obtuse request for two trays of sushi and sashimi.

I questioned the wisdom of feeding such high-risk foods to the already compromised passenger, whereby the flight attendant told me that it was their favourite food, as it was the pilot's!

Horrified at the prospective impact that a food poisoning outbreak implicating sushi could have if both an immunocompromised passenger and the pilot in charge became sick, I set about the job in hand with trepidation. Having failed to convince the flight attendant of the lack of wisdom with which she had devised the menu, I had no choice but to do as I had been told. Had I refused, it would have resulted in the sushi and sashimi having been purchased elsewhere and then served alongside all of the other catering supplied by myself. I preferred to keep control of the whole order for fear of reprisals implicating my food. As it was, we did not manufacture the sushi and sashimi products ourselves; we outsourced them to a specialist supplier in the manufacturing sector who could verify product safety with shelf-life in excess of 24 hours.

It was a glorious sunny July day when I arrived at the aircraft in my refrigerated truck and began to offload my gorgeous cargo of diligently and immaculately prepared goodies. Slowly but surely every available area of chilled storage space quickly became overburdened: ice drawers were packed with sodas and the gasper air cabinet was crammed with soup and sandwiches. Finally, I carefully carried the two trays of stunning looking sushi and sashimi out to the aircraft. The flight attendant took them from me and then peered around the galley looking for an appropriate place to store them. High and low she craned and strained as if expecting some miracle of ergonomics to occur.

Eventually, accepting that all appropriate food storage areas were, by this time, fully laden, she threw open the front lavatory door and placed the two sushi trays on the toilet seat! My mouth fell open, I stumbled and spluttered trying to make utterances in protest. She merely waved an immaculately manicured hand at me and dismissed my concerns: 'Honey, I've been doing this job 30 years and I ain't killed no one yet. It will be fine.'

With that I left. My mind was clouded with a vision of screaming front-page headlines, 'First man of film dead – killed by tray of in-flight sushi'!

That day changed my life, and my obsession with advancing the issue of aviation-specific, recurrent, food hygiene training for all cabin crew began.

Seven years later very little has changed. There remains no mandatory industry requirement to train crew in food hygiene despite legislative compliance demands in many countries including Europe to suggest that:

'all food handlers be trained in hygiene matters commensurate with their work activity.'

The aviation industry will not recognise hygiene training as part of the same compulsory training agenda as fire and egress training, medical emergency and standard emergency procedures (SEP) training and crew resource management (CRM). The perspective remains that the primary function of crew in the commercial sector is to fulfil a safety role in terms of an emergency situation. They are not openly acknowledged as food handlers.

Conversely, in the business and corporate aviation arenas the situation is the reverse. With no mandatory requirement for any third crew member to be in evidence on private aircraft with less than 19 seats[81], the corporate flight attendant is perceived more in terms of their service function than their safety one. The common denominator in both cases is that the need for food hygiene training is not acknowledged and fully understood by either the commercial or corporate aviation sectors.

We look later at the impact of the absence of training on passenger and crew safety and on aviation safety issues generally, but before then the second of my stories.

In 2004 my partner and I took a short break early in the season to Crete. The availability of scheduled flights was restricted to travelling via Athens so for ease of purpose we booked a flight-only deal with one of the UK-based charter airlines.

The outbound departure was early and as we left London Gatwick we waited in anticipation of what culinary breakfast delights we might be offered. As the tray arrived I scanned the contents looking for verification that all labelling and ingredients declarations were indeed missing from my 'special meal'. Having satisfied myself that the industry had once again surpassed itself in its inability to meet both its statutory and moral obligations in all labelling arenas, I cast the contents of the tray to one side. I refused all further enquiries from the crew during the endless buy on board drinks rounds and made notes for my own reference.

On arrival in Crete some four and a half hours later, I watched in horror as the crew covertly attempted to remove the hot meals for the return flight to Gatwick from behind blankets and cushions in the overhead lockers. Box after box came down from the lockers as the crew hurried up and down the aisles in readiness to load the meals into the oven racks.

Once in the arrival hall, I checked on the board to discover that the outbound flight back to Gatwick had incurred a three-hour delay. My first thought was the return flight meals on board, sitting in oven racks with the heat of the Greek day only just beginning.

The following week we boarded our flight back to London. Shortly after taking off we were approached with the inevitable request from the crew: 'Would you like some lunch?'

I called over the chief flight attendant and politely enquired as to whether my lunch had been stored during the sector from London earlier that morning in the overhead locker. Her face dropped and the endless excuses began to flow: 'All the meals are frozen.' 'It would cost the airline too much money to stow them in the hold and then have to pay to have them brought up.' 'Environmental health have said it is OK.' 'We haven't killed anyone yet.'

I suggested an analogy. I enquired whether, if she went to a restaurant on the ground and found the chef taking a meal out of the boot of his car and reheating it for her consumption, she would deem that acceptable. Silence at last pervaded.

I continued my line of enquiry. 'Have you ever received any hygiene training? Did you take a test? Is it recurrent? What did it involve? Do you take the temperature of the food at all? How do you monitor the integrity of the food?' All enquiries were met with the same responses; 'No', 'I don't know' or 'I can't remember'.

The purpose of recounting this story is to illustrate effectively the reason why crew hygiene training is such a non-issue for the industry. As one hygiene officer, working for one of the UK's most media-prominent scheduled carriers, once remarked to me: 'If we empowered our crew with food safety knowledge we would never get our planes off the ground.'

It is against this backdrop then that we look at the issue of cabin crew hygiene training in terms of its impact on food safety. In-flight food safety management has to begin and end with the crew and therefore their ability to understand the critical food safety issues that affect both passengers and crew is paramount.

Cabin crew food service role explained – chefs, or merely waiters in the sky?

Whilst I am not going to attempt to rewrite the history of aviation food service evolution (I will leave that to the experts like George Banks at British Airways) it is important to understand how the ever changing nature of in-flight food service concepts has seen a simultaneous transformation in the connection that the cabin crew have with the food on board.

Historically, the crew role has fluctuated hugely from the early days when they cooked and prepared everything from scratch on board and the aircraft galley was located in the hold; to the evolution of non-exclusive air travel and box or tray meals for the masses; to the establishment of premium and economy cabin differentials, whereby the food at the front was prepared on board and the food at the back retained box meal status.

To fully understand the role that cabin crew play in connection with food on board, it is impossible to generalise. Every airline will have its own food service itinerary, menu format and service protocols. The main focus for attention in terms of the crew involvement with the food has to be in the premium business and first class cabins. However, that is not to negate their connection and responsibilities in the food safety arena with the food and drink served on low cost and buy on board services. Ultimately, whatever the airline and whatever the service protocols, front end or economy cabins, they will all be required to undertake beverage and drink services. For the purposes of liability and the definition of food in terms of the law, both water and ice constitute food.

For many years cabin crew have fought against their trolley dolly or steward/stewardess labels, in an attempt to have themselves more positively viewed by both the public and their peers as cabin service and safety professionals.

Food service techniques are all based around what happens in the cabin in the full view of the passengers, with very little attention paid by the airlines to galley service protocols. Many of my crew friends are happy to recount tales of their inappropriate behaviour in the galley. Whilst it is true to say that this type of picture could likely be viewed from the other side of any restaurant door, in many circumstances the inappropriateness of the behaviour is not acknowledged as such by the crew themselves due to a fundamental misunderstanding of the safety issues. The same cannot be said of trained chefs and kitchen staff, who are well aware of the rules even if they do not necessarily abide by them.

Often crew will be actively encouraged by their supervisors or trainers to do whatever it takes to get the job done, with little regard for food safety. The best example of this that I can think of is the common practice of scrambling raw shell eggs for premium cabin breakfast service in the ice bucket. In the absence of crew training and in the absence of having anything more suitable in which to carry out this task, the practice is commonplace on many international airlines.

To put the crew connection with food into perspective, the first thing we have to do is break down the nature of their job in its entirety. Table 9.1a illustrates this.

A glance at Table 9.1a shows the complete cross-section of tasks potentially undertaken by the average cabin crew member. Whilst the table is my own breakdown and labelling of tasks, not necessarily that of the industry, it is clear that

Table 9.1a Spectrum of cabin crew responsibilities

Job aspects	Skill category	Training given	Training recurrent
First aid	Medical	Yes	Yes
Cleaner	Service	No	No
Pest control	Duty	Yes/No	No
Fire fighter	Safety	Yes	Yes
Evacuation	Safety	Yes	Yes
Retail sales	Duty	Yes	Yes
Customer service	Service	Yes	Yes
In-flight police	Duty	Yes	Yes
Food service	Service	Yes	No
Food safety	Safety	No	No
Food auditor	Safety	No	No

the crew are empowered with the necessary skills set required to undertake these tasks only when regulatory compliance is an issue, or where there may be some perceived impact on revenue being generated, i.e. on-board sales techniques.

Despite Table 9.1a being slightly generalised for the purposes of this exercise, it is obvious that in the service arena the incidences of both initial and recurrent training fall below those encountered in the duty and safety arenas. By categorising the crew job description in this way it is interesting to see not only how many areas of responsibility the crew remit encompasses, but also how safety in relation to food is not perceived in the same way as safety in any other aspect of the job. Safety-related undertakings are mandated by the aviation industry where crew are involved in performing these tasks. The service areas are only regulated in terms of the impact that service styles may have on health and safety, but it remains in the airline's interest to lay down service training requirements in order to present a good image and gain a service driven reputation. In terms of the safety impact of any food or drink-related activity, the acknowledgement is non-existent.

The actual tasks undertaken in connection with the catering are broken down into the steps shown in Table 9.1b, which may or may not apply depending on the sector length and service protocols. This figure illustrates the operational aspects of the crew management of catering products on board.

The list of tasks is generalised, to encompass a vast range of airline service styles, and will be hugely influenced by sector length and status of the individual

Table 9.1b Cabin crew responsibilities in connection with food and beverage service on board

1	Crew receive catering stores from the catering provider on board (chilled, ambient, frozen
2	Crew check catering stores and stow in designated stowage
3	Crew undertake preboard drinks service in premium cabins
4	Crew undertake post take-off drinks service in all cabins
5	Crew reheat meals for hot meals service (if applicable)
6	Crew serve special diet meals including restricted diets, babies and children
7	Crew prepare premium cabin appetisers by plating and presenting or silver serving
8	Crew conduct tray set meal service in some or all cabins (hot or cold)
9	Crew plate and present premium cabin entrées or silver serve from trays having been transferred by them
10	Crew plate and present dessert and cheese or silver serve from pre-prepared trays
11	Crew serve in-flight snacks from pre-prepared tray sets or prepare themselves from bulk-packed items in the galley
12	Crew conduct 'buy on board' drinks and chilled and ambient snack service
13	Crew serve ice cream in economy and premium cabins. Economy is prepacked in tubs or bars, premium cabins may be served in a bowl or as a side accompaniment. Ice cream dessert may be served from a trolley
14	Crew serve tea and coffee in all cabins
15	Crew serve water and juice in all cabins
16	Crew prepare and serve the cockpit crew

airline. It is, however, pretty comprehensive in taking an overview of what cabin crew may perceivably be asked to undertake in connection with food and drink preparation and service, on board.

Let us not lose sight of the fact that all crew will be expected to undertake all tasks potentially simultaneously, and whilst some will be designated to operate the galley and some the cabin end of the service, this will vary from flight to flight. If one considers the nature of the rest of the tasks that fall within their remit (Table 9.1a) and the possible combination of any of these in connection with food service and preparation, the case for hygiene training solutions for crew that deal with the multirole, multirisk issues grows ever stronger:

- Cleaning – sick bags, toilet areas, floor spillages, chemical usage+food prep and service.
- Pest control – spraying insecticide aerosols for disinsection procedures+food prep and service.
- First aid – potential contact and exposure to bodily fluids+food prep and service.
- Retail sales – handling of a variety of products whose tertiary packaging will be contaminated: bottles, packets, newspapers and magazines+food prep and service.

It is essential not only that crew receive in-depth, certified and recurrent hygiene training but that it is generic to their specific issues. I have always maintained that there is no benefit in putting crew through a basic level qualification in food hygiene such as those mandated for kitchen staff, when their food safety and hygiene issues are so peculiar to them.

It is essential that protocols such as those covered in the medical emergency training and general policy directives laid down by Port Health Regulations, in connection with the identification and treatment of sickness on board, are brought together so that the connections are made between the risk factors inherent in undertaking multi-roles. To expect crew to devolve the information from all risk-related aspects of the job and make these connections for themselves, with only a basic knowledge or no knowledge at all, is unrealistic and naive to the potential hazards.

Cabin crew as in-flight auditors

I have always believed that the most critical role that crew have to play, and one which probably presents the best reason why they should receive mandatory hygiene training, is that of in-flight auditor.

For the most part, airlines will carry out hygiene audits on their catering suppliers around the world at various intervals. Many will also have key personnel stationed down route who routinely check the catering content and quality against specifications. Depending on the size and scale of the airline operation they may or may not have dedicated personnel who have food safety specifically as their remit. More often than not, even with some of the largest international airlines, this most critical aspect of the business will be designated as an adjunct to an existing role. On many occasions I have been referred to 'Cabin Service' or 'Catering Managers' at airlines' HQ, when I have made a food safety-related enquiry.

With these factors in mind, we start to build a case for the role of the cabin crew as in-flight food safety and quality auditors. Despite their lack of training in this role, it is the crew who are charged with the overall responsibility to both accept and stow all the catering supplies at the commencement to the flight. They sign to acknowledge acceptance of goods received and record any shortages or anomalies. It is they who are at the sharp end in terms of being in a position to judge whether the safety and quality parameters of the catering supplies are being completely satisfied. If the food arrives on board out of temperature control or in poor condition, it is they who are left to make critical decisions as to how to deal with it. The consequences of the decisions they make at this stage of the process, without sound knowledge and guidance being set, can be disastrous.

Historically, airlines are reluctant to acknowledge this critical role undertaken by the crew on a daily basis for fear of empowering them with the authority to reject catering items that are not fit for purpose and consequently delaying the flight. In aviation terms delayed flight status, particularly for scheduled carriers, is the most expensive incident to befall an airline outside of physical aircraft damage or full-blown accident. In the absence of hygiene training the crew remain ignorant to many of the risks posed to in-flight catering by mishandling and temperature abuse.

Often, when I have battled the issue of mandatory hygiene training for crew with my industry colleagues, they have argued that the crew's handling connection with the food on board is so minimal for it not to be an issue. In the case of tray set up or boxed meals, there may be some substance to that; however, in terms of the critical role they have to play at the aircraft door the industry case must surely be dismissed.

Much of the acceptance paperwork carried out at the commencement to the flight is filled out by the representatives from the catering suppliers themselves. Not empowering crew with the tools and training to check arrival temperatures and record non-compliances, allows non-performances by catering suppliers to go unrecorded outside of scheduled audits and the accompanying paperwork may well prove not to reflect the true picture in terms of the catering provider's performance.

In terms of ensuring that the best practice protocols established in the industry guidelines are upheld on the aircraft as they are intended to be on the ground, there remains no case not to acknowledge the food safety training requirements of the crew. The in-flight monitoring of the product integrity, quality and safety undeniably rests in their hands; to suggest otherwise is a nonsense. Having established that it is critical in any supply chain to monitor product integrity and safety at every stage, it is unrealistic to assume that there is any satisfactory, defensible procedure for the transfer of goods from supplier to aircraft that does not implicate the crew in a quality assurance and auditing capacity.

It is interesting to note in guidelines issued by both ICAO and IATA, that the cabin crew role is defined in this capacity by the documentation of recommended cabin crew training content.

Chapter 12 of Annex 6[82] to the convention on ICAO, 'Operation of aircraft', contains in paragraph 12.4 the following standard:

'an Operator (airline) shall establish and maintain a training programme approved by the State of the Operator, to be completed by all persons before being assigned as cabin crew member. Cabin crew shall complete a recurrent training programme annually.'

This standard is supported by the ICAO Training Manual cabin attendant's safety training, Part E-1 (Doc 7192–Second Edition 1996), which contains in Chapter 8, section 8.2, 'Training Objectives', a performance requirement that flight attendants after training:

'will be able to describe medical aspects related to air transport operation, identify the basis of transmissible diseases and protect themselves and their passenger from such diseases.'

This is followed in section 8.3 by a list of 'required knowledge, skill and attitude', which includes:

'transmissible disease; risks posed by drinking water, milk, ice, fruit, salads and raw vegetables, meat and fish and perishable foods.'

To acknowledge that the food chain quality and safety monitoring protocols should extend on board and into the flight, would invariably implicate the crew and automatically necessitate and validate the requirement for the specific types of training recommended by the industry guidance standards. As I see it, what the current practice actually does is invalidate all other procedures.

The complexities of why or why not the industry chooses to acknowledge the crew role in food safety issues on board is part of a much larger debate. It revolves around the issue of whether the airline catering suppliers can legitimately be classified as such, thus being subject to food service/catering sector systems management protocols, or whether it is the airline itself that remains in this sector, with the food service environment actually being the aircraft. In this instance supply to it of ready prepared meals would render airline caterers actually airline food manufacturers. To acknowledge and embrace this distinction would dictate that crew acceptance and possession of the food would be the dividing line.

Table 9.2 shows the inherent risks attributable to the crew connection with the food, in a documented format that should be transferred into a full blown in-flight HACCP plan. If one takes the earlier breakdown of functions associated with the crew role, then the devolved risks are easy to quantify and the necessary controls and monitoring procedures are vital to establish. As in any plan, certain aspects of the crew food function will prove critical to food safety. Table 9.2 illustrates the crew role in the continuation of the in-flight supply chain and associated hazards.

Whilst the list in Table 9.2 is not a consummate example, it serves to illustrate the critical and ongoing control that crew have with the food chain on board. In the absence of any job-specific hygiene training that highlights the risks and amalgamates training examples into on-board SOPs, many critical aspects of the supply chain are left in the hands of uninitiated in-flight personnel.

Table 9.2 also illustrates the extent of the audit role forced upon the crew to monitor temperatures at point of receipt, during stowage and at point of service and to monitor and control pack and product integrity. Not only must the crew be given the training necessary to carry out these duties effectively,

Table 9.2 Crew's role in the product supply chain in-flight and hazards they need to document

Crew function	Hazards	Monitor	Control
Receive catering	Out of temp	Check centre temps	Keep chilled
	Packaging damaged	Check pack integrity	Accept/reject
	Not to spec	Check against spec	Accept/reject
Stow catering	Microbial growth	Check temps of chiller	Keep chilled
	Physical contamination	Check pack integrity	Keep covered
			Galley free from debris
Preparing/plating	Contamination	Check in date	Personal hygiene/handwash
	Physical	Check for contaminates	
	Chemical	Check usage and storage	Galley free from debris
	Biological		No spraying of insecticide
Reheat meals	Microbial survival	Check centre temps	Heat to +75 °C
Serve SPMLs	Not to spec	Check against spec	Accept/reject
Ice cream service	Microbial growth	Check temps −18 °C	Accept/reject

they must also be equipped with the necessary tools to do so. Calibrated probe thermometers must be in evidence, as must systems and documentation of temperature and product integrity monitoring. Ultimately it is vital that critical limits are set by the airline in conjunction with their catering providers, and that any deviation from these is documented and reported.

Any failure by the industry, and the catering community that serves it, to acknowledge the vital role that cabin crew have to play in this regard, has a major impact on the undermining of all other food safety management protocols which may have been established prior to the catering reaching the aircraft door.

Cabin crew health and the risks posed to food safety

Having established that the role cabin crew play in the ongoing food safety management supply chain is critical, and having established that their connection with the food and drink handling and service on board is extensive, it is necessary to analyse inherent factors which may impact on their overall suitability as professional food handlers.

It is well documented in the safety management protocols of all food businesses that there is an obligation to monitor and control any food handler's state of health and establish a strong sickness reporting policy and culture within the business. It is also critical that those persons engaged in a food handling activity who travel to high risk destinations on holiday or during the course of their business activity, *must* prove themselves fit to return to work so that any incidence of food-related illness can be identified, documented and dealt with before they return to work. The risks posed to food safety by those displaying healthy carrier status in a food environment are well documented; all the more critical then to ensure that those who travel and handle food for a living have their fitness to work carefully monitored and controlled.

Cabin crew sickness reporting

In terms of the role that crew have to play in the food chain, it is vital that they are made aware of the impact that handling food whilst sick can have on the food safety of both the passengers and fellow crew members they serve. If one takes into account the increased probability that these most unique of food handlers will perceivably prove far more likely to be exposed to incidences of food-related illness and display symptoms of these, due in part to the travel nature of their profession, then the requirement to monitor closely their state of health and control their food handling activity is absolutely critical.

Many airlines' sickness reporting policies have far more to do with the ability of the crew to carry out their safety function than their food safety function. The rules vary from airline to airline; however, many crew who report themselves sick, either during a trip or while on duty down route, will be expected to remain down route until they can work home fit.

Whilst on the surface of it this may seem an adequate way of dealing with the situation, if the illness is food or drink-related the symptoms can vary hugely in magnitude and duration, and without proper sampling it will be difficult to diagnose the precise nature of the complaint and therefore impossible to be sure of the potential risks posed to food safety of a crew member returning to work.

In the probable absence of any indoctrinated crew awareness of the potential risks a bout of sickness and diahorrea can have on the food safety status of those they may serve, it is essential for the airlines to ensure that those crew members displaying food-related illness symptoms refrain from food handling, as they would on the ground, and are not permitted to return to work until they have proved their biological fitness to handle food; far better then to ensure that those crew members who are fit to travel are passengered home, not left down route at all. The temptation among cabin crew members in these circumstances may be to avoid reporting bouts of food-related illness at all, for fear of being left down route, and unless the symptoms are so debilitating that they cannot move, to take some over-the-counter medicine to temporarily quell the symptoms of the complaint and work home.

For the educated food safety professional and indeed for those with an interest in the global transit of food-borne disease, the situation of cabin crew/food handlers travelling sick and handling food is extremely disturbing. Earlier I suggested that these food handlers were part of a unique group. What other profession where a connection with food and drink preparation and service is so prevalent, is simultaneously combined with a situation whereby the food handler has the capacity to eat and drink in a global cross-section of as many as ten different countries a month? One has to question these individuals' suitability as food handlers in terms of the standards of food safety protocol, established and mandated in food businesses on the ground.

At this point we begin to make strong connections between the real impact of the non-existence of mandatory, job-specific food hygiene training for crew and the potential magnitude of the risks posed to food safety and public health. Later in the chapter we look at the specifics of what the training content needs to be to satisfy the risk factors really posed by travelling food handlers, but for now we need to continue to make connections between the crew role as food handlers and the potential impact of their compromised health status on passenger and fellow crew safety.

Cabin crew role in monitoring food standards

Whilst discussing the role of the crew as in-flight auditors it was possible to witness the extent to which the crew connection with the catering supplied and consumed on board had an impact on the integrity of the final stages of the overall supply chain. What was not considered, however, was the possibility that the crew also had a significant role to play in the monitoring of the quality and integrity of the in-flight food supplied, by virtue of the fact that they are the ones most often consuming it.

Even for the most extensive and frequent travellers, exposure to in-flight food could in no way compare in terms of frequency and proliferation with that of the cabin, and indeed flight-deck crews, who are consuming it at a variety of destinations every day of their working lives. The real evidence of the state of its safety and integrity lies not with the fare paying public, but with those who take to the skies every working day.

The fact is that the crew are the best sources of research material in this regard. They are the undeniable in-flight auditors of product safety and integrity, not only through incident reporting but through a system of monitoring their own personal state of health following the consistent consumption of in-flight food and drink products provided to them from a variety of out stations around the world.

It is important when considering the wider aspects of any study undertaken into this subject to consider the inclusion of any food or drink products consumed on the ground while down route. This includes food and drink consumption in the audit-approved crew accommodation on the ground as well as the potential impact posed by eating outside of crew accommodation in high-risk destinations.

I have always asserted that it is the responsibility of the airlines to ensure that all crew, including the flight-deck crew, only eat and drink at approved hotels and restaurants down route to ensure not just food safety, but in the case of the flight-deck, aviation safety. The crew, whether flight-deck or cabin, when down route are, to all intents and purposes on duty, not on holiday. Therefore like any responsible company charged with ensuring that health and safety considerations are fully met, the airline has an obligation to ensure that crew accommodation is safe and secure. This includes any impact posed to health by inadequate food and water supply. Having established a list of audit-approved hotels all over the world, why run the risk that crew may become unfit to fly by allowing those on duty to eat and drink at non-approved premises? In the absence of any directives along these lines, the statistical divide monitoring in-flight food integrity is coloured by the consumption of food and drink by crew from non-approved sources on the ground.

To put the situation into context, one only has to look at the best practice directives initiated by the food manufacturing sector, in which any food handler's state of health in the workplace is not allowed to be compromised by bringing food and drink in from outside with the potential to contaminate the food chain. Many food manufacturing environments will go further and provide subsidised canteen facilities and vending products. Whilst any food handler on the ground has the capacity to compromise their own food safety and integrity by mishandling in the home and by making poor choices when eating out, let us not forget that the crew down route are not on their own time; they are on duty and as such should be expected to abide by the rules to ensure food safety. In the absence of any formal food hygiene training having been undertaken by the crew, it is unlikely, however, that they will fully appreciate their personal capacity to impact on the integrity of the food they are handling in-flight and the risks posed by them as infected food handlers.

I believe it is essential for all airlines to acknowledge and accept the very direct connection that the health status of the crew has on the quality and integrity of the food handled by them on board, and to embrace the benefits of using their state of health indicators as an effective and relevant way in which to monitor the true picture with regard to the safety of the food supplied around the world. By adopting a proactive approach and by achieving a situation in which the crew culture of health reporting has more to do with highlighting potential problems with the catering supply chain than apportioning blame, a far truer picture will emerge and the reactive strategies currently employed when the airline is faced with an allegation of food-associated illness will prove redundant.

Whether the industry accepts it or not, the undeniably critical role that crew have to play in monitoring and controlling the product safety and integrity, in tandem with the potential risks that their own state of health can pose, has to be recognised and the issues formalised into effective training strategies.

Food hygiene training protocols for cabin crew

Having spent the earlier parts of this chapter building the case for the crew training requirement, it is now time to look at the potential training content which needs to be employed in order to cover all of the hygiene and food safety aspects of the cabin crew role. Let us not lose sight of the fact that any crew training in this regard has to have a direct connection with the nature of the food safety protocols established on the ground at point of production and during transit. The systems management protocols that have been mandated and established at ground level will have a direct bearing on what is required and how it is to be validated and verified by the crew on board.

It is critical that the training requirement, whatever it may be, evolves year on year if it is to remain credible and viable, in exactly the same way as other aviation-focused training programmes. The training solutions must also be airline and product-specific, with generic aspects kept to a minimum.

It would be impossible to describe in detail all of the potential training solutions required by every airline without an in-depth knowledge of each individual airline's operation, service styles and techniques. For the purposes of this exercise we will divide airlines into generic groups, which may each have a variety of catering options on board but at least will be categorised by sector length and their likely type of food provision. Table 9.3 illustrates the methods that can be employed to determine food safety training parameters for crew.

In order to determine the factors that influence whether all the cabin crew or just key members of the crew require training, the dividing line probably needs to be drawn between the provision and service of high risk or low risk foods. In the event that the service provision includes high risk foods in the menu and service portfolio, all crew should automatically be trained. In an ideal world, all crew should be trained regardless of the service type or provision, but the reality may prove easier to achieve if a distinction is made based on an assessment of risks associated with the products provided, served or sold on board, and the potential handling requirement of the crew.

We will look at each potential service provision and application in turn.

Table 9.3 Methods of determining food safety training parameters for cabin crew

Flt classification	Sector length	Service	Training required
Long haul scheduled	Plus 6 hours	Full	All crew
Short haul scheduled (Domestic only)	<6 hours	Snacks and drinks (ambient foods)	Key crew
Short haul scheduled (Non-domestic)	<6 hours	TSU/hot and cold (snacks and drinks)	All crew
Long haul charter	Plus 6 hours	BOB/TSU (high-risk foods)	All crew
Short haul charter	<6 hours	BOB/TSU (high-risk foods)	All crew
Low cost short haul	<5 hours	BOB (snacks and drinks) (ambient foods)	Key crew

Long haul – scheduled

Potential service requirement

- Tray set up in economy cabins comprising hot meal, and accompaniments such as salad, fruit or dessert. Additional snack service may occur which may comprise a sandwich or cake item. Where the hot main is lunch/dinner, the snack item may be breakfast oriented. Where the hot main is breakfast, the snack service may be lunch or dinner oriented.
- Hot and cold beverage service.
- In premium cabins, the requirement will be a full service selection of starters, hot and cold entrées, salads, sandwiches, desserts and cheese.
- Snack service comprising sandwich provision will also be required. Depending on sector length, full service hot and cold breakfast is also likely to be available.
- SPML service in all cabins.

High risk food potential

Salads, meat and fish served hot or cold, desserts, cheese, sandwiches, cold entrées, prepared meals, SPMLs, fresh squeezed juice, egg dishes, sushi, canapés, milk and dairy including ice cream.

Crew service/handling requirement

- Food and drink receipt and appropriate storage – ambient, chilled or frozen.
- Reheating and serving.
- Plating and presenting.
- Cooking from raw, i.e. scrambled egg for breakfast in premium cabins.
- Ice handling and stowage.
- SPML service.
- Flight-deck meal service.
- Hot beverage service.
- Cold beverage service.

Crew training template

- Personal hygiene and hand-wash protocols.
- Food handlers' fitness to work.
- Temperature control – receiving, storing, reheating and serving.
- Methods of contamination, and controls and monitoring appropriate to in-flight.

- In-flight auditing tasks – raw material QA, temp monitoring, peers' health status, galley cleanliness and pest activity, crew personal hygiene and hand-wash.
- Galley cleaning techniques appropriate to in-flight chemical usage (Control of Substances Hazardous to Health – COSHH).
- SPML comprehension and allergen information.
- In-flight HACCP training and documentation familiarisation.
- Microbiology and food poisoning.
- Port Health Regulations – waste disposal of food down route and impact on global food chain.
- Ice hazards.
- Knowledge tested by competency examination.
- Training recurrent annually.

Short haul scheduled – domestic only

Potential service requirement

- Ambient snacks served – crisps, peanuts, pretzels, etc.
- Hot and cold beverage service.

High risk food potential

Only in relation to allergens.

Crew service/handling requirement

- Food and drink receipt and appropriate stowage – ambient or chilled (drinks only).
- Ice handling and stowage.
- Flight-deck meal service.
- Hot beverage service.
- Cold beverage service.

Crew training template

- Personal hygiene and hand-wash protocols.
- Food handlers' fitness to work.
- In-flight auditing tasks – stock rotation and risks posed by out-of-date foods including ambient snacks and beverages.
- Methods of contamination, and controls and monitoring appropriate to in-flight.

- In-flight auditing tasks – raw material QA, peers' health status, galley cleanliness and pest activity, crew personal hygiene and hand-wash.
- Galley cleaning appropriate to in-flight chemical usage (COSHH).
- SPML comprehension and allergen information.
- Ice hazards.

Short haul scheduled – non-domestic

Potential service requirement

- Tray set ups in all cabins – comprising hot and cold meal service with accompaniments, salad, dessert, cheese, etc.
- Hot and cold beverage service.
- Sandwich or snack service if full meal not available.
- SPML service.

High risk food potential

Pre-prepared and plated cold meat and fish, salads, cheese, desserts and prepared fruit, SPMLs, sandwiches, milk and dairy.

Crew service/handling requirement

- Food and drink receipt and appropriate storage – ambient, chilled, frozen.
- Serving cold from tray set ups (TSUs).
- Ice handling and stowage.
- SPML service.
- Flight-deck meal service.
- Hot beverage service.
- Cold beverage service.

Crew training template

- Personal hygiene and hand-wash protocols.
- Food handlers' fitness to work.
- Temperature control – receiving, storing, reheating and serving.
- Methods of contamination, and controls and monitoring appropriate to in-flight.
- In-flight auditing tasks – raw material QA, temperature monitoring, peers' health status, galley cleanliness and pest activity, crew personal hygiene and hand-wash.
- Galley cleaning techniques appropriate to in-flight chemical usage (COSHH).

- SPML comprehension and allergen information.
- In-flight HACCP – training and documentation familiarisation.
- Microbiology and food poisoning.
- Ice hazards.
- Knowledge tested by competency examination.
- Training recurrent annually.

Long haul – charter

Potential service requirement

- Tray set-up meals in all cabins – hot and cold service.
- Additional snack service may translate into breakfast or lunch depending on sector length.
- Hot and cold beverage service.
- SPML service.
- Buy on board – snacks ambient and chilled; drinks – hot or cold.

High-risk food potential

Prepared meals chilled and frozen, sandwiches, salads, desserts, cheese, prepared fruits, SPMLs, fresh juices, milk and dairy including ice cream.

Crew service/handling requirement

- Food and drink receipt and appropriate storage – ambient, chilled, frozen.
- Reheating and serving.
- Ice handling and stowage.
- SPML service.
- Flight-deck meal service.
- Hot beverage service.
- Cold beverage service.
- Retail service of high-risk, pre-prepared foods.
- Retail service of low-risk ambient foods.
- Return catering of held frozen prepared high-risk foods – defrosting and reheating.

Crew training template

- Personal hygiene and hand-wash protocols.
- Food handlers' fitness to work.
- Temperature control – receiving, storing, defrosting, reheating and serving.

- Methods of contamination, and controls and monitoring appropriate to in-flight.
- In-flight auditing tasks – raw material QA, temperature monitoring, peers' health status, galley cleanliness and pest activity, crew personal hygiene and hand-wash.
- Galley cleaning techniques appropriate to in-flight chemical usage (COSHH).
- SPML comprehension and allergen information.
- In-flight HACCP – training and documentation familiarisation.
- Microbiology and food poisoning.
- Ambient stock rotation and risk posed by out-of-date foods including ambient snacks and beverages.
- Labelling compliance directives for retailed food and drink products.
- Ice hazards.
- Knowledge tested by competency examination.
- Training recurrent annually.

Short haul – charter

Potential service requirement

- Tray set-up meals in all cabins – hot and cold service.
- Hot and cold beverage service.
- SPML service.
- Buy on board – snacks ambient or chilled; drinks hot or cold.

High-risk food potential

Prepared meals chilled or frozen, sandwiches, salads, prepared fruit, desserts, SPMLs, fresh juices, cheese, milk and dairy including ice cream.

Crew service/handling requirement

- Food and drink receipt and appropriate storage – ambient, chilled, frozen.
- Reheating and serving.
- Ice handling and stowage.
- SPML service.
- Flight-deck meal service.
- Hot beverage service.
- Cold beverage service.
- Retail service of high-risk, pre-prepared foods.
- Retail service of low-risk ambient foods.
- Return catering of held, frozen prepared high-risk foods – defrosting and reheating.

Crew training template

- Personal hygiene and hand-wash protocols.
- Food handlers' fitness to work.
- Temperature control – receiving, storing, defrosting, reheating and serving.
- Methods of contamination, and controls and monitoring appropriate to in-flight.
- In-flight auditing tasks – raw material QA, temperature monitoring, peers' health status, galley cleanliness and pest activity, crew personal hygiene and hand-wash.
- Galley cleaning techniques appropriate to in-flight chemical usage (COSHH).
- SPML comprehension and allergen information.
- In-flight HACCP – training information and documentation familiarisation.
- Microbiology and food poisoning.
- Ambient stock rotation and risks posed by out-of-date snacks and beverages.
- Labelling compliance directives for retailed food and drink products.
- Ice hazards.
- Knowledge tested by competency examination.
- Training recurrent annually.

Low cost – short haul

Potential service requirement

- Selection of buy on board ambient snacks.
- Hot and cold beverage service.

High-risk food potential

Only in relation to allergens.

Crew service/handling requirement

- Food and drink receipt and appropriate storage – ambient or chilled (drinks only).
- Ice handling and stowage.
- Flight-deck meal service.
- Hot beverage service.
- Cold beverage service.
- Retail service of low-risk ambient foods.

Crew training template

- Personal hygiene and hand-wash protocols.
- Food handlers' fitness to work.

- Methods of contamination, and controls and monitoring appropriate to in-flight.
- In-flight auditing tasks – stock rotation and risks posed by out-of-date foods including ambient snacks and beverages.
- Galley cleaning techniques appropriate to in-flight chemical usage (COSHH).
- In-flight auditing tasks – raw material QA, peers' health status, galley cleanliness and pest activity, crew personal hygiene and hand-wash.
- SPML comprehension and allergen information.
- Labelling compliance directives for retailed food and drink products.
- Ice hazards.

By examining the catering requirement of each type of operation in tandem with the training templates required, a pattern begins to emerge between the sector length and potential type of meal provision. In any airline operation where food and/or drink form an integral part, whether the offer is traditional full service, buy on board or a combination of both, an acknowledgement and clear understanding of the specific issues that face the crew are critical to ensure that the appropriate training templates are adopted.

The food safety and hygiene issues for crew are impacted by several factors. These in turn will determine the training requirement specifics:

- Sector length.
- High-risk food provision.
- Reheated food provision.
- Crew plating and presenting application.
- Retail sales or integral full service provision.
- SPML provision.
- Flight-deck food provision.
- Return/round trip catering mechanisms.
- Chilled storage available on board.

It is also crucial to take an overview of the entire catering picture airline-to-airline in order to determine, in less general terms than those in the examples above, exactly what is required. For example, the non-availability of chilled stowage on board when the catering provision is high risk results in a situation where specific training will need to be given to crew to underwrite the safety protocols established on the ground. Specific temperature and time-related HACCP would need to be devised, and an understanding of the systems management required to monitor and control the risks would need to be proven by the crew themselves.

It is interesting to examine exactly how great the crew safety impact and connection with the food and drink on board really is. It is only when the aspects are broken down, as previously documented, and examined in the light of the fashion of the catering provision and service styles, that this becomes apparent.

It is crucial that before any training remit can be fulfilled, this exercise is undertaken and reviewed to encompass any changes in catering provision or service activity elsewhere or down route.

It is important that this most overburdened of professions does not fall foul of inappropriate training mandates in this arena. The information and its relevance must be predetermined and predefined, taking account of the nature of the entire production and supply process, in order that the mandates are of maximum effect. Poor training in this area will ultimately have a massive impact on the overall success and integrity of the catering supply chain, whatever it may be.

In operations where buy on board is the preferred service option, adequate and effective training will have a major influence on the efficiency of stock rotation systems and ultimately product safety. This in turn will have a bearing on stock losses and costs. Whether it is the airline or catering provider who is underwriting such losses, the role the crew have to play will prove instrumental in determining the fiscal success or failure of such a system. In this area of the industry where airline operational costs are potentially partially underwritten by buy on board sales revenue, the requirement for crew food safety and hygiene training takes on a different meaning and the ramifications of crew failures in this area are potentially less onerous to detect.

It is important to continue to stress that without appropriate food safety systems management protocols being established on board by the airline, the impact of crew training in this area will have limited potential. It is critical that the training solutions operate in tandem with in-flight HACCP procedures and documentation. There is little benefit in training crew to take centre temperatures of received or reheated foods and then not provide them with a mechanism and reporting protocols to document and control. Conversely it is irresponsible to regiment crew in the significance to food hygiene of hand-washing protocols, and then fail to provide adequate and appropriate hand-washing facilities and apparel.

The indisputable link between cabin crew training and ensuring in-flight food safety has to result in a fundamental reassessment of the overall picture of airline catering HACCP and systems management. We examine the role the crew have to play in this in the final part of this chapter.

Cabin crew role in ensuring effective food safety management

Having spent this chapter understanding the nature and proliferation of the role that crew have to play in the food chain, it is now for us to define the specifics of what has to occur in tandem with crew training, to ensure that the integrity of the systems are upheld.

To achieve this effectively each aspect of the supply chain needs to be scrutinised and the variable applications defined and applied. It is important to remember that the crew role is more about a continuation and further assurance of effective systems management on the ground, than a satellite-style operation of unrelated aspects in the air.

Catering supply goods receipt

Cabin crew checks

- Check pack integrity – monitoring of contamination in transit.
- Check product against specification – monitoring of spec compliance.
- Check date marking and labelling – monitoring of date marking.
- Check centre temperatures – monitoring of temperature control.

Cabin crew action

- Pack integrity intact? Accept or reject document action.
- Product to spec? Accept or reject document non-compliances and action taken.
- Product in date? Accept or reject document non-compliances and action taken.
- Product within temperature? Accept or reject document non-compliances, deviations from critical limits and action taken.

Airline catering SOPs verified and sustained by above

- Verification of contamination in transit. Verification of caterers' despatch documentation.
- Verification of product specification compliance. Quality and safety parameters determined.
- Verification of correct application of date coding and product coding.
- Verification of temperature control procedures at despatch and during transit. Verification of caterers' despatch documentation.

Implementation of airline documentation required

- Crew goods receipt document referencing:
  - pack intact
  - product to spec
  - product in date/correctly labelled (SPMLs)
  - temperatures of high-risk foods.

- Product specification documentation or menu specs on board.
- Non-conformity reporting documentation; product reject sheet.

Catering supply reheat and service

Cabin crew checks

- Check product integrity – monitoring of raw material QA.
- Check service temperatures – monitoring of chilling and heating equipment efficiency.
- Check product content and quality against spec – monitoring of specification compliance.

Cabin crew action

- Product integrity intact? Accept or reject, document non-compliances and action taken.
- Centre temperatures achieved during heating or sustained during storage? Accept or reject document non-compliances and action taken.
- Specification compliance achieved and product attributes confirmed? Accept or reject, document non-compliances and action taken.

Airline catering SOPs verified and sustained by above

- Verification of product integrity. Verification of effectiveness of caterers QA systems management.
- Verification of safe service temperatures being achieved and sustained. Verification of effective equipment maintenance and accuracy of cooking times.
- Verification of quality and safety parameters being achieved. Verification of product specification compliance.

Implementation of airline documentation required

- Crew food preparation and reheat documentation referencing:
  - service product QA checks
  - centre temperature service checks, hot and cold.
- Product specification documentation or menus on board.
- Non-conformity reporting documentation; product reject sheet.
- Equipment maintenance reporting.

Catering supply prerequisite issues

Cabin crew checks

- Check cabin cleanliness and galley hygiene – monitoring of in-flight environmental hygiene issues.
- Check pest activity – monitoring of effective disinsection and disinfestation protocols.
- Check food handlers' fitness to work – monitoring the state of health of fellow crew.
- Check hand-wash facilities and protocols being upheld – monitoring hand-wash facilities adequate and fully serviceable.

Cabin crew action

- Cabin clean and galley hygienic environment for food preparation and service? Accept or reject, document non-compliances and action taken.
- Presence of pests or evidence of pest activity? Confirm or deny, document non-compliances and action taken.
- All food handlers fit to work? Confirm or deny, document any non-compliances and action taken.
- Hand-wash facilities and apparel serviceable and available? Confirm or deny, document non-compliances and action taken.

Airline catering SOPs verified and sustained by above

- Verification of hygiene standards being maintained. Verification of cleaning contractors' obligations being met.
- Verification of effectiveness of pest control regimes. Verification of presence of pests and requirement for disinsection or disinfestation to be carried out.
- Verification of food handlers' health status. Verification of sickness reporting systems.
- Verification of adequate hand-washing facilities. Verification of crew personal hygiene procedures.

Implementation of airline documentation required

- Crew environmental hygiene documentation referencing:
    - cleanliness of galley work tops
    - food stowage areas
    - ice drawers
    - food service utensils

- ovens
- trolleys and carts.

- Pest control effectiveness and presence or activity reporting documentation.
- Food handlers' fitness to work confirmation and reporting documentation.
- Serviceable and equipped hand-wash facility documentation.

On the surface of it, the documentation attributable to the crew aspects of the food chain seems immense. However, by amalgamating catering product or environmental non-compliances into existing systems of incidence reporting, it will make the requirements much more achievable. In any food business activity, the monitoring and control at each step of the supply chain has to be achieved and the relevant reporting and documentation processes put in place in order to totally assure the process controls throughout. It is not acceptable to verify only the processes on the ground with no consideration being given to the food service environment and activity in the air.

All of the reporting and documenting procedures have to happen in tandem with appropriate crew training and by empowering crew with the same food handling obligations and responsibilities in the air as those undertaken by individuals in the same activity on the ground. In this way, the effectiveness and sustainability of food safety systems management, at point of production and during transit, can be verified by those operating at the consumer-focused end of the supply chain. If systems failures remain undetected by those who form the missing link between supplier and consumer, the entire supply chain remains unverified and ultimately vulnerable to a speculative rather than robustly documented interpretation of events in the light of a food complaint.

As the nature of airline food service styles and provision grows ever more eclectic, the crew connection with the products grows ever more critical. By communicating the food safety issues through effective crew training templates, the evolution of quality assured, appropriately handled, safe food throughout the entire supply chain can at last begin.

10 Managing aircraft water

I aim in this chapter to bring together the whole spectrum of issues inherent in the global management of aircraft water, and to formulate an accurate future resource for this most complex and diverse of subjects. The confronting considerations have less to do with a limit to the science-based data and information available and more to do with the plethora of emotive material and speculative comment on the subject matter.

Whilst at the time of writing there are no specific global regulations or guidelines governing the microbiological quality of drinking water on aircraft, there is a quantitative requirement for the qualitative and equitable replication of standards of potable water supply on board, on a global scale. Whilst the parameters governing potable water on the ground are defined nationally and firmly in terms of microbiological integrity, there has seemed little point in using these standards as a yardstick by which to quantify water safety and quality on board aircraft for fear of setting the industry up for immediate failure.

By its nature the process of water outsourcing and supply to aircraft of any size and scale is extremely complicated, and the requirement for global replication of potable standards renders the entire process fraught with difficulties. Every aircraft operator has an obligation to assure the integrity and potability of the water on board their aircraft and to ensure that standards are maintained, regardless of the port of call. Throughout this chapter we will look at current standards of regulation and guidelines as they appertain to aircraft and their operators, as well as the risk factors inherent in the aircraft water supply chain and the methods and relative success of implementing safety standards on a global scale.

With the recent aviation history requirement to link passenger levels of water consumption in-flight with a variety of other aviation-related health issues, the quality and integrity of the in-flight water supply chain has never come under such intense scrutiny in terms of the success of its qualitative assurances.

Current standards

Having already established that there are no specific enforceable or defined standards for the quality and safety of water on board aircraft, it will come as no surprise to discover that there are no international standards for drinking water quality that mandate the integrity of supplies on the ground at source.

Whilst there is no doubt that diseases related to the contamination of drinking water constitute a major burden on human health, and interventions to improve the quality of drinking water provide significant health benefits, guidance documents such as the WHO *Guidelines For Drinking Water Quality* (GDWQ)[83] are designed merely to assist national agencies in establishing water management strategies of their own, and not to mandate regulatory compliance standards for potable water supply.

The basis on which the GDWQ is designed to work is to provide support and advice in the development and implementation of risk management strategies that will ensure the safety of drinking water supplies through the control of hazardous constituents in water. What the guidelines do make provision for, however, are descriptions of the

'reasonable minimum requirements of safe practice to protect the health of consumers and/or derive guideline values for constituents of water or indicators of water quality.'

Neither the minimum safe practices nor the numeric guidelines values are mandatory limits, however. It is the opinion of advisory bodies such as the WHO that the definition of such limits relies on consideration being given to the context of local or national environmental, social, economic and cultural conditions before appropriate standards can be set.

In terms of the impact of this situation on managing aircraft water where the requirement for airlines with operations internationally is to outsource water from many different outstations all operating under variable national and regional water safety directives, there are two concerns that should be considered by aircraft operators:

- The quality and safety parameters by which potable water uptake is governed on the ground are likely to be subject to huge variation from nation to nation and therefore from port to port.
- The non-definition of specific safety and quality parameters for aircraft water supply leaves the industry operators subject to the surveillance of national inspectors operating within regional regulations enforcing fluctuating standards of water integrity.

As we will see later, the major factor to govern the safety and quality of in-flight water supplies is the relative integrity of the initial source of supply. Attempting to assure international compliance to non-enforceable guidelines remains an almost insurmountable challenge to the industry and is unique to the requirement to take up, transport, store and serve water from a multitude of global locations.

Whilst the WHO GDWQ do not mandate, only recommend, water quality standards, section 14.2 of the International Health Regulations (IHR)[44] states:

> 'Every port and airport shall be provided with pure drinking water and wholesome food supplied from sources approved by the health administration for public use and consumption on the premises or on board ships or aircraft. The drinking water and food shall be stored and handled in such a manner as to ensure their protection against contamination. The health authority shall conduct periodic inspections of equipment, installations and premises and shall collect samples of water and food for laboratory examinations to verify the observances of this Article. For this purpose and for other sanitary measures, the principles and recommendations set forth in the guides on these subjects published by the organisation shall be applied as far as practicable in fulfilling the requirements of these Regulations.'

For aircraft operators this has the effect of applying international legal provision to the quality of water supply to airports by making reference to the standards laid down in the WHO GDWQ. The standards of supply to airports then are based on best practice infrastructure but are given legal enforcement stature by the IHR provision. The problem still remains, however, that in the absence of any formalised safety indicators in the WHO GDWQ, interpretation is influenced by local and regional considerations.

That said, there is no doubt that most nations subscribe to the theory that preventative management of the water supply chain is the preferred approach to drinking water safety and should take account of the WHO recommendations to:

> 'take account of the characteristics of the drinking water supply from catchment to consumer.'[83]

In terms of the source of supply to aircraft, many aspects of drinking water quality management are outside the direct responsibility of the aircraft operator and end supplier. It is essential therefore that a collaborative approach involving the definition and assessment of multiagency responsibility port to port is established by the aircraft operator so that an adequate water safety plan can be established that adequately takes account of all risk factors inherent in the supply chain specifics.

The collaborative infrastructure of responsibilities in terms of water source supply needs to be understood by those charged with the responsibility of outsourcing water on behalf of airline operators and looks something like this:

- National agencies – provide a framework of targets, standards and legislation to enable and obligate suppliers to meet defined quality and safety parameters.

- Supply agencies – required to ensure and verify that the systems that they administer are capable of delivering safe water and that they achieve this on a regular and consistent basis.
- Surveillance agencies – an independent external surveillance agency responsible for periodic audit of all aspects of safety and verification testing.

It is clear in the directives laid down by the GDWQ that the two functions of surveillance and quality control are best performed by separate and independent agencies because of the conflict of interest that arises when the two are combined. If this were to be true of the surveillance of the quality of aircraft water, it would require that samples taken from the aircraft for testing were not derived from or handled by the airline or their agents.

In most countries the agency responsibility for the surveillance of drinking water supply is the Ministry of Health and its regional or departmental offices. In some countries it may be an environmental protection agency, in others the environmental health departments of local government may have some jurisdiction.

Whilst it is common for the industry suppliers of water on the ground to come under the surveillance of independent agencies such as the Environmental Protection Agency (EPA) and Food and Drug Administration (FDA) in the USA, and Association of Port Health Authorities (APHA) in the UK, the transient nature of aviation means that often the water sampling responsibility and therefore the surveillance activity of in-flight water supplies from on-board source, fall into the hands of the airlines and/or their agents contrary to WHO GDWQ recommendations.

Public health authorities

In order to be effective in the protection of public health, a national entity with jurisdiction over public health issues will normally act in four areas:

(1) Directly establish critical limits for drinking water safety – national public health authorities have the primary responsibility for establishing norms and standards for drinking water supply. This may include setting water quality targets (WQTs), performance and safety targets and directly specified treatments. Jurisdiction may extend to specifying the chemical and treatments permitted for use in the production and distribution of safe drinking water. This is an ongoing organic process, with the evolution of standards occurring in tandem with epidemiological data and surveillance results and reporting.
(2) Surveillance of health status and trends – outbreak detection and investigation either directly or through a regional body.

(3) Direct action – by providing guidance to local and/or regional governments in the surveillance of drinking water supplies. This role may vary hugely and be determined by public health policy and governmental infrastructure.
(4) Representing health concerns in wider policy development – health policy and integrated water resource management. A supportive role in resource allocation and conflict resolution.

Public health surveillance teams traditionally function at national, regional and local level and contribute hugely to verifying drinking water safety. Their work can be enhanced in a variety of ways to identify possible water-borne outbreaks or deterioration of water quality. Epidemiological investigations include outbreak investigation, intervention studies to evaluate intervention options, and cohort studies to evaluate the role of water as a risk factor in disease.

Routine public health surveillance incorporates the following roles:

- Ongoing monitoring of reportable disease which may be related to water-borne pathogens.
- Outbreak detection.
- Long-term trend analysis.
- Geographic and demographic analysis.
- Feedback to water authorities.

It is essential not only to comprehend the jurisdiction and remit of public health surveillance teams but also to be clear that they cannot be relied on to provide information in a timely enough fashion to enable short-term operational response to control water-borne disease. These limitations include:

- Outbreaks of non-reportable disease.
- Time delay between exposure and illness.
- Time delay between illness and reporting.
- Difficulties in identifying causative pathogens and sources.

Local authorities

Local environmental health authorities often play an important role in managing water resources and drinking water supplies at source. This may include any one of a number of the following aspects[83]:

- Water hygiene awareness training.
- Basic technical training in drinking water supply transfer and management.
- Motivation, mobilisation and social marketing activity.
- Consideration of and approaches to overcoming sociocultural barriers to acceptable water quality interventions.

Local environmental health authorities will give specific guidance to communities or individuals in designing and implementing community drinking water systems and deficiencies and may also be responsible for surveillance of the quality of community and household supplies.

Aircraft operator responsibilities in water safety management

It is essential that in the context of responsibilities, aircraft operators understand the specifics of the manner in which regional, national and local water management and surveillance agencies operate, not just from their home domicile but at each nation to which they fly (see Table 10.1). This type of information devolution and strategy responsibility will highlight any endemic water quality management and control issues at source and can be written into the water safety plan (WSP) devised by the airline operator for that region.

Whilst the WHO GDWQ do not lay down any specific parameters by which aircraft water quality should be determined microbiologically, they do make the point that the quality and safety of water supplies on board aircraft should be such that it is of potable quality and as such conform to all the aspects of water safety management laid down in the GDWQ. To this extent the unique aspects of the mechanics of water supply to aircraft are given special reference in the WHO *Guide To Good Hygiene and Sanitation In Aviation*, which is currently under review. These mechanical considerations are looked at later in the chapter when we examine the problems inherent in verifying the aircraft water supply chain on a global scale.

In terms of the current standards of regulation and surveillance that affect the operational activities of water supply to aircraft, airline operators are left to design and manage their own internal WSPs, which may be more generic and less specific than is actually required if one takes the water safety framework provisions of the GDWQ as gospel. Whilst these provisions are designed to underwrite the water supply management strategies of piped water distribution and supply mechanisms on the ground, the principles should remain the same in the air.

There can be many risk factors inherent in the supply chain, depending on a host of factors that may fall largely outside the aircraft operators' control. To this extent the case for the specific development and implementation of WSPs for every destination from which an aircraft operator derives supply is a strong one, and in the light of the requirement to comply with the IHR may well assist in the avoidance of outbreaks of water-borne disease rather than having to deal with them in an historically reactive fashion.

Table 10.1 General responsibility for safety of aircraft water supply

Agency responsible	Area of responsibility
Public Health Administration	Approval of source of supply and compliance with IHR Article 14.2
Health Authority	Surveillance and regular inspection of installations
	Water and ice sampling from airport source
	Water and ice sampling from aircraft in accordance with WHO GDWQ best practice surveillance v quality control
	Dealing with implementation of outbreak remedial action – removal and safe disposal of contaminates on board (IHR Article 63.1)
Airport Authority	Provision of suitable (pure) source of supply to airport buildings, airport catering outlets and for aircraft supply
Airlines	Ensuring WSP is outstation-specific and based on risk assessment
	Verifying source of supply and determining and surveying water transfer and loading protocols
	Ensuring all water is treated to contain residual chlorine
	Determining and implementing programme for aircraft water tank servicing and disinfection
	Verifying health status of water servicing personnel
	Water sampling programme definition and implementation based on risk; sampling to be carried out on aircraft and from water servicing vehicles
	Collation of sampling data and sharing of results with interrelated agencies on a regional, national and international basis
	Ensuring all water and ice served on board are of potable quality
Water service providers	Compliance to all attributes of the WHO GDWQ
	Surveillance and verification of the 'portable' source of supply
	Training and supervision of all water servicing personnel
	Risk-based assessments of mechanisms of water service supply to aircraft consumers
	Compliance with the codes of practice laid down in the GDWQ with regard to the specifics of water supply to aircraft

Risk factors in aircraft water supply

The capacity for any travel-related vector, but particularly aircraft, to be implicated in the spread of endemic disease that may have a global impact, is well documented. In general terms, the biggest risks are posed by the microbial contamination of aircraft water by animal or human excreta. With the 21st century potential for the relative masses to have access to air travel on a global scale, come the added problems encountered by aircraft operators who transit both into and out of endemic disease-affected areas, or areas which demonstrate variable standards of general hygiene and sanitation. Water-borne diseases that are still being transmitted in many parts of the world include cholera, enteric fevers (*Salmonella*), bacillary and amoebic dysentery and other enteric infections.

It is often not possible for aircraft undertaking long haul travel to carry sufficient potable water supplies to last for the duration of the flight, regardless of the aircraft size or flying capability. If the initial potable source is compromised in any way, it is clear that disease can be readily spread through the medium of aircraft water.

Despite the relative quality and integrity of the initial potable source of supply, there are numerous other factors for consideration in the aircraft water supply chain that may affect the safety of the aircraft derived end product. To this end, it is critical that a risk-based assessment of supply chain issues is conducted, if necessary to take account of the entire global supply chain where risk factor intensity and qualification will vary from port to port.

Whilst Article 14.2 of the IHR allows for the availability and provision of potable water at every airport and port facility, it is essential not to allow reliance on potable water standards at source to influence the implementation or otherwise of adequate and effective measures of control during transfer, storage or distribution in aircraft.

Many of the risk-based control measures that are employed to assure the quality and the safety of aircraft water are based on the assumption that microbial contamination is likely to cause the greatest hazard to human health and subsequent spread of disease. It is important to note also, however, in the case of water supplies specific to global travel vectors, the relative problems posed by the presence of small animals in the potable water distribution system.

Whilst supplies of potable water on the ground remain traceable and consistent in developed nations, the global outsourcing requirement of the aviation industry allows for a persistent taking up of supplies from many different outstations all over the world. The WHO *Guidelines For Drinking Water Quality* (third edition 2004)[83] lay down the parameters by which potable water standards should be attained, and under Article 14.2 of the IHR every airport location worldwide must have facilities to accommodate pure drinking water. However, it is the responsibility of every aircraft operator to ensure that these standards are being upheld, not just in terms of quality of source of supply on the ground but also in terms of the level of quality assurance provided by the tanked systems in-flight.

Bearing all of this in mind, it is with particular interest then that we look at some of the less well-documented risk factors inherent in water supply to aircraft. Whilst microbiological safety and integrity are an extremely important issue, it is not the only one, and to ensure that safe water is consistently achievable in-flight there are interrelated factors that need to be examined and certainly need to form part of any water safety plan developed by the airlines.

Small animals in the water supply system

Invertebrate animals are naturally present in many water resources used as sources for the supply of drinking water. Small numbers of adults and their larvae have the capacity to pass through the water treatment works, especially at works where the defensive mechanisms put in place to particulate matter are not 100% effective. Many of these animals can survive and even reproduce within the supply network by deriving their food source from the micro-organisms and organic matter in the water, or in deposits on pipe and tank surfaces. It is perhaps surprising to understand how prolific populations of small animals are in treated water distribution systems around the world, with reports suggesting that few if any continents have water distribution systems that are completely free from animals[84].

The composition, population and density of animals in treated water systems varies hugely from infestations of visible species which invariably prove abhorrent to consumers, to the less concentrated occurrences of microscopic species. Whilst the presence of small animals in treated water systems has widely been regarded as an aesthetic problem (due to the association between levels of small animal presence in treated water and water discolouration) there has been some suggestion in recent years that animal presence may affect the microbiological quality of water.

In temperate countries there has been no evidence to suggest that any of the metazoan animals found in water distribution systems are injurious to human health; however, in tropical and subtropical climates, specific species of aquatic invertebrate can act as intermediate hosts for parasites. The parasitic nematode *Dracunculus medinensis*, or guinea worm as it is more commonly known, is currently found only in Sub-Saharan Africa but has been found to be historically endemic in North Africa, the Middle East and the Indian subcontinent. It is transmitted solely by water consumption; its intermediate host is the copepod *Cyclops* and human infection results when water containing infected *Cyclops* is ingested.

The five species of parasitic flatworm *Schistosoma* that cause schistosomiasis have been found in many countries throughout Central and South America, Africa, Asia and the Western Pacific[85]. By virtue of the complexities of their aquatic lifestyle they use snails as intermediate hosts. Eggs released by infected human beings are infective to the snails and they develop and release sporocysts, which in turn develop into cercariae, which are infective to other human beings.

In natural waters bacteria are found on the surfaces and in the gut of various invertebrates and this has led to some speculation that if the same were found to be true of invertebrates found in water supplies, then this would pose some significant sanitary risks[84].

A variety of studies have suggested that invertebrates found in potable water distribution systems could harbour micro-organisms in their gut, thus protecting them from the disinfection process. Chang and colleagues[86] conducted laboratory experiments using two species of nematode (round-worms) isolated from potable water in the USA and exposed to suspensions of micro-organisms. It was demonstrated that the nematodes would ingest both-*Salmonella* and *Shigella* bacteria and Coxsackie virus and echo-virus. It was found that a small percentage (1%) of these micro-organisms survived in the gut of the nematodes for 48 hours. The nematodes were also found to be highly resistant to chlorination and as a result viable micro-organisms were isolated from the gut after the nematodes were subject to chlorination. What was not demonstrated by Chang and colleagues, however, was a demonstration of the excretion of viable pathogens. Smerda and colleagues[87] did later show that viable salmonella might be excreted by a nematode.

These studies, along with another carried out by Levy and colleagues[88], have demonstrated the possibility that invertebrate presence in treated water systems may protect micro-organisms from disinfection. In theory this mechanism would only present a significant risk if pathogens were already present in the distribution system and were protected from the levels of disinfectant carried through distribution. Those most likely to be protected in this fashion are those present in biofilms and sediments which themselves offer protection from disinfection. It has been argued that grazing animals present in the water distribution system allow more penetrative disinfectant effect by reducing the amount of organic matter present in biofilms and sediments. It must be stressed at this point that this notion is purely theoretical and is not based on any sound scientific data; therefore the emphasis must remain on minimising the formation of deposits and biofilms via appropriate treatment and routine maintenance.

Control methods

The methods available for the control of existing infestations in water mains include both physical and chemical applications. When assessing the risks of small animal presence in the water distribution system, it is essential that any intermediate supplier to aircraft via water bowser or vehicular tanking, can produce evidence of the methods of small animal control from source. This will allow the operator to take a view on likely types of animal presence and aspects of chemical control measures that may have been employed. Any conflict with potable water treatment, particularly in terms of chemical usage employed at source, can be assessed by the operator and integrated into the water safety plan specific to that outstation.

Chemical methods

- Chlorine – the concentrations of chlorine usually found in water leaving treatment works that are acceptable to consumers, are not particularly effective against most of the types of small animals found in the distribution systems.
- Pyrethroids – natural pyrethrins and a synthetic analogue, permethrin, have been used very successfully to control Asellus and other crustaceans such as Gammarus[89–92]. In spite of the fact that permethrin is chemically distinct from pyrethrins, it shares a number of similar properties that are crucial in its use for controlling animals in the water mains. It must be noted at this point that there is a vast difference in concentration between the dose effective in killing a range of aquatic mammals and the dose that would prove toxic to mammals when drunk. The dose commonly used is $10\,\mu g/l$, which is half that of the safe recommended level in the third edition of the WHO *Guidelines For Drinking Water Quality*[83].

It needs to be considered, however, that the effective dose for controlling animals in water mains is highly toxic to fish and for this reason the addition of pesticides to drinking water is now prohibited in many countries.

So, with an established the connection between animal presence in piped water systems and the potential impact on end product integrity, any airline water safety plan has to include provisions for the monitoring and control of animal presence both at source of supply and in the tanked end product on board.

Physical methods

- Systematic unidirectional flushing – removes most freely swimming animals provided that adequate flows are available. In smooth pipes it will also remove loose deposits and animals burrowing within them, but higher flows are required to achieve good results. The solid particles transported by the water move more slowly than the water itself, so at least twice the nominal volume of water in the section of the main should be flushed[93].
- Swabbing – may be used where only moderate flows are available and it is generally effective at removing loose deposits and burrowing animals. However, swabbing is not generally effective in badly encrusted mains.

- Air sourcing – may be used where only moderate pressures are available and is less effective on encrusted pipe walls than foam swabbing. However, air sourcing is usually restricted to mains up to 200 mm in diameter and it may exacerbate corrosion in iron mains[94].

Having established that there are measures that can be employed to control small animal infestations at the source of water supply, it is essential that any airline operator examines the water safety schematics specific to the region of uptake. By far the best solutions to controlling small animal infestations in the water distribution system are those long-term principles which will be evident in the make-up of national water safety plans. These are recommended to prevent animals reaching nuisance levels or following disinfestations to prevent recurrence of problems. The principal objectives are to deny the animals a food supply and to restrict their entry into the distribution system in the first place.

Controlling microbial risks in aircraft water supply

In order to secure the microbial safety of drinking water supply, the use of multiple barriers from catchment to consumer is essential to prevent contamination or to reduce contamination to levels that will not prove injurious to health. Invariably with the introduction of multiple barriers, safety is increased[83].

Whilst some of the barrier steps to safe water provision remain directly in the hands of the aircraft operator, many of the initial steps are subject to the enforcement of potable water standards at source on the ground. The breakdown of responsibilities looks something like the list below, with some interrelated aspects becoming the responsibility of both supplier and aircraft operator at different stages of the supply chain.

Supplier responsibility/regional enforcement responsibility

- Protection of water resources.
- Management of water distribution systems.
- Maintenance and protection of treated water quality.
- Emphasis on pathogen prevention in water source rather than reliance on removal treatment processes.

Aircraft operator responsibility

- Water safety plan implementation – risk-based and location of supply specific.
- Proactive on-board systems maintenance programme.
- Management of water distribution from source of supply to aircraft.

- Barrier or pore filtration at every outlet.
- Proactive sampling regime – nature and frequency determined by risk-based strategy.

In general terms the greatest microbial risks to water supply are associated with the ingestion of water that is contaminated with human or animal faeces, including bird faeces. Such faeces can be a source of pathogenic bacteria, viruses, protozoa and helminths.

The traditional approaches adopted by the aviation industry to assure water safety on aircraft have relied on end product sampling. Microbial water quality has the capacity to vary rapidly and over a wide rage. Intermittent peaks in pathogen concentration have the capacity to increase disease risks and trigger potential outbreaks of water-borne disease. To this extent, if reliance is placed solely on end product water sampling, then by the time microbial contamination is detected many people may have been exposed. Therefore, although sampling can verify that the water is safe, it is not suitable for early warning detection purposes.

Traditionally the detection of contaminants in both source water and the end tanked product delivered to passengers and crew is often slow, complex and costly. It is therefore essential that a risk assessment-based HACCP-style approach to water outsourcing and maintenance is adopted and a proactive as opposed to reactive culture pervades in management approaches to water safety in-flight.

Water safety plan

In order to implement an effective approach to water safety and quality management, the development of a detailed and outstation-specific WSP is critical and this forms part of the guidance criteria in the WHO GDWQ. Whilst the guidance is not specific in terms of what should be included in the plan, it suggests breaking the approach down into three general strategies:

- Health risk assessment along the water supply chain from airport to aircraft.
- System risk assessment to determine whether the water supply chain as a whole can deliver water of a quality that meets the above criteria, i.e. controls hazards to meet defined targets.
- Setting of control measures, management and monitoring of control measures and corrective action.

We look more at WSP inclusions and implementation in the next section, but for now let us return to microbial risk control measures that ultimately form aspects of the WSP systems management process.

The growth of micro-organisms following water treatment can occur particularly in protracted systems of water distribution or, as in the case of aviation supply, during stored or tanked supplies. The problem of regrowth can become endemic in aircraft water supplies as the tanks are topped up from a variety of water sources.

In addition to faecal derived pathogens, other microbial hazards, e.g. guinea worm, toxic cyanobacteria and *Legionella*, may prove of public health importance under specific circumstances. Public health concern with regards to cyanobacteria relates to their potential to produce a variety of toxins known as cyanotoxins, which in contrast to pathogenic bacteria proliferate only in the aquatic environment before intake. Toxic peptides, e.g. microcystins, are usually contained within the cells and may be largely eliminated by filtration; toxic alkaloids such as cylindrospermopsin and neurotoxins are also released into the water and may break through filtration systems.

Regrowth is typically measured and reflected in terms of increasing heterotrophic plate counts (HPC). Micro-organisms will normally grow in water and on water contact surfaces as biofilms and can cause nuisance through the generation of a variety of tastes and odours as well as discoloration of water supplies. Elevated levels of HPC occur especially in stagnant aspects of piped systems and can be aided by a rise in temperature, availability of nutrients and lack of residual disinfectant.

HPC testing has a long history of use in water microbiology in the 19th century, with HPC testing used as an indicator of the successful functioning of processes and therefore as direct indicators of water safety. With the advent of specific testing for faecal indicators, HPC usage declined; however, it continues to figure in water regulations guidelines in many countries. HPC measurements are used:

- to indicate the effectiveness of water treatment processes as an indicator of levels of pathogen removal;
- as a measure of numbers of regrowth organisms that may or may not have a sanitary significance; and
- as a measure of possible interference with coliform measurements in lactose-based culture methods. This application is of declining value, as lactose-based culture media are being replaced by alternative methods that are lactose-free.

Whilst it has been accepted that HPC testing alone is not satisfactory to assess the health risks posed to the piped water distribution systems, in aviation environments it is an extremely useful indicator of in-flight distribution system conditions. HPC counts will arise from stored water stagnation, deficiencies in residual disinfection, high levels of assimilable organic carbon, higher water temperature and availability of particular nutrients. Therefore, obtaining a high HPC count from tanked aircraft water may indicate the necessity to examine procedures for taking on water, maintenance of the system and disinfection protocol and procedures.

Having established the risk factors and then the scientific methods of performance indication through sampling, it is now for us to look at the methods employed to control the risk factors in the supply of water to aircraft.

Disinfection

Disinfection is of unquestionable importance in the supply of safe drinking water and is utilised in the destruction of microbial pathogens. The utilisation of reactive chemical agents such as chlorine are most commonly used in this process.

Chemical disinfection of a drinking water supply that has faecal contamination, whilst reducing the overall risk of disease may not necessarily render the supply safe. To this extent it is crucial to consider a whole series of factors when formulating a WSP that is systematically designed to eliminate pathogenic contamination and/or regrowth in the tanked system. Chlorine disinfection of drinking water has limitations against the protozoan pathogens, in particular *Cryptosporidium* and some viruses. An overall management strategy is needed incorporating multiple barriers, including water source protection and filtration as well as protection and treatment during storage and disinfection.

The use of chemical disinfectants in water treatment usually results in the formation of chemical by-products; however, the health effects associated with these by-products are considered to be small in comparison to the risks posed by microbial contamination. Disinfectants such as chlorine can be easily monitored and controlled, and frequent monitoring of usage levels and presence in end product is recommended. Acceptable levels of chlorination specific to aircraft supply are laid down in the WHO *Guide To Hygiene and Sanitation In Aviation*, which is currently under revision, and whilst not mandated they are generally considered the acceptable level to be administered to control general levels of risk.

Aircraft water systems

The modern requirement for aircraft water storage dictates that all water be stored in tanks on board and that all water on board should meet potable standards, with no distinction between quality of supply at drinking outlets and at toilet outlets.

All components that combine to create the physical make-up of the system should be corrosion resistant and suitable for use with hyperchlorinated water. As part of a proactive approach to water safety, it is recommended that barrier method filtration is fitted as well as filters designed to neutralise the chlorine content of the water. Great care must be taken to ensure that these filters, and indeed any filter system, are regularly maintained and filters changed to prevent disintegration and subsequent contamination of the very supply they are designed to assure. The tanks themselves should be impervious in construction and designed to drain completely with a single fill-point if tanks are located together.

Aircraft water sampling

The nature and frequency of water sampling should be documented in the WSP and should be based on risk. Sampling regimes fall into four categories:

(1) Water sampling from airport source of supply.
(2) Water sampling from aircraft servicing vehicles.
(3) Water sampling from incoming aircraft.
(4) Water sampling from departing aircraft.

Whilst the jurisdiction for sampling the water source supplied to the airport should be shared between local health authorities and the airlines, aircraft operators must ensure as part of their WSP that they have carried out appropriate independent sampling among their due diligence processes; this applies to water servicing vehicles also.

The frequency of sampling airport water supplies should be considered in the overall framework of the port-specific WSP. It will be determined by such factors as the quality of source, the risk of contamination based on historical data as well as demographic factors, the complexity and length of the distribution system, the possibility of the spread of endemics and the size of the population to be serviced.

The nature and frequency of samples from the aircraft themselves also need to be considered in the WSP and firmly based on assessment of risk. As we have already seen, sampling has historically been part of a reactive culture that has done little or nothing to prevent the passenger exposure to contaminated water risk. The effectiveness of sampling as a preventative measure can only be evident in increased sampling activity to highlight issues as they arise, and an interrelated programme of source, vehicle and end supply sampling whereby data is immediate and shared aspects are documented as one instead of in isolation.

Aircraft water risk assessment inclusions

When conducting an audit of the water supply and loading protocols employed by the water servicing companies contracted around the world, it is essential that all of the risk factors inherent in the supply chain have been identified. It is for the aircraft operator to ensure this process is undertaken as part an initial risk assessment process and that it is carried out at every outstation.

It is critical that potable water integrity at source is not taken for granted, particularly in disease endemic areas but also in perceivably disease-free nations. The supply chain mechanisms involved in getting the potable supply of water from source to aircraft storage need also to be assessed. This process cannot afford to be generic for all ports and must take into account the prevailing safety risk factors from outstation to outstation. Historical sampling data results collated from different outstations can be an effective tool in predetermining the nature and frequency of the sampling regime and which suppliers pose a bigger risk to water supply integrity than others.

The risk factors inherent in the aviation water supply chain can be broken down into the following steps:

- Potable source of supply – outstation categorised in terms of risk, microbiological and chemical safety of water verified, and animal presence identified and quantified. Chemical and physical treatment processes verified as meeting international standards for drinking water safety.
- Filtration systems and treatment systems on board aircraft – filter purpose determined as extra precautionary and not as a replacement for chlorination. Regular cartridge and filter maintenance to prevent disintegration and contamination. Filters fitted to every outlet. Best practice installation of pore-style filtration downstream of main tanks.
- Design and construction of water servicing vehicles – designed and maintained so that water in transit between fill point and aircraft storage mechanism cannot come into contact with any external matter or be affected by handling.
- Water loading techniques – hoses durable and impervious, nozzles protected from contamination when not in use. Before loading, hoses should be flushed through by pumping a small quantity of water through them.
- Design and construction of storage tanks and pipes – tanks constructed of welded steel or fibreglass and designed to drain completely. Exclusive to the purpose of water supply. Fill points separated from toilet servicing panels. All components must be corrosion resistant and suitable for use with hyperchlorinated water.
- Backflow prevention – distribution lines not cross-connected with distribution lines for any non-potable system.

It is evident that the supply chain issues inherent in the supply of potable water to aircraft are fraught with risk. Every step in the process must be clearly established and documented in line with international guidance standards. Water safety audits must be all encompassing and must be undertaken by those with specialist knowledge of the subject to ensure that all attributable risk elements have been considered in the context of disease colonisation, demographics and local issues, i.e. natural disaster or war zone status, that may impact on the integrity of water supply quality.

The continuous vigilant assessment of drinking water supplies is essential to ensure that every component of the system – source, treatment, storage and distribution – operates without risk of failure[73].

Implementing water safety standards on a global scale

Having looked at all the potential applications that need to be considered in the supply of aircraft water, in terms of integrity of source and then the distribution and extended supply chain, it is now time to consider theoretical guidance for realising and consistently achieving the quality and safety attributes of the end product on a global scale.

The difficulties inherent in this task are often underestimated, not only by the aircraft operators charged with the responsibility of implementation but also in terms of the regulatory and advisory bodies who attempt to offer guidance and advice. As every aircraft operation will be unique, it is for every aircraft operator to develop their own water safety plan against a framework of considered risks specific to their global outsourcing requirements. Just as one would not expect to trust the efficiency of a generic global HACCP for airline food production, the same must be true of generic WSPs when the outsourcing requirements span every continent. Defining the quality and safety parameters of the end source of supply is crucial in working out the specifics of the WSP that will deliver the defined end result. Guidelines on microbiological integrity issued by local agencies are a good place to start; however, as they will vary from port to port, the evidence of fluctuating enforcement standards should not be used as an excuse for fluctuating standards of compliance or a lowering of the safety bar.

As with any global outsourcing requirement, the issues that dominate the entire supply chain will need to be considered and documented, and the GDWQ emphasis on the supply chain in terms of the focus of safety assurances being 'from catchment to consumer'[83] should be considered in terms of the overall chain of supply.

As part of the water safety plan, every aircraft operator should develop their own specific guidelines to ensure that they not only have an understanding of the issues unique to aircraft supply but can also demonstrate knowledge of the practical mechanisms and steps that need to be taken by the individuals and companies involved in the source of supply.

In order to audit the systems and procedures of those companies supplying aircraft water, these types of operation-specific guidelines will prove invaluable.

Having undertaken a risk-based assessment of the practical considerations for water supply, it is then for the WSP to provide guidance on the management of the procedures involved in getting the water supply on board. This process involves several key aspects and can be documented in terms of the procedural considerations as laid out here.

Airport water source of supply

The risk is assessed and evaluated port to port in terms of any demographic or historical concerns. It should conform to IHR Article 14.2 and WHO GDWQ for water quality at source. The microbial quality is surveyed and scientifically verified at periodic intervals. Surveillance and sampling from source schedules are determined by category of risk, but should be a minimum of four times annually. The water can be obtained direct from the mains supply or transported in water servicing vehicles.

The mains supply point from which aircraft water is derived from source should be above ground level and under cover to protect it from contamination. As far as possible each airline should have its own supply point and be responsible for its maintenance and hygiene. If the supply point is shared by a number of different aircraft operators and/or servicing companies, then control and maintenance jurisdiction should fall to the airport authority. The supply point should be used exclusively for aircraft drinking water and should be situated at least 30 metres away from the supply point for toilet servicing vehicles. The hydrant hose should have a self-sealing, non-return valve coupling. The diameter of the hose should be different from that of the hose supplying water to toilet servicing vehicles.

Water servicing vehicles

Ensure that tanks are constructed of smooth, strong corrosion-resistant material and that they do retain sediment after full drainage. All corners should be rounded. Covers should be provided which will permit full access to the interior so that the tank can be cleaned and maintained. Ensure a tap is fitted to allow for sampling. Inlet and outlet valves should be self-sealing, non-return and quick release and have caps installed that should be fitted when not in use.

All water servicing vehicles should be cleansed and disinfected according to a schedule laid down in the WSP and the procedure determined and documented, e.g. fill tanks with 50 mg/l of residual chlorine, to remain in tanks for minimum of 30 minutes. The vehicle is then emptied through the delivery hose as opposed to the drain plug. After draining the drain plug is removed and the tank flushed out with potable water through the valve coupling. The interior of a tanked water vehicle should be scoured with hypochlorite solution or jet steamed, to remove deposits, on a schedule to be determined in the WSP based on risk, but no less than once a month. Instructions for the cleansing and disinfecting of water vehicle tanks should be affixed to the vehicle as a reminder of the protocols to water servicing personnel. Dates for treatment should be documented and cross-referenced with the WSP at the point of audit.

Transfer of water from airport supply to aircraft

Sanitary safeguards must be in place during the transfer from mains supply or water servicing vehicles. These should include connections to the aircraft system and through the aircraft system at each outlet to prevent contamination and pollution of the water. Dedicated water servicing staff should be in place, who are fully trained and not engaged in any toilet servicing activity.

Storage tanks and pipes on board

Regular servicing and maintenance of chlorine neutralising filters are essential to prevent disintegration and contamination. Filters must be fitted at every outlet to ensure that chlorine removal occurs at the end of the system to avoid bacterial introduction downstream of the filters.

Treatment of aircraft water

Regardless of the integrity of the potable water supply, it is necessary to introduce extra precautions to prevent contamination during transfer and in the aircraft water system itself. Prior to loading, all water should be treated to maintain the level of residual chlorine at 0.3 mg/l. At the point of water conveyance, dosing can be carried out automatically or by using portable or fixed chlorinating units at the supply points. Verification methods for checking that the water has been treated and for approximating the amount should be available to ground engineers, cabin crew and catering officers, all of whom should be responsible for testing at the appropriate times, i.e. at point of supply and prior to service.

Cleaning and disinfection of tanks

A risk-based and documented procedure is needed to determine the nature and frequency of tank disinfection. Where chlorination of the supply chain can be verified categorically at all ports, tank disinfection can occur less frequently but not less than every four weeks. Where the integrity of the supply chain cannot be verified, disinfection must occur much more frequently.

Disinfection protocol should involve filling the tank with water containing residual chlorine at a level of 50 mg/l, and leaving it for a minimum of 30 minutes. Alternatively a 200 mg/l solution may be left in the tanks for between 3 and 55 minutes. The tanks should then be drained, flushed out completely with potable water to ensure that hypochlorite solution is completely removed and then refilled with treated water as above.

Maintenance protocols

These protocols should include not only authorised, supervised and documented works to water servicing equipment on board, but also the training and cleanliness monitoring of personnel engaged in water servicing activity.

Operational monitoring

The aircraft operator is responsible for monitoring the relative success of water safety management procedures and protocols. Monitoring allows the success of the management systems to be verified. The scheme for monitoring will be controlled by the determinants identified in the risk-assessed WSP and will depend on the types of control measures applied by the management systems. These should include quality of water source; hygiene and maintenance of hydrants and hoses; disinfectant residuals and pH backflow preventers; and filters and microbial monitoring, particularly following maintenance.

Verification programmes

The frequency of implementation of water sampling programmes from airport source, water vehicles and on-board tanked supply is determined by assessment of risk laid down in port-specific WSPs. Unsatisfactory sampling results are to be investigated, reported to the relevant authorities where necessary, and remedial action implemented pending a review of control measures.

Airport health surveillance

Independent assessment and surveying of airport water source supply integrity is needed, including periodic audit, review and approval or otherwise of water safety plans.

So, having established the logistical requirements and management strategies for the equitable replication of aircraft water provision on a global scale, it is possible to gain some kind of perspective on exactly what it takes to achieve it. The process steps are long and variable, the development of the WSP will need to take account of the attributable risk strategies of a whole source of different supply chain mechanisms and the management protocols on board will need to be established within a framework of alerting procedures capable of highlighting issues early on before they develop into an insurmountable problem.

It is essential, however, that the water safety issues are understood in the context of what a possible breakdown of protocol and procedure could mean to public health as well as to passenger and crew safety.

Water consumption in-flight and its critical link to cabin health issues

How safe is aircraft water?

Invariably, with the advent of a new generation of cabin health concerns, the profile of the quality of water supplies to aircraft has come under intense scrutiny. Traditionally links have been made to the prevention of DVT and to dehydration risk factors by the increased consumption of water in-flight. The aircraft operators' capacity to distribute only bottled supplies throughout the duration of the flight is impacted upon by not only fiscal but space constraints. To this end, it is more essential than ever that the potable quality of aircraft water is verified and assured.

With increased consumption comes increased public awareness and concern over the safety and quality of the product in the in-flight environment, and as a response to that concern several studies to investigate the safety and quality attributes of aircraft water have been conducted both officially and unofficially.

In 2000 the results were published of a six-month joint study investigating the microbiological quality of potable water supplied to and stored on commercial aircraft, conducted by the Public Health Laboratory Service (PHLS) environmental surveillance unit and APHA. The study was carried out for a six-month period at each of 13 major airports in the mainland UK as well as the Isle of Man between July 1998 and March 1999 and was designed to compare the microbiological quality of potable water at different points in the supply chain. A total of 850 water samples were obtained from airport mains supply points, from bowsers during transport to the aircraft and from water taps on board the aircraft. Hygiene practices associated with the water supply chain within airports and on board aircraft were examined by requesting airports, airlines and bowser companies involved to complete a questionnaire.

Current UK water regulations state that the bacterial indicators faecal coliforms (*E. coli*) should be absent from 100 ml samples of water and total coliforms should be absent from 100 ml samples of water, in 95% or more of samples. In this study samples were considered to have failed if there were any *E. coli*, other coliforms or faecal streptococci in the water. The study showed that 8.7% of samples failed, with the highest percentage of failed samples taken from the aircraft drinking water fountain (15.8%), followed by bowsers (10.2%), aircraft galley tap (7.9%) and then mains supply points (6.8%).

Water quality varied greatly between airports, with just over a third of all failed samples having been taken from one airport, while over half were from a combination of two airports. Whilst few samples were positive for *E.coli* (3/845), in the two positive samples taken from bowsers the *E.coli* counts were high (264 and >20 organisms per 1000 ml respectively.)

On the other side of the Atlantic the Environmental Protection Agency (EPA) in the USA published results in 2004 of a survey conducted on board 158 randomly selected passenger airlines. The survey showed that 12.6% of domestic and international passenger aircraft tested at US airports carried water that did not meet EPA standards. Initial testing of the on-board water supply revealed 20 aircraft with positive results for total coliform bacteria; two of these also tested positive for *E. coli*. In repeat testing on 11 aircraft the agency confirmed that the water from 8 still did not meet the EPA's water safety standards.

On the basis of this study the EPA issued guidance to passengers recommending that those with suppressed immune systems should request bottled or canned beverages only on board and should avoid drinking the tea and coffee also.

In June 2003 the US FDA published the contents of a warning letter sent to the owner of an aircraft water servicing business in Puerto Rico, following an inspection of the premises by an FDA inspector. The deviations noted constituted violations of several pieces of US public health legislation including the Interstate Conveyance Sanitation Regulations, at title 21, and the Control of Communicable Diseases Regulations, Part 1240[95].

The deviations reported on the potable water bowser were as follows:

- The water inlet line that fed into the tank was comprised of a piece of garden hose.
- The nozzle of the water inlet line was not equipped with a protective cap to prevent it from becoming contaminated by pests or bacteria.
- There was no evidence to verify that the bowser and its ancillary equipment used to deliver potable water to aircraft were cleaned and sanitised regularly.
- The mobile tank was not identified with a sign to differentiate it from the mobile waste tanks.

In another FDA-reported incident, the President of a major American airline was issued with a violation warning in July 2003. During a routine inspection of the airline point and service area at an international airport, the inspector found 'significant potable water backflow deficiencies as well as poor employee practices and unsanitary conditions'. Other offences noted at the time of inspection included:

- The potable water hose was allowed to be placed in the floor drain and used to flush blocked drains.
- Potable water and lavatory carts were stored adjacent to one another on the ramp.
- The hose used to mix and fill deodoriser in the lavatory carts was housed in the same room as the potable water source, as was the hose used for de-icing.
- The employee smoking break room was located in the same room as the potable water source, with the room itself being soiled and unkempt with miscellaneous debris, standing water and cracked flooring.

In addition to the numerous surveys and inspection data published by official government and public health surveillance agencies, there have been a number of covert reports undertaken by the media in recent years. The culture of concern that governs all aspects of aviation-related health matters appears to have gained momentum in recent years, making the industry ample fodder for the media wishing to expose the 'truth' with regards to food and water quality on board aircraft.

In November 2002 the *Wall Street Journal*[96] published a detailed account of the results of their own on-board survey where the microbiological safety and integrity of water samples from 14 different airlines were tested. The samples were collected not only from the galley taps but also from the lavatory and revealed that in all but two cases the bacteria levels exceeded maximum legislative values set by the US government. In September 2004 a similar article was run by the *USA Today* newspaper, documenting in detail the findings of the previously mentioned EPA survey.

There is no doubt that the heightened public and media awareness in relation to the integrity of aircraft water supplies leaves the industry with an even bigger challenge in its endeavours to assure water quality and safety on a global scale. It is critical that the adoption of a risk-based approach results in documented, trained and enforced procedures throughout the extended supply chain, and that documented protocol is evaluated and reviewed in terms of the results of the verification data available.

The biggest issues, as I see them, appear to be the difficulties inherent in making WSPs and verification testing schedules port-specific and not airline-generic. It is difficult to break with the traditionally accepted methodologies and inaugurate change when verification systems are historically designed to deliver blanket results. Airlines must not be afraid to face the realities of what improved data collation studies could mean in terms of procedural methodology and poor results reporting. If indeed standards of water safety and integrity are being compromised by aspects of the operational and/or logistical supply chain, then it is essential to establish verification and sampling programmes to alert operators to that fact at the earliest possible opportunity, so remedial action can effect immediate consumer safety.

One thing is for certain: the debate over the quality and integrity of aircraft water is unlikely to assuage. It is therefore in the interests of consumer satisfaction that the risk factors inherent in water supply to aircraft are identified, managed and controlled.

11 Aircraft disinsection and pest management

The main purpose of this chapter is to identify the issues appertaining to aircraft disinsection and pest management and draw some parallels between employed methods ideology and possible links to aircraft food safety.

Throughout the time I have worked in the industry, I have been aware of some very sharp distinctions being drawn between the utilisation of pest control methods in this environment and their connection with aircraft hygiene and sanitation, where food and drink ultimately become an integral part.

The industry ideology is that pest control and disinsection issues fall directly against the backdrop of a wider range of cabin health concerns that include DVT, air quality and other vector and airborne communicable disease issues such as Severe Acute Respiratory Syndrome (SARS).

Meanwhile food hygiene and safety concerns are painted against an acutely different canvas, occupying a different threshold of priorities in terms of their perceived impact on the global food chain. My perspective remains that food hygiene and sanitation issues, particularly in terms of international airline travel, pose just as much a threat to the spread of disease and global health as their designated cabin health counterparts.

During the course of this chapter we focus on the issues surrounding vector-borne diseases and analyse methods of control in the aviation environment. Later on in the chapter we look at pest management in terms of its connection and impact on food safety, particularly in terms of chemical usage and contamination.

Vector-borne disease – the case for disinsection

A 'vector' is the term used to describe 'a vehicle for pathogen transmission between hosts'[97].

Invariably, in the case of international air travel and transport, it is essential that passengers, crews, and indeed endemic populations, are protected against diseases spread by insects. Given the speed and distance capability of modern aircraft, the emphasis has to be focused on early detection and rapid destruction of any responsible vectors.

The control of vectors is a joint responsibility at any airport location between adequate barring and detection methods on the ground (particularly in the case of airport catering facilities which will provide an attractive harbourage for pests) and the aircraft operators themselves who have the capacity to transport vectors and their diseases around the globe on a daily basis.

It is noted by the WHO 1995[98], that vector-borne disease has the capacity to infect half the world's population at any one time and cases of malaria, leishmaniasis and dengue-fever are already on the increase.

Following the WHO informal consultation on aircraft disinsection[98] great concern was shown regarding the upsurge of arthropod-borne disease and its geographical transmission and spread. Arthropod-borne viruses are transmitted between hosts by insect vectors and are classified in a group known as arboviruses. A large proportion of these have mosquito vectors and occur mainly in tropical regions. They fall into four distinct groups[99]:

(1) togaviridae
(2) flaviviridae
(3) bunyaviridae
(4) reoviridae.

The following vector-borne diseases fall into the arboviruses category:

- malaria
- yellow fever
- dengue-fever
- Japanese encephalitis
- Rift Valley fever.

Other types of diseases spread by vectors include the following:

- bubonic plague (spread by flea bacteria)
- leishmaniasis (spread by sand fly parasites).

Other vectors notable for their capacity to infiltrate aircraft or aircraft cargo and transmit disease, include rats and cockroaches. We will look at these two later in terms of the disinfestation and fumigation methods employed by the industry to deal with them.

Having established and accepted that the importation of vectors by both sea and air leads to outbreaks and upsurges of disease in non-endemic areas, the remaining burden is for the industry and regulators to agree on the appropriate treatment and prevention methods that need to be employed, in order to tackle the problem safely and effectively.

As in any aspect of the industry that requires global standards cohesion, the aviation and health directorates should come together to produce one standard based on sound scientific evidence and data, that satisfies the criteria. Historically in the case of aviation this has not been so. Just as in many areas of food safety and security, cohesion of standards and systems even at national level proves almost impossible to achieve outside any international requirement for compliance. From airline to airline, systems and standards vary hugely and much of this is governed by budget and in turn by aircraft time on the ground constraints also linked to cost management. Later, when we look at types of disinsection methods available, the influence of budget constraints on these choices becomes more apparent.

Overall, in terms of the issues impacting on vector disease control and in turn disinsection standards, the key factors revolve around the following generic considerations being satisfied:

- Disinsection, fumigation and disinfestation practices must be carried out in such a manner that passengers and crew do not undergo any discomfort or suffer any injury to health.
- No damage must be done to the structure or operating equipment of the aircraft.
- All risk of fire must be avoided.

These conditions formed the initial basis on which the IHR[100] stipulated the parameters by which vector controls on aircraft should be developed and approved.

I find it ironic that even from the outset of the IHR guidelines, no specific reference was made to the protection of food or drink on board, nor to the protection of food or drink stowage areas during residual disinsection. As we will see later, many of the processes employed to satisfy disinsection and disinfestation criteria in confined galley and stowage areas, have the capacity for a large amount of residual insecticide to remain on both food and drink receptacles as well as the products themselves. There are vague references to covering food during residual spraying, in the cabin crew guidance documents of some airlines, but nothing that refers to the monitoring of insecticide residual build-up on galley surfaces and food service equipment.

In my view, the initial avoidance of the IHR to make any specific reference to the protection of the integrity of food and drink stowage, receptacles and galley areas, in their initial recommendations, contributed significantly to the devolution of food safety concerns across the board. Aviation food safety has remained since the early 1970s part of an alternative debate, far removed from the consistently higher profile cabin health debate and even further removed from any connection to the primary debate, which is the aviation industry's impact on global health.

What is disinsection?

Disinsection is defined as 'the destruction of epidemiologically significant anthropods, which are vectors of disease or which cause national economic damage'[101].

Disinsection is defined as 'just one process that contributes to the overall management of aircraft pest control. Residual disinsection is the process or processes employed whereby the aircraft is either sprayed with aerosols or treated with an approved residual insecticide on a scheduled basis'[101].

Health authorities in many countries are becoming increasingly aware about the potential deadly risks of vector-borne disease being carried into their counties or territories via aircraft.

Whilst disinsection techniques are various, the objective remains the same: to destroy epidemiologically significant arthropods that are vectors of disease and that subsequently have the capacity to impact on the spread of global disease whilst simultaneously posing a threat to animals and plants.

Disinsection is not required on all routes currently and over the years since disinsection of aircraft became a matter for public health debate, the routes to which disinsection has been a mandatory requirement have varied also. The general rule has always been, however, that it is appropriate for disinsection to be carried out when an aircraft is leaving an area endemic in yellow fever or malaria, or where protection against crop pests is necessary.

The Australian Quarantine and Inspection Service (AQIS) however, mandate that all flights entering Australia and New Zealand are subject to residual disinsection or aerosol disinsection. This is just one example of a variation in international standards influenced by national rather than international legislation and outside international guidance standards. We examine the debate thrown up by the variation in disinsection application later.

The threat posed to global health by vectors

There has always been much evidence to support the theory that the importation of vectors by both sea and air transport mechanisms results in cases of disease in non-endemic areas.

The specific issue of insect importation between nations was highlighted in a study undertaken at two Tokyo airports in 1984. Insect vectors were found to prevail on 40.6% of flights that had already been disinsected[102].

Malaria has not always been considered a tropical illness, with many vector species present in Northern Europe and the UK until the mid-1890s. Anopheline mosquitoes were recorded as far north as the Scottish Highlands during the 18th century and there are many historical references to ague (the former term for malaria).

The eradication of malaria from most of Europe by the mid-1900s coincided with an upturn in the numbers of people travelling by both sea and air. Over the past century outbreaks of previously eradicated vector-borne disease have been in evidence among non-endemic communities and many of these are directly linked to proximity to airports of overseas travel experiences. Generally in areas where there are not enough carriers of the disease to support a large outbreak, incidences can remain isolated; however, despite widespread vector control schemes aimed at reducing mosquito populations and in turn eliminating disease, the situation in areas such as Africa and India is worsening. Many factors have determined this but they fundamentally lie in the fact that endemic nations and immune populations lack efficiency, funding and interest, and that many of the mosquito species are showing resistance to both drugs and insecticides.

Imported malaria is a real threat, which continues to increase throughout Europe and North America. Total reported cases in Europe rose from 6480 in 1985 to 7244 in 1995[103]. There remains increasing concern that malaria vectors will be imported into the Western Pacific[104].

Much of the recent research into cases of imported malaria has been undertaken in France. It has given rise to the term 'airport malaria', which is a slight mistranslation of the phrase adopted by French airport workers, paludisme de l'aeroport, which actually means malarial disease from mosquito species imported by aircraft. It refers to cases where the patient has no travel history and it does not include malaria contracted on stop-overs in endemic areas.

Historically, outbreaks among airport workers are reasonably common where an outbreak has been identified; nonetheless the vectors can be transported further into the wider area, via passenger luggage or in vehicles from the airport, and infect the wider population. In 1984 vectors transported in the vehicles of air crew from Gatwick airport in the UK resulted in two cases of malaria, the first in a pub visited by the staff and the second affecting the spouse of an airline worker over 15 km from the site.

Australia provides the best examples of vector imports, with over half of the insect pest species present in Australia having been introduced. The Australian government has had to take steps to control the spread of several diseases including Japanese encephalitis and dengue-fever. Various serotypes of dengue-fever occurring in Northern Queensland have the potential to spread to southern temperate coastal regions via vector importation.

In terms of the impact of vector-borne disease on the global economy, it is not just issues associated with human health that prevail. The governments of the world must also consider the protection of crops, animals and ultimately the food chain from pests and the diseases they give rise to.

As early as 1951 the 6th International Plant Protection Convention was held. This conference aimed to address the issue of vector importation and it was made clear that a great risk to crops and animal species was posed by the airborne vectors associated with air travel. Many plant viruses are known to be carried by insect vectors, and huge economic losses have been sustained as a result of outbreaks, e.g. *Aphis (Doralis) fabae* is the known vector of the pea mosaic virus[105].

Incidences of global importance associated with vector-borne diseases include Dutch elm disease, where the vector is a beetle, and various cholera outbreaks associated with the transportation of domestic flies. Whilst there remain strict rules in place in most countries controlling the importation of animals and plants, there are no such rules to protect against the importation of insect vectors. Appropriate penalties should be applied to nations who do not put measures in place at ports and airports to inhibit the potential for vector proliferation and transportation.

The history of disinsection

The control of unwanted anthropods has taken place since the early 1940s on board aircraft and prior to that on cargo ships. In terms of disease-carrying vectors such as mosquitoes, disinsection techniques have been employed in varying techniques and formulations, and in the case of cockroaches a routine disinfestation technique has been employed.

The WHO has always taken an active role in overseeing disinsection protocols since their inception and has been instrumental in giving the issue global importance. The 1949 WHO expert Committee On Plague first set the precedent by expressing concern and recommendations that control of vectors both by air and sea be carried out, citing disinsection as part of these measures.

In 1959 a review by the WHO Expert Committee On Hygiene and Sanitation In Aviation broached the subject of insect control, not only in the aircraft themselves by within airports, and as a result reviewed methods of disinsection. Two years later vector control within countries was redressed, resulting in the establishment and identification of malaria-free zones.

In 1969 the IHR[43] were adopted by the 22nd World Health Assembly with the aim of:

'Ensuring the maximum security against the spread of diseases with a minimum interference with world traffic.'[100]

Member states were, and continue to be, urged to adopt the recommendations of the IHR and adhere strongly to the procedures and recommended practices contained in them. These regulations covered all of the perceived measures that needed to be taken to underwrite the protection of public health both at sea and airports.

Article 19 of the IHR refers to the control of vectors around ports and perimeter areas, including larval and adult stages of growth of both mosquitoes and associated insect vectors.

Article 25 of the IHR sets out guidelines regarding general pest control measures in international traffic, taking into account the key features of passenger and crew health and safety; structural damage and fire hazards.

Article 83 has a direct link to Article 25 and gives details of the circumstances in which aircraft disinsection shall be carried out:

- Upon leaving an airport in a malaria-free zone
- Upon leaving an airport where resistant mosquitoes are present
- Upon leaving an airport where mosquito species are present that have been eradicated at the destination.

The later chapters go on to identify the disease covered by the regulations, i.e. plague, yellow fever and cholera.

The WHO has continued to amend many of the directives of the IHR and disease issues have changed either due to demographic and global warming effects, or as a result of the success of vaccination programmes such as smallpox. Disinsection is no exception to that rule. It was under review in 1995 and again in 1998, with the latest debates held at a conference on cabin health in Geneva in June 2004.

The WHO regulations are reinforced by ICAO, also part of the United Nations in the same fashion as the WHO. The collaboration is necessary to bridge the gap between issues of global health significance and their devolution into the aviation arena.

WHO member states have also to consider their own legislation governing Port Health. The legislative powers of each member state allow for a variation in both the method of disinsection applied and the extent to which it is mandatory. From state to state, the provisions laid down by governments in the regulations preventing and controlling the spread of disease vary hugely despite WHO guidelines.

The best example of member states setting their own guidelines via their own legislative processes is the New Zealand MAFF Quarantine Service (MQS) and the Australian Quarantine Inspection Service (AQIS). These regulations mandate the disinsection of all aircraft coming into their territory, not just those aircraft hailing from areas prescribed by international organisations such as the WHO as high risk.

Methods of disinsection defined

There are several methods of disinsection available, most of which are employed in one way or another by airlines around the world. The decision as to which method to use can be determined by a variety of factors:

- Potential impact on crew and passenger health.
- Fiscal cost.
- Availability of approved contractors.
- Level of risk of vector-borne disease.

Current aviation practices for disinsection currently fall into two groups:

- Spraying either before or during the flight with aerosols.
- Residual treatment.

Within either of these two categories there are also a variety of methods of application with varying levels of effectiveness and suitability being demonstrated from airline to airline.

The blocks away method

The blocks away method has been popular since the early 1970s and refers to the process by which the aircraft is sprayed at the beginning of the flight. The term refers to the shift in aircraft wheel blocks in readiness for take off.

The system is based on the principle that if there are vectors present at the commencement of the flight, it is logical to destroy them before they pose a hazard to passengers or crew. The primary advantage of adopting this method is that it has to be administered postembarkation and as the aircraft begins to taxi, so the crew have to undertake the task, as opposed to other methods that require specialist contractors, therefore satisfying a primary contributory factor for any airline: cost control.

Once the ventilation has been switched off, a single use aerosol container is activated until all of the contents have been discharged. A valve cap prevents the container from being discharged accidentally. Bizarrely, these chemical canisters are transported to the aircraft with the catering supplies and stored post usage in the galley, and made available for inspection by Port Health Authorities upon arrival at destination.

Cabin crew must proceed down the aisles slowly spraying the insecticide in the direction of the overhead lockers, away from the passengers who are advised by the crew to cover their nose and mouth. The cans contain insecticide and discharge at a rate of 1 g per second. The number of cans discharged will vary according to the size of the aircraft.

When spraying has been completed, the disinsection certificate in the Aircraft General Declaration of Health must be completed by senior cabin crew. Every aircraft carries its own registration number which corresponds to the registration number on the Aircraft General Declaration of Health. POH officers check this information on arrival and have the power to insist on further disinsection measures being undertaken if they are not satisfied that these procedures have been carried out effectively and in line with the regulations.

The ground staff are responsible for disinsecting the hold, but as an article in the medical journal *The Lancet* attested in 1990, the cargo section is 'largely ignored'.

Effectiveness

The blocks away method has the advantage of speed in terms of the swiftness by which insect species are knocked down following administration of the prescribed dosage. This type of spraying can also be used to combat insects in an emergency situation. Whilst this method has good immediate effect, its low residue results in frequent re-application being required, which may not always be carried out properly.

In 1995 the efficiency of this method of disinsection, was studied on Airbus 310 flights from the Ivory Coast, using caged mosquitoes (*Anopheles gambiae* and *Culex quinquefasciatus*). The cages were placed at different sites throughout the interior and levels of mortality were recorded after 24 hours: 100% for non-resistant strains of *A. gambiae* and 70–100% for *C. quinquefasciatus* in open spaces. In concealed areas, however, particularly in the overhead lockers, figures were considerably lower. In the case of insecticide resistant *Culex* species, a mortality rate of only 34% was recorded.

The popularity of this method of disinsection is without doubt based on its ease of application and budget benefits. Despite a growing lobby of critics who claim that this type of aerosol cabin spraying is detrimental to human health, particularly during long periods of exposure (cabin crew), there remains much industry evidence to the contrary. Based on the effectiveness of this method versus the potential contamination risks, it would appear that other residual type methods do undoubtedly prove more effective. The WHO stipulated in 1995[106] that:

> *'Financial considerations should not preclude any of the preferred methods of aircraft disinsection.'*

Despite this, airline budget restraints and the cost of keeping aircraft on the ground while they are residually disinsected by approved contractors in controlled environments, is extremely high. Invariably cost factors cannot fail to have a major bearing on the type of techniques employed from airline to airline.

The top of descent method

This method fundamentally operates in two stages:

(1) Pre-flight residual spraying by approved staff.
(2) A respray by cabin crew as the aircraft begins its descent.

Pioneered and ultimately adopted in 1989 by Australian airlines, following detailed trials, the system did not gain WHO approval until later.

Certification is required as part of the Aircraft General Declaration Of Health for both sections of the treatment; and in the same fashion as the blocks away method, aerosol cans must be retained for inspection. In the same way also, a cabin announcement will be made to alert passengers that spraying is about to commence, and air recirculating systems are shut down.

Effectiveness

The top of descent method is widely considered to be more effective in terms of its ultimate purpose, i.e. to destroy insect vectors[107], and during the early trials in Australia in the late 1980s 100% mortality rates for *Culex quinquefasciatus* were observed[108]. The suggestion is that the pre-flight residual treatment combined with the in-flight spraying is more effective overall. The employment of this technique among airlines, however, is less prolific due to the extra cost implications which result from two separate treatments, one carried out by specialist personnel.

The other issue that impacts on the overall effectiveness of both of the above methods is the fact that both result in restricted access and coverage being achieved in the overhead lockers and toilet areas, as during both taxi and descent these areas are closed.

In both of the previously described methods the presence of passengers and crew has given rise to concerns over the balance between vector control and chemical safety. The effectiveness of the cabin crew-derived techniques involving aerosols is often influenced by customer complaints and poor training methods. Ultimately successful application relies in both cases on the crew carrying out the procedures as efficiently as possible. The wisdom of placing the critical responsibility for in-flight vector control in the hands of an already overburdened and consumer-sensitive profession has to be questioned in terms of its acceptance as an approved industry SOP primarily in place for reasons of cost effectiveness and efficiency.

The on arrival method

With the on arrival method disinsection is carried out at the destination airport. Quarantine officers (QOs) board the plane and treat the cabin with a swift knockdown of insecticide. Passengers and crew are not allowed to disembark until the treatment is complete and the QO has signed a certificate of practique[109]. Cargo and luggage may not be removed until an inspection has been carried out and the officer is satisfied that adequate measures have been taken. The actual process of disinsection does not vary from those of 'blocks away' or 'top of descent'; the aircraft is sprayed with aerosols of the same type and concentration in the same way.

Effectiveness

The effectiveness of this method can be measured in the same way as for 'blocks away' and 'top of descent'. The major factor here is not the nature of the application, more the timing of the application. No pre-flight or mid-flight disinsection renders passengers and crew susceptible to risk from vectors present throughout the duration of the flight. More often than not the 'on arrival' method is employed in conjunction with one of the other methods as an extra precaution or when it is believed that the application of one of the pre-flight or in-flight methods has proved unsuccessful in any way. This is usually determined by the obvious prolonged presence of airborne vectors throughout the duration of the flight, or may prove necessary should all of the in-flight disinsection records prove incomplete. The fiscal impact on an airline employing this technique in tandem with one of the others renders its usage rare as part of a two-step process. More often than not it is a method adopted in isolation and carried out as described at the conclusion to the flight.

The pre-embarkation method

The pre-embarkation method was developed in Australia primarily in response to consumer concerns about the health effects of aerosol spraying in-flight. Currently still not a WHO approved method of disinsection, it is nonetheless employed by some airlines in an effort to employ a technique that accommodates disinsection in the absence of passengers.

The rationale of the pre-embarkation technique is to treat the plane after all the catering is loaded, just before the passengers board and within an hour of departure or closure of the main entrance to the door. The spray is carried out like a normal on arrival treatment. The aim of the spray is to kill soft-bodied insects that may be present inside the cabin at the time of disinsection. As in all other proven methods of spray treatment, the air conditioning or recirculating systems have to be turned off.

The chemical component released will result in a light and patchy coating of insecticide residue, designed to knock down and kill any vectors that may enter the cabin during the time between disinsection and aircraft departure. Moreover the chemical residues will continue to prove lethal to stow away insects during the flight.

The main advantages of this method are:

- It does not inconvenience passengers or cause delay.
- It is easy to carry out by crew or airline staff and does not require complicated training.
- It is a simple, inexpensive method that can easily be audited by the authorities.
- It uses relatively safe insecticides recommended by the WHO for use in aircraft.

Effectiveness

In insect mortality trials carried out in Australia and New Zealand in 1995, the pre-embarkation method was utilised with live house flies, which are generally considered to be more robust than mosquitoes. To monitor and measure the efficiency and effectiveness of the pre-embarkation spray, several different trials were undertaken, some involving netted and caged flies and others involving the release of flies into the cabin areas. The health status of the insects was measured systematically over a 24 and 48-hour period and categorised in terms of knockdown and dead effect.

It was concluded overall that in the absence of air movement as a result of air recirculation and air conditioners being turned off, the spray dosage reached the furthest places underneath the seats; however, only a 74.2% knockdown or kill occurred. It was suggested that the small and scattered amounts of residual insecticide which remained on wall and contact surfaces would ultimately mortally affect any flies not initially knocked down, when they settled and came into contact with the chemical residue. In view of the growing opposition to cabin spraying in the presence of passengers, this seemed a worthy alternative. However, with no capacity to measure the chemical residue build-up on catering equipment, galley areas and food contact surfaces, consideration might need to be given to the chemical contamination of galley areas and its potential impact on food safety.

The economic cost of this method in terms of each airline having to submit themselves for audit before it can be adopted, and the training required to carry it out, effectively makes it for the most part cost prohibitive; however, it remains a popular method in Australasia.

In-flight vapour disinsection

This method was devised in the early 1970s and approved for use in aircraft in 1977[110]. There were two primary reasons for developing this alternative method at the time:

(1) To overcome the efficiency and accuracy issues connected with aircraft disinsection procedures conducted by crew.
(2) To allow the hold to be treated in-flight and thus ensure that this is carried out effectively and without fail.

This semi-automatic system was activated by a crew member who simply pressed a button. Dichlorvos (now a banned substance in this application) vapour was then forced through a tubing system by a compressor into all areas of the aircraft. After 30 minutes it closed down automatically. This method was approved by WHO member states following airworthiness trials ICAO during the early 1970s. Despite some concerns by certain aircraft manufacturers that the Dichlorvos vapour might result in long-term corrosion issues, none of the trials conducted between 1970 and 1976 showed this to be true.

Effectiveness

There are several major advantages of this technique over some of the other methods already looked at. Primarily there is far less risk of error, as the system is automated and not subject to human error. The convection currents produced result in a greater penetration of potential insect harbourage sites[111] and the formulation used is of low mammalian toxicity, odourless and leaves no visible residue. Despite the recommendations from both the WHO and ICAO in favour of this system and their assertion that:

> 'Financial considerations should not preclude any of the preferred methods of aircraft disinsection, being undertaken,'

this system was never adopted by the airlines as, despite its obvious benefits and effectiveness, the installation and maintenance costs were deemed cost prohibitive by the industry.

The residual disinsection method

The residual disinsection method was developed in the early 1980s in New Zealand[112] and is gaining popularity among airlines, particularly among those which fly regularly to and from high risk destinations. It is undertaken by professional contractors when the aircraft is routinely grounded for maintenance. These specially trained personnel are equipped with protective clothing such as dust masks or canister-type respirators.

The insecticide is applied with compressed air spray guns or pressure retaining garden sprayers. Meantime aerosols are used to treat electrically sensitive areas. To ensure optimum effectiveness it is critical that all surfaces are evenly coated with an insecticidal film, and in areas where cleaning occurs more frequently, like the galley, additional spraying with aerosols will be required as well as touch-up respraying. This process of residual treatment is undertaken every eight weeks with a recommendation for fortnightly touch-up spraying. A certificate must be completed for each aircraft and included in the Aircraft General Declaration of Health[109].

Effectiveness

This method is likely to be applied more efficiently as it is carried out by professional contractors, not the crew. As the passengers and crew are not present during treatment, it eliminates discomfort to them and the risks of chemical hazard exposure associated with in-flight spraying of aerosols. Whilst the principal benefits of residual treatment remain greater passenger comfort and a reduced risk of insecticide exposure, this method is not favoured by many airlines due to the cost incurred. While it was intended that this treatment would be part of routine maintenance, to avoid aircraft downtime, modern aircraft fleets are less likely to require maintenance within the eight-week time-frame than those used during the early 1980s, when this type of treatment had its trials.

It has been estimated that residual disinsection costs are 2.6 times greater than the costs associated with the use of aerosols[113], with aircraft down time and the amount of chemical required for effective treatment being the major contributory factors.

It has been suggested that this method produces a lower insect knockdown rate as it relies on insects landing on surfaces and the immediate knockdown effect is not demonstrated, leaving passengers susceptible to 'stowaways' for at least a proportion of the flight.

An alternative approach to residual disinsection is the ultra low voltage (ULV), which was first used against locusts in the 1950s. An electrical generator is required to rapidly diffuse the insecticide aerosol droplets. Air currents carry the droplets, which can penetrate harbourage sites and produce a flushing effect. A lower volume of insecticide is required and it can be applied via a hand-held applicator without specialist training. All of these factors render this method more economical than normal residual disinsection.

In 1995 a study in Cameroon recorded a 100% mortality rate for *Anopheles gambiae*[114]; however, the major downside to this type of treatment remains the presence of kerosene in the solvent used. This has led to some aircraft manufacturers not approving it for use because of the fire risk posed by the solvent[101]. Paragraph 1 of Article 25 of the IHR states that:

'disinsection operations must not present a fire risk.'

Disinfestation

Disinfestation in this context is the term usually applied to the treatment of cockroaches in the aircraft environment. Aircraft, by their nature, provide abundant harbourage opportunities and are a natural attraction for cockroaches. While these are considered second-rate vectors for disease, the risks posed to human health when they come into contact with food or food preparation and storage areas are well documented.

Galley areas in particular on aircraft provide perfect harbourage and breeding grounds. The favoured warm environments of two of the most common species, *Blatella germanica* (German cockroach) and *Periplaneta americana* (American cockroach) can be found in abundance among the warm spaces between and behind beverage makers and around the water pipes. The evidence of food scraps in and around galley areas, trolleys and carts will also encourage nesting and proliferation.

It is currently mandatory that all long haul aircraft be routinely disinfested. In line with European guidelines on pest control practices around food service environments, this must be carried out by specialist, licensed pest control contractors or specifically trained airline staff. In the case of the airlines, this situation is not as carefully monitored and enforced as it should be, and in the case of many nations it is not mandated as it is in Europe, by the provisions of the food safety legislation. In this scenario, disinfestation procedures involving the spraying of insecticides, not only in the cabin and hold areas but particularly in the galley and food stowage areas, are open to potential abuse and misuse by the application of non-trained personnel who may have no perception of the inherent risks of chemical contamination of food.

On short haul aircraft the disinfestation schedules are not mandated and are left to the discretion of the airline. For reasons of cost and time on the ground, airlines will usually only treat as a reaction to a report of a sighting on board by the passengers or crew.

Fumigation

Fumigation refers to the procedures employed in an emergency situation on board aircraft whereby rodents or other hazardous species may be observed as present or loose in the cabin. In this scenario, the aircraft is immediately grounded and fumigation with a lethal gas is carried out. Only licensed contractors are able to undertake this work with the obvious risks of aircraft and chemical safety high on the agenda.

Chemical safety and public health

Whilst it has for a long time been accepted that there remains an undeniable necessity to disinsect aircraft and therefore inhibit vector transportation and spread of disease, there is a growing body of concern among many that both short-term and long-term exposure to the chemical substances, used in residual or aerosol form, poses numerous hazardous to human health.

The occupational health debate has continued to rage despite the numerous changes made over the years to the types of chemicals used and approved for use on aircraft. Currently, the chemicals approved for usage in the residual market are limited to three. The WHO lay down strict guidelines as to the appropriate percentages to be used to provide for optimum rates of knockdown and eradication efficiency; however, various health and safety governing bodies will stipulate different formulations from country to country. AQIS have their own standards governing the percentage formulations to be used, which fall in with WHO guidelines; however, in the UK the Health and Safety Executive (HSE) is not alone in not allowing certification for percentage doses as high as those recommended by the WHO and insisted upon by AQIS.

Bearing in mind that the rather critical issues of which chemical and how much chemical are subject to such variations of interpretation, it is easy to understand the growing lobby of opinion among those who believe that they have suffered health ill effects as a result of occupational exposure to them. Coupled with all of these factors there is the ongoing issue of who actually administers the residual and disinfestation treatment-trained pest control contractors or untrained airline staff or engineers.

Most of the cases in which adverse health effects have been documented as a result of exposure to disinsection procedures, have cited the aerosol spraying techniques particularly.

Long-term exposure issues have been alluded to by ex-crew members such as Diana Fairechild[115] who was grounded permanently after 21 years' service with an American airline after developing multiple chemical sensitivity (MCS) and sited as the root cause exposure to residual insecticides not only via aerosol spraying but also through the residual effects of crew blankets and pillows.

Overall, the industry view is that the health effects posed by vector-borne disease far outweigh any cause for concern over the occupational exposure risks posed by insecticides. Whilst there is some truth in that, let us not forget the IHR stipulation that:

'All disinsection, disinfestations and fumgiation practices must be carried out in such a manner that passengers and crew do not undergo any discomfort or suffer any injury to health.'

Bearing in mind that there is so little cohesion over techniques, application and chemical usage, despite international recommendations, the issue of airline self-regulation to a standard commensurate with cost control and not best practice rears its ugly head again. Just as in food safety management protocol, disinsection issues are subject to a host of interpretations, determined less by optimum safety concerns and more by what has always been the status quo, despite the high profile nature and global impact of the issue.

Disinsection and disinfestation techniques and their link to food safety

There have always been two main thrusts to my concerns over disinsection and disinfestation techniques and how they impact on aviation food safety.

The first has always been the traditional industry desire to compartmentalise fundamentally integrated issues such as food safety and pest control into different arenas and have them subject to different areas of debate and concern.

Earlier in this chapter I expressed my opinion that if only food safety issues in aviation had been given the same priority and presence on the global stage as disinsection and pest control, the industry would be much improved. The integration of directly connected aspects of cabin safety and occupational health must surely require that food and water supply and pest management concerns become part of the same debate.

The level of research and international involvement in disinsection has been phenomenal. The statistical data have constantly evolved historically in a way that food quality and safety verification in the same arena can only dream of. The level and spectrum of international involvement in the issue of disinsection is astounding. Study after study, trial after trial, yet still no cohesive, collective decision on what, when or how.

The second issue I have felt has been the fact that the impact on chemical food safety of disinsection and disinfestation chemical exposure, over either prolonged or even short periods of time, has never been assessed to my knowledge, nor as far as I am aware have any toxicological tests been carried out to measure residual build-up of chemicals on galley surfaces or on food receptacles and containers.

Out of respect for those who have assisted with my research in this chapter, I pledged not to detail the specifics of any chemical formulations rendered suitable for use on aircraft, rather to refer the reader to my colleagues in the industry for verification. At the same time I would like to add that in terms of their suitability for use in food service or manufacturing environments, they would not be approved as safe for use with food.

I would challenge the theory, also, that following residual treatment of aircraft, including galley areas and surfaces, all that is required to remove the chemical is a wipe over with a damp cloth. If the assertion is that following residual treatment a deep clean occurs, then what is the chemical suitability of residual disinsectent in conjunction with cleaning chemicals? Furthermore, if the disinsectant is cleaned away, what impact is there on airborne vectors?

The WHO recommends that:

'all exposed food be covered during the spraying of aerosols.'

But what impact is there on food safety if that doesn't happen? We already know what an incredibly overburdened profession cabin crew are. So far we have examined their food service and safety roles, which are in addition to their nursing and policing roles, and now we have also discovered that they are pest control contractors too. I would imagine that the chances of approved in-flight disinsection protocols floundering in the hands of crew, whether in terms of food safety or the wider issues of successful application of the aerosols, are pretty high.

12   Special meals – special hazards

Before starting to write this chapter of the book, I spent a considerable amount of time attempting to distance myself from the topic entirely, in a very real effort to view the subject matter as a novice to both the concept and its manifestation within the aviation environment.

I began by typing the words 'special meals' into my internet search engine. The results were astounding: 168 000 references to the term, a massive percentage of which were entrenched in the plethora of international and domestic airline websites. For hour upon hour I looked through these sites, devouring the information, not as a seasoned airline traveller and special meal manufacturer, but as an ordinary statistic, a member of the fare-paying public.

For many, the concept of being able to request an in-flight meal which falls outside the ritual delights contained within the standard tray set, is remarkable in itself. Many of my business traveller colleagues are astonished to discover that almost every airline will offer a selection of dietary preference meals, which can be requested instead of the standard menu items. Special meal booking policies and procedures vary hugely from airline to airline, as does the range of meal types available; however, the general requirement is that any non-standard meal request should be placed at the time of booking and with a minimum of 24 hours' notice prior to departure.

Having discovered that a non-standard meal request is a possibility, one sets about building perceptions of what one should come to expect in the provision of such a meal. The assertion that the industry does not discriminate, by making such a vast array of meal options available to such a huge cross-section of ethnic, medically challenged and preference groups, is accurate and as we will see later, is entrenched in air transport protocols and procedures that in reality have little or nothing to do with food. Nonetheless the need is met and the requirement satisfied.

Or is it?

The most telling reflection of industry attitudes towards the responsibility of producing meals that make such a vast array of medically and religious-based claims, is illustrated in the IFCA/IFSA World Food Safety Guidelines. Despite the fact that their introduction attests to the complexity of the requirement, there is no further reference or commitment to special meal provision protocols in any other area of the document, least of all on the

HACCP example. The assumption is, therefore, that the industry believes that general standards of product safety suffice in this area with no concession made to the incredible complex nature of the component development, production and appropriate and legal labelling standards.

Throughout this chapter we will look at where special meal provision within the aviation industry really needs to go, in order to fully satisfy consumer expectations. We will draw all necessary parallels with the food manufacturing sector, to establish a best practice model for this most critical area of all aviation catering provision, and consider its impact on consumer safety and satisfaction statistics.

What are special meals?

Having established that special meals are available and offered on almost every airline, it is important to understand the distinct categories into which these meals fall. The International Air Transport Association (IATA) developed the original special meal (SPML) coding system in the early 1970s and it continues to be the model against which all SPML codes are attributed.

The meal request is denoted at the time of booking by the four letter coding SPML and the categories are divided between preference meals, religious meals and medical meals. Within each of these categories are specific meal types, each with its own attributable coding. Meal requests which fall outside any standardised coding, or which may be a combination of several coding profiles, are simply classified SPML.

The meals are denoted as 'special' on board, usually by a meal labelling system generated by the airline or airline caterer, which will contain such information as the passenger name, flight number, etc. Service protocols vary from flight to flight but for the most part SPMLs are served in advance of the main meal service in economy cabins and often in premium cabins also.

The tray set or menu profile may contain some generic items from the main menus, such as fruit and bread, depending on the classification of the SPML, and we examine the impact on overall product safety of this practice later. Most often, however the meal components will not resemble the standard set-up in any way.

Most airlines set no parameters or restrictions on SPML ordering, and when the meal falls within IATA coding guidelines they require no formal verification by the passenger of the necessity for the request. In the case of SPML non-generic meal requests, some advance consultation with the passenger or passenger representatives may be required and may occasionally result in the meal request being refused. For example, in October 2002 British Airways, who had previously attempted to accommodate all non-standard meal requests from 'nut-free' and 'no onions or 'no mushrooms' to 'diabetic vegan with a gluten intolerance', ceased to offer anything other than meals which fell within strict IATA coding guidelines. This could have been viewed by some as a discriminatory action against the dietary afflicted; however, what is clear from an examination of the SPML systems management protocols within their catering establishments, is that they do not allow for such specific allergen intolerances to be accommodated effectively and therefore the decision to withdraw the service was a sound one.

It is ironic, I feel, that the aviation food service sector is burdened with the broadest possible spectrum of dietary meal requests to fulfil, and boldly ventures where the food manufacturing sector do not dare to tread without the most stringent protocols being established, yet in my opinion they are fundamentally less well equipped to deal with them in every sense. The vast spectrum of meal categories, the necessity to produce in a unit which is conducting non-special meal manufacture also, the relatively small volume requirement, the component nature of the meal, the non-technology-based specifications, the labelling requirements, the list goes on and on.

There is no better demonstration of the huge divide between appropriate food safety management systems in the manufacturing sector and in the aviation sector than in the special meal arena. The component nature of the meal, which requires that every product going to make up the tray set must individually and collectively satisfy the specific SPML criteria, renders the establishment of appropriate systems to control and monitor fraught with difficulty. Here we examine what protocols and procedures are required to conduct the development, manufacture and assembly of special meals safely and effectively with optimum diligence.

Preference meals

Preference meals denote a meal request that is a preferred option rather than one which is critical to sustain health. Typical meal types which fall into this category are:

- Vegan meal – VGML.
- Lacto ovo vegetarian meal – VLML.
- Seafood meal – SFML.
- Oriental meal – ORML.
- Child meal – CHML.
- Raw vegetarian meal – RVML.
- Asian vegetarian meal – AVML.
- Fruit plate meal – FPML.

There is a very close association between the preference and religious meal choices in terms of the types of products that are likely to be excluded, the difference being the style of the menu. The necessity to ensure that the component parts of these types of preference meals satisfy the criteria exactly is less about compromising safety and more about offending the passenger with foods they have chosen not to consume. That is not to suggest, however, that people may not revert to a vegetarian, seafood or low protein derived diet to assist medical conditions or improve well-being; however, preference meal options tend not to be restrictively developed in the same way as medical meals.

At the other end of the preference meal spectrum are child and toddler meals, which as I see it form the most fascinating aspect of the preference meal group. Historically, the products within this meal category have been developed with assumptions about the types of menu options typically enjoyed by this group.

Traditionally, CHML choices will bring forth a host of salt and sugar-laden products, sweets and chocolate bars, fun size sugary drinks and fried entrées. Whilst this category should be about aesthetically appealing to younger travellers, it should also consider the fact that the majority of these SPML requests are going to be consumed by a large, high-risk group. Here we draw comparisons with the regulation of the manufacturing sector which, certainly within the EU, is being urged to reduce the salt, fat and sugar content of foods marketed directly at children.

I am in no way down playing the requirement by the aviation sector to provide a CHML which is fun and enjoyable; however, serious consideration has to be given to the likely food intolerances of an already immunocompromised group and not bowing to the fiscal and often parental pressure to provide meals which are fundamentally nutritionally unsound. The integrity of these products must be maintained, not only in terms of the strictest standards of hygienic manufacture but also in terms of the responsible nature of their nutritional development.

Religious meals

Religious meals are those meal categories that combine a preference meal with a religious requirement. The typical meal types that fall into this category are:

- Moslem meal – MOML.
- Hindu meal – HNML.
- Kosher meal – KSML.
- Strict Indian vegetarian/jain meal – JAML.

These meals are based on strict religious preferences and have to be developed with a sound understanding of the particular procedures and protocols that need to be developed to satisfy the criteria exclusively. The best example of this is the KSML, where specialist religious supervision has to be employed which goes way beyond the food and beverage components to encompass the raw material outsourcing, the tray set equipment, packaging and washing, as well as in-flight service protocols. In some ways this is the best example of the necessity to employ a specialist manufacturer who can meet the exclusion requirements without conflict with other operational processes within the catering unit infrastructure.

Much of the development of religious meals should focus not only on the food groups and combinations of foods to be excluded, but also the typical styles, flavours and combinations of flavours generic to the ethnic group in question. Attention must be paid not only to 'banned' foods but also to those that would be expected to form an integral part of every meal. Once again, the multicomponent nature of the meal requires a detailed analysis of every product from salad to bread, individually and then collectively to ensure that all components blend and complement. Ultimately both the preference and quality criteria of the meal have to be satisfied simultaneously.

In this category the appropriate raw material outsourcing can also pose difficulties. With restrictions on the numbers and types of Halaal and Kosher raw material suppliers, it is increasingly difficult for aviation caterers to commit to the necessary supply chain even if it is available. Most raw material suppliers of religious meat and fish product raw materials will be food service/catering licensed and therefore will not meet the food manufacture licensing standard requirements. This poses traceability issues and a conflict with the successful implementation of GMP standards in the production of religious meals, unless animal by-products can be eliminated altogether.

Medical meals

Medical meals are those requested as a result of the direct necessity to consume food in a particular manner or format with the restriction or exclusion of one or a number of food groups. Mainly, these requests are a result of food intolerances and/or allergies, but they may be derived from a variety of related medical conditions or ailments or form part of a rehabilitative programme pre- or post-surgery. Other meal types in this group indicate a possible combination of necessity and preference choices, e.g. low fat, or low calorie. The typical meal categories found in the medical meal group are:

- Low fat meal – LFML.
- Low salt meal – LSML.

- Low purine meal – PRML.
- Low calorie meal – LCML.
- Diabetic meal – DBML.
- Gluten-free meal –GFML.
- Non-lactose meal – NLML.
- Low cholesterol meal – LFML.
- Bland meal – BLML.
- High fibre meal – HFML.
- Low protein meal – LPML.

As one can see, the list of meal types attributable to the medical meal category far outweighs that of the preference and religious meals. As such, the food safety management protocols employed in the development and manufacture of these most critical of meal groups should be the same in all categories. In this way the process flows allow for allergen and prohibited product exclusion across the board so that the level of dietary claims which is capable of being made, can remain in line with retail manufacturing standards and is not compromised overall by the catering/food service ethic of the products.

Also found in this group are the broader base of medical requests which are just denoted by SPML. These are the specific requests that British Airways decided to refuse to accommodate, as mentioned earlier, and they come in many guises: liquid meal, nut-free, no onions, low salicylate, basically anything that falls outside defined meal code parameters.

To suggest that these most specific and highly delicate dietary requests can be successfully accommodated within an average airline catering unit, without engaging food technology type protocols, GMPs, specifications and expertise, is sheer folly. To attempt to make a nut-free claim in an environment which would most certainly process huge volumes and types of nuts is insane. These claims are often shied away from by all major food manufacturers who, even with the most highly developed and strict food technology protocols and expertise, know that they cannot meet the criteria that the claim demands in the truest and legal sense of the word. In the meantime these claims are embraced by the aviation catering community on a daily basis.

Since the IATA special meal coding system was devised nearly 30 years ago, food labelling, allergen identification and the level of the requirement for dietary meals has surged ahead at a rapid rate. The focus of attention, particularly in the area of food labelling and the attributable minefield that is medical claims, should have rendered the airline practice of catering manufacture in the SPML arena redundant at least 10 years ago. Even with the growing acceptance among airlines that allergen requests outside of the coding system cannot be accommodated, still the specific coded medical meals require a greater level of understanding and expertise applied to both the development and processing.

Having studied many airline special meal policies, it is obvious that many airlines are confused as to exactly what it is reasonable to expect from them and what constitutes an unsafe request that they and their catering partners are ill equipped with facilities and expertise to satisfy.

The issue has far more to do with the component nature of the airline meal than anything else. In defence of the aviation industry, food manufacture has only to contend with making claims about products in isolation. The 'product' in the aviation case can be up to as many as six or eight items, to which the dietary claims have to stick collectively as well as individually. The technical specifications required to achieve this accurately are complex and outside of the catering box. We look later at the specification development and systems management skills and tools required.

It would be naive of me to suggest that every airline caterer should cease producing SPMLs on the basis that fundamentally they are not logistically or technically capable, when many outsource at least one of a number of components from manufacturers anyway. However, where the requirement is that they are made in-house, the information laid out in this chapter should most certainly be implemented at the earliest opportunity to ensure successful product development, accurate allergen labelling and controls, customised nutritional data to establish both the collective and individual component composition, and above all the quality assurance and product safety protocols which guarantee optimum levels of safety and diligence.

Special meal menu planning and development

In advance of the menu development of any special meal menu suitable for aircraft application, one has to consider all of the necessary production and quality management protocols first. This concept itself is alien within the aviation catering environment where, as in all food service environments, it is the product that drives the concept, with process flow and product safety considerations following behind.

Before beginning a special meal development plan it is useful to complete a list of 'banned' ingredients in order to cross-reference every product and component as it is short-listed for inclusion. Never lose sight of the fact that the meal is not just one but a host of components that must come together, collectively as well as individually, to satisfy the specific criteria. Remember that the more restrictive the product list and recipe development across all categories, the easier it will be to maintain the appropriate diligence during production, and the less complicated the systems will need to be.

Tables 12.1a, 12.1b and 12.1c illustrate the manner in which certain food groups are attributed to everyday ingredients, products or product components. It is useful to examine the relationship between product groupings and allergen-based products and witness the attributable meal coding criteria. Utopia would be a system of menu development which excluded all raw materials and finished products that appear on this list.

Is it possible? Well, it is possible to create menus to exclude all of the products and raw materials that appear in Table 12.2, but in order to satisfy the quality, regional and taste expectations which form just as important a part of special meal development, some of them have to be utilised. For airlines that still offer the SPML request service, it is critical in terms of systems management and diligence that as many as possible of the products referred to in Table 12.2 are restricted.

Table 12.1a Attribution of food groups to everyday ingredients, components and product

Banned food group	Potential sources	Attributable code/codes	Meal type
Milk + milk products	Cream soups Butter sauces Yoghurt Cheese Dairy-based dressing Custard Mousses Smoothies Bread Cakes Pastries Confectionery	NLML VGML AVML HNML	Medical Preference Preference Religious

Wheat + wheat products
Gluten + gluten products
	Bread – pastries
	        cakes
	        biscuits + cheese biscuits
	Sausages/processed meats
	Breadcrumbs – kievs
	        fish fingers
	        chicken nuggets
	Pasta cereals + porridge
	Stock + gravies + sauces
	Salad dressing
	Beer + malted drinks + gin/whisky
	Snacks – pretzels/crisps
	Soy sauce
		GFML
		SPML
			Medical
			Medical

Table 12.1b Attribution of food groups to everyday ingredients, components and product

Banned food group	Potential sources	Attributable code/codes	Meal type
Nuts + nut products	Packaged nuts Nut oils in sauces Satay sauce Confectionery M&Ms + cakes Muesli + muesli bars Breakfast cereals	SPML CHML	Medical Preference
Sesame + seed products	Sesame oil Sesame seed – Chinese style foods Muesli + muesli bars Bread toppings Salad dressings Savoury snacks	SPML	Medical
Eggs + egg products	Egg mayonnaise Hard boiled eggs Scrambled egg Omelettes Custard Quiche/flans/tartes Cakes Pastries Glazes Dessert mousses Pasta Meringue Icing	SPML HNML AVML VGML	Medical Religious Preference Preference

Table 12.1c Attribution of food groups to everyday ingredients, components and product

Banned food group	Potential sources	Attributable code/codes	Meal type
Animal products	All meat + fish inc. shellfish All meat + fish derivatives Stocks + animal byproduct Sauces Gravies Suet Honey Eggs Cheese Gelatine mousses	AVML HNML VGML SPML	Preference Religious Preference Medical

(Continued)

Table 12.1c (Continued)

Banned food group	Potential sources	Attributable code/codes	Meal type
Salt/sodium	Additives – MSG	LSML	Medical
	Stocks	DBML	Medical
	Gravies	LFML	Medical
	Added salt	BLML	Medical
	Vegetables naturally high in sodium, e.g. carrots	PRML	Medical
		CHML	Preference
	Cured meats		
	Smoked fish		
	Cheese		
	Bread		
	Salad dressing		
	Tinned/canned foods		
	Salted butter		

Table 12.2 Guide to appropriate and inappropriate group considerations for each IATA-coded special meal

Code	Prohibited	Allowed
AVML	Meat/poultry/fish/shellfish/egg	Veg/starches/margarine/nuts/pulses/fruit/pasta/rice/potatoes/sugar and preserves/soya product herb and spices
BBML/CHML	Take notice of allergens	Appropriate foods to aid digestion
BLML	Mustard/pickles/garlic/pulses/nuts/fatty food	Lean meat/fish poached/low fat food/chicken/veg/maize/potatoes (boiled)/pasta/fruit/rice/white flour
DBML	Sugar/syrup/jam/cake/chocolate/sweet/fat food	Veg/lean meat/pasta/milk/fish/brown bread/rice/baked + boiled potatoes/low fat/fruit/low fat cheese/eggs/pulses
FPML	Meat/fish/veg/dairy/breads	Fruit fresh + dried
GFML	Flour/oats/soup/sauces/pastry/sausages/pasta	Meat/fish/rice/soya/fruit/veg/maize/herb/spices/eggs/potatoes/cheese/dairy product/sugar/specialist gluten free product
HNML	Beef/eggs/veal meat extracts	Lamb/pork/eggs/fish/chix/rice/fruit/veg/starches/maize/dairy product/herb/spices/pulses
KSML	Pork/sausages/cured meat	Poultry/beef/lamb/liver/sweetbreads/eggs/dairy/flour/fruit/veg/sugar/potatoes/rice/fish (scaled)
LCML	Fat food/sugar pate/full cream cake/sweets/dressing	Fish/white meat/eggs/low fat/lean meat/fruit/veg/skimmed milk/rice/low fat cheese/potatoes (baked/ boiled)/brown bread/pasta
LFML/LPML	Fat foods/paté/sausages/nuts/cheese/egg yolk/shell fish/offal	Lean meat/fish (white/oily)/poultry/veg/fruit/rice/sunflower/olive oil/margarine/skimmed milk/starches/low fat yoghurt

Code	Avoid	Include
LSML	Salt/garlic/tinned cured meat/ gravies/pickle/ cheese/sausage	Meat/fish/cream/unsalted butter/cream cheese/ fruit/veg/milk/potatoes/oils/rice/yoghurt/herbs/ spices/pasta/maize/vinegar/sugars
HFML	White flour product	Meat/fish/nuts/pulses/fruit/veg/dairy/brown bread/maize/pasta/wholegrain
PRML	Gravies/offal/ shell fish/oily fish roe	Dairy/yoghurt/pasta/fruit/veg/fish white/meat/ sugar/oils
MOML	Pork/sausages/ alcohol/eel/fat animal/shell fish	Fish/beef/lamb/rice/poultry/pulses/yoghurt/nuts/ fruit/veg/maize/pasta/eggs/herb/spices/dairy/all meat must be halaal
NLML	Dairy/yoghurt/ sauces/soups/ choc/sweets/ cake/batters	Meat/fruit/veg/fish/eggs/pulses/sugars/potatoes/ rice/pasta/bread/nuts
ORML	Western food i.e. pastry, pie, etc.	Appropriate Asian-style foods, rice, stir fry, Chinese or Thai style
RVML		Appropriate raw veg
SFML	Meats/dairy	Fish/shellfish
VLML	Meat/fish + shell	Dairy/eggs/nuts/rice/fruit/pasta/sugars/pulses
VGML	Meat/fish/honey/ dairy	Veg/fruit/pulses/nuts/sugars/oils/non-meat/pasta
SPML	Requirement not covered by specific code	

The next fundamental question to be answered is whether the special meal development and product range are capable of spanning all classes, or whether the dictates of the logistics and methods of manufacture require that variable and separate product development occurs in each class. Obviously, if the latter is true, the restrictive picture of the product range looks set to be less successful in its application. In order to achieve a greater range and style of products, with variable budgets and set-up specifications, undoubtedly a greater range of raw materials will need to be utilised.

Factors influencing special meal development strategies

In all aviation product development, the wider complications appertaining to tray set equipment types and styles, budgets, aircraft and oven size, galley constraints and sector length, which impact on all aviation menu selections, cannot be forgotten during the development of special meals. It is more crucial than ever that a firm understanding of all of these factors is gained in advance of any serious product development being undertaken.

A prepacked breakfast fruit product that comes in an air exclusion disposable may perceivably have the capacity to pass muster on an economy/coach tray set up or snack service, but will it translate as effectively if the standard equipment is ceramic or glass? Probably not is the answer, so whilst the product intended for the same purpose in the premium cabins may well still be fruit, its style and application will be vastly different. This problem tends to be less of an issue with the hot part of the set up, where as long as the source of supply is consistent, the same meal components can be packed larger or smaller in the equipment. For example, chicken curry and rice would be a perfectly acceptable Moslem meal offer in both economy and premium cabins; all that would need to vary would be the size of the foil or entrée dish, which would equate to a larger or smaller portion of the same meal selection in all classes.

It has become increasingly common practice for airlines that offer the whole spectrum of special meal types, to roll as many meal types as conceivably possible into one or two products. This is a useful way of cutting down on menu development time and costs, but in my experience can be a dangerous practice unless it is incredibly carefully scrutinised and controlled.

Getting back to our chicken curry, if we made it vegetable then would it not pass muster for all AVML/VGML/VLML/HNML/strict Indian vegetarian and MOML? It may, but only if the recipe was incredibly restrictive. All meat and dairy derivatives would have to be excluded, as would garlic, onions and onion derivatives, bulbous and root vegetables. Here we see an example of a solution to a menu development issue that actually results in an end product which has little to offer most meal groups in terms of quality or aesthetic appeal. Along the same lines, and looking at the wider component aspects of the job, would it not be appropriate to put a fruit cocktail on all set ups for breakfast in all classes? Whilst fruit may hold universal appeal and affords the ease of global replication, it may not meet all the carbohydrate restrictions of the DBML and LCML, depending on what the other carbohydrate components and counts are on the rest of the tray.

The key to devolving meal categories into one group is to work with the broader scope meals together and then look at the more restrictive in isolation. To cross cultural divides by mixing meal groups that combine both ethnic and medical or preference restrictions may also lead to quality complaints. Not every low fat or low salt request would necessarily welcome a vegetarian or curry choice. Whilst there is certainly an element of not being able to please all the people all of the time, which is true of all menu development in some ways, it is important to give some consideration to the quality issues in tandem with assuring that the content meets the necessary restrictive criteria.

Previously I have stressed the point that SPML development must consider every component on the tray that forms the meal, in isolation as well as collectively. The aviation-style special meal is not afforded the same luxury as the retail-style product, where one item in isolation, albeit a main course/entrée, snack or dessert component, is subject to nutritional scrutiny. In terms of what constitutes the airline brand special meal, as many as six or seven components must come together to meet the overall criteria. This renders the blanket use of seemingly suitable products quite dangerous unless they are placed directly into context at the development stage alongside the other components that will make up the tray. Is it unreasonable to render the sum total of all the tray components the meal? Why can we not assume that the nutritional data of each individual component will suffice? Because we have to assume that the passenger will consume everything on the tray and if the total calorie, sugar, fat, carbohydrate, gluten, lactose or salt count of all products combined exceeds industry or legislative guidelines, the suitability of the meal as fit for purpose is eradicated and may result in ill effects in the passenger who consumes it.

Summary of factors to be considered in special meal development

To summarise, the following list shows the primary factors that must be considered before any specific development can commence:

- Compile a list of banned and/or restricted foods and food groups, highlighting allergens specifically.
- Attribute potential meal categories that may apply (see Tables 12.1a, 12.1b and 12.1c).
- Decide conclusively which foods/raw materials will be banned and make a list.
- Decide conclusively which foods/raw materials will be restricted and make a list. Attribute the special meal types from which the foods in the restricted group will be omitted.
- Compile a list of foods/raw materials which may have universal usage opportunities across all meal types, e.g. fruit and or fruit products, rice, certain vegetables; and highlight any meal types where their nutritional composition and inclusion will impact on the overall picture of the meal in terms of suitability nutritionally and/or in terms of the legality of claims made about the meal, e.g. low calorie, low salt, diabetic, low fat, high fibre, gluten/lactose-free.
- It is also useful at this stage to compile a list of products which will appear generically on every tray set, as an accompaniment to every meal, and which will require replication in different guises in every category.

An example of the list mentioned in the last point is given here.

Milk

- Soya required in AVML/VGML/HNML/strict Indian vegetarian.
- Skim required in LFML/DBML/LCML.
- Whole required in CHML.
- Lactose-free required in NLML.

Butter

- Soya spread required in AVML/VGML/HNML/strict Indian vegetarian.
- Low fat spread required in LFML/LCML/DBML.
- Salt-free butter required in LSML.

Bread

- Dairy-free required in AVML/VGML/HNML/strict Indian vegetarian/NLML.
- Salt-free required in LSML.
- Gluten-free required in GFML.
- Low fat required in LFML/DBML.
- Low calorie required in LCML.

Salad dressing

- Dairy-free required in AVML/VGML/HNML/strict Indian vegetarian/NLML.
- Salt-free required in LSML.
- Gluten-free required in GFML.
- Low fat required in LFML/DBML.
- Low calorie required in LCML.

Condiments

- No salt or pepper in cutlery packs or on the tray in any of the following:
  - LSML/DBML/LFML/PRML/BLML/strict Indian vegetarian.
- No sugar sweetener substitute in cutlery packs or on the tray in any of the following:
  - DBML/LCML.

Once all the above basic considerations have been redressed, it is possible to begin the fine detail and recipe formulation aspect of the special meal development. This process involves looking in detail at the types of meals which will be required sector-by-sector, in tandem with an equipment analysis to ensure that whatever the products evolve into, they can be accommodated by the available equipment for each class and for each airline and aircraft type that is being catered.

Table 12.2 clearly illustrates the appropriate and inappropriate food group considerations for each IATA-coded SPML category.

Menu development

Translating the restricted and non-restricted raw materials into recipes and menu items is a much more difficult process and a very alien process in comparison to the usual practices employed. In normal circumstances four major factors dictate the standard menu development considerations:

- budget
- equipment
- class of travel
- route.

In special meal development, these four factors play a part but only after:

- banned and restricted foods
- allergen controls
- nutritional compliance
- labelling criteria.

To attempt to make existing products which are being utilised outside the SPML arena is tempting, to ensure that menu development time and costs are kept to a minimum. However, as we will see later in the chapter, the food safety management protocols which will need to employed cause conflict in this area.

Therefore, whilst it is perfectly acceptable to attempt to develop mainstream recipe ideas with consideration given to the requirement to restrict or prohibit certain mainstream raw materials – salt, sugar, butter, milk, etc – it is not advisable to adapt existing products to fit the criteria because the true picture of what needs to be done to ensure that they fit the SPML requirement may be much more than is first envisaged. Figure 12.1 gives some examples.

From the examples given in Figure 12.1 it is clear that the perception that certain common foods may have universal application in the SPML arena tends to be misplaced. A thorough focus on the banned and restricted food

Mainstream product: cauliflower cheese Restricted product: cauliflower cheese	
Likely ingredients Cauliflower, hard cheese, milk or cream butter or margarine, salt and pepper, mustard, wheat flour	**Amendment requirements** Non-dairy milk, cheese, butter or spread No added salt, low salt cheese Low fat milk, cheese, butter or spread Non-wheat flour Non-white flour
Perceived categories suitable for in mainstream state VLML, CHML, HFML, BLML, PRML, DBML	
Actual categories suitable for in mainstream state VLML, CHML	
Mainstream product: roast chicken and gravy **Restricted product: roast chicken and gravy**	
Likely ingredients Whole chicken carved or chicken breasts, fats or oils, salt, alcohol, packet or dried stock or gravy, meat juices, flavour enhancers	**Amendment requirements** No added salt Non-wheat or yeast-based sauce No additives or flavour enhancers such as MSG No added fat or oil No other meat derived extracts such as beef or pork No alcohol
Perceived categories suitable for in mainstream state CHML, DBML, HFML, BLML, LFML, GFML, NLML	
Actual categories suitable for in mainstream state NLML	
Mainstream product: mashed potato **Restricted product: mashed potato**	
Likely ingredients Potato – fresh or dried, butter, milk or cream, salt, other seasoning	**Amendment requirements**: Use of fresh potato only No added salt or seasoning Non-dairy fat Non-dairy milk or cream Low fat butter Low fat milk
Perceived categories suitable for in mainstream state CHML, VLML, LFML, DBML, MOML, GFML, BLML, PRML	
Actual categories suitable for in mainstream state VLML	

Figure 12.1 Considerations for converting mainstream menu items for special meals.

groups must preclude any decisions about the inclusion of mainstream items in the menu. My previous assertion that this is not a good idea, in terms of assuring the complete nutritional integrity of the product, is borne out by Table 12.2 and Figure 12.1 and will be re-affirmed when we examine special meal manufacturing protocols in the next section.

Having focused primary development issues around banned and restricted raw materials and food groups, the next area of focus has to be allergen control. Even if the airline directive is to make no specific claims in terms of allergens, my experience is that many SPML policies still make reference to them in terms of an awareness of their inclusion or not in certain meal groups. It is not uncommon to find policies that state a partial inclusion of allergens such as nuts or sesame in some meal categories, whilst simultaneously suggesting an avoidance of them in others. This practice is both confusing to the consumer and impossible for the caterer or special meal provider to police.

The only way to make allergen exclusion claims effectively and genuinely is to develop menus that derive no raw materials or finished products from the banned list and are manufactured by dedicated personnel in dedicated areas which see absolutely no allergen sensitive product throughput. Generally, in the aviation catering environment, this is impossible due to the catering not manufacturing style application of most facilities. Whilst a good practice avoidance of allergens is still an excellent idea in order to underwrite better safety practice, making specific 'free from' claims must be avoided. Where does this leave airlines and their catering providers then, if the policy of the airline remains to cater the SPML medical diet request? With no choice but to develop all products with a total dedication to using neutral raw materials and finished products only (see Table 12.2).

The potential impact of airline code-sharing agreements in the special meal arena is significant but for the most part remains unconsidered, save for brief statements to exempt airlines from the special meal policies of their code-share partners. It would be sensible to encourage code-share partners to formalise their SPML policies into one cohesive document so that not only were all meal categories on offer the same, but the menu development occurred in consultation with one another. Ultimately this would ensure that the fare-paying passenger was not subject to one special meal policy at the time of booking, only to find a different one in operation at the point of travel. One would imagine that this is a reasonably critical issue if allergen sensitivity or medical necessity is at the heart of the special meal request.

The sense is also that airlines develop one complete range of special meals for all applications in all classes and suitable for all sectors, and then devolve that development data into food technology-type specifications which could then be either centrally produced and distributed to all outstations in a frozen or ambient format, or replicated exactly from precise data and recipe solutions. I have made this assertion to many of the airlines that I have either provided meals for, or for whom I have undertaken consultancy. Their issues revolve around the variable standards of production and facilities available and varying availability of raw materials. I have noticed, however, that when mainstream product replication is at stake, this argument appears to hold no water! My conclusion then is that, for the most part, special meal development and implementation practices hold far less interest than mainstream products, particularly in front-end cabins, and there is a fundamental reluctance by the airlines and the airline caterers to accept that these products *must* be cohesively developed and replicated, or at the very least manufactured by dedicated professionals outside the catering environment who understand the parameters of safe production and legal compliance.

Labelling as a menu development consideration

Applying the nutritional data to every product which comprises the meal in the special meal arena may seem a little overzealous bearing in mind the food service style and application of the product. However, food service style aside, let us not forget the dietary claims issues attached to these products. If one was to consume food items in any other area and in any other food service environment which purported to provide such restricted diet foods as gluten and lactose-free as well as low fat, low purine and vegan, would one not require some verification of the dietary complexity and nutritional make-up? At the very least a full ingredient listing with percentage composition and source of protein derivatives would be expected.

The decision to include in the development make-up of the meal as many retailer branded components as possible that satisfy such claims, is ideal. Despite the fact that they may prove a more expensive option in the long term, they restrict development costs in the initial stages and build confidence in the consumer, who views their inclusion as a positive endorsement and acknowledgement that the needs of the consumer group have been fully recognised and understood. It has widely been recognised by the airline industry that the provision of branded food items as part of the overall meal experience greatly enhances the perceived quality and suitability of the product by the consumer, particularly if the destination or outstation from which they are served is geographically located far from the home port. If retailer branded items are not an option, than even a simple gluten-free or dairy-free label will do much to enhance consumer confidence, assuming of course that the menu item bears out the claim.

In summary, it is vitally important when developing special meal menus that a full and all-encompassing consideration is given to the following, in advance of any detailed development:

- The SPML categories that are to be catered.
- The decision to accommodate or not accommodate allergen and medical requests outside of IATA standard codes.
- The types of meal to be served, i.e. long or short sector provision or a combination of both.
- A detailed analysis of suitability of production facilities and the availability of dedicated, trained personnel.
- A detailed HACCP plan documenting the specific risks and focusing on allergen controls.
- A detailed listing of banned and restricted food groups available for product development.
- A detailed inventory of branded, nutritionally compliant finished goods or raw materials, which may be utilised as part of the meal.
- A consideration of whether any mainstream meal items will transfer into SPML development criteria and if so how the contamination hazards will be monitored and controlled outside a dedicated special meal production environment.
- A full and detailed consideration given to nutritional analysis and compliance issues, particularly in the 'low' or 'free from' arenas.
- Consideration given to the SPML policies of code-sharing partners and an acknowledgement of responsibilities in this regard.
- Allergen consideration given to the composition of the food product packaging with an awareness of allergen sensitivity to some packaging materials in conjunction with certain food groups.

Special meal manufacturing protocols and specifications

Having established that there are clearly two types of special meal policies in existence, it would seem possible on the surface to differentiate between the protocols required in both.

More often, medical and allergen meals which fall outside standard IATA codes are being shunned by the airlines and their catering providers. However, there are some airlines that do still accept allergy-based or medical-based requests. Again, the variation in the policies from airline to airline is confusing.

One such airline clearly states:

'We do not guarantee to provide nut or allergy-free meals. Passengers will be asked to select from the nearest alternative IATA-coded meal selection.'

The same airline in its caterer's notes then states:

'Non-standard meals must be approved by in-flight services and an approval code attributed.'

Another writes:

'The use of nuts, nut derivatives, nut oils or nut products in the preparation of bakery goods and in the preparation of Special meals is strictly forbidden and must be enforced at all times with the exception of the VLML and the VGML.'

For the avoidance of doubt, we need to progress along the lines that the requirement to produce non-standard meal requests is necessary for at least a small proportion of the time. The protocols that govern both standard and non-standard meal production have to be united under a single system of both GMPs and HACCP.

Regardless of whether any claims are being made about specific allergens like nuts, allergens such as lactose and gluten pose equal danger to an intolerant and should be treated with the same level of caution. The assumption that one's systems management cannot control the flow and appearance of one type of allergen, but can control the flow and apparance of another, is foolhardy. The resulting statement, 'Cannot guarantee that this product is free from nuts', is a nonsense when the equally allergy restrictive claims of 'fat-free', 'salt-free' and 'gluten and lactose-free' continue to form part of the standard IATA-coded offering without hesitation.

Whatever happens, all systems management and production processes must be based on the requirement to produce or outsource products that have been made in a restrictive environment with due consideration given to the high risk nature of the product. Even if the SPML being produced is of a 'religious' or 'preference' nature, the requirement for certain ingredients and raw materials to be excluded from the production environments and process remains the same.

Current industry codes of practice in the SPML arena are at best vague and at worst non-existent. As an example, the industry food standards bible, IFCA/IFSA World Food Safety Guidelines, reveals no reference to any specific safety protocols that should be employed in the development, production or outsourcing of SPMLs and their raw materials or components. What is more concerning is that there is no reference or guidance standards in terms of the types of audit criteria that SPML manufacturers should meet, nor is there any reference to appropriate standards of labelling and nutritional analysis.

In the light of these apparent oversights, in a document that is designed to give multilateral guidance on all issues appertaining to the production and distribution of safe airline food, it is even more critical that airline catering providers assume their own standards in this regard and embrace their obligations in the supply and manufacture of IATA-coded special meals.

Producing special meals using manufacturing protocols

In any food manufacturing process, the initial focus of attention in tandem with the menu or recipe development has to be the critical source of supply. If the product is going to be manufactured in-house from a variety of ingredients that are procured elsewhere, then attention must be given to breaking the ingredients down in terms of their sources of restrictive and non-restrictive supply (see Figure 12.2).

Recipe dish: vegetable curry and rice
SPML categories suitable for: **AVML, HNML, MOML, GFML, NLML, VGML**
Ingredients Onion, carrots, swede, parsnip, turnip, garlic, cumin powder, cinnamon, tomato purée, chick peas, paprika powder, garam masala, basmati rice **Supplier classification restricted or non-restricted**: Vegetables – pre-prepared onions, carrots, parsnip, turnip **Restricted – no dairy, meat or fish, no prepared meals or sandwiches on site or on transportation, no nuts or nut products** Vegetables unprepared – swede **Non-restricted** Dry goods tinned – chick peas, tomato purée **Non-restricted** Dry goods powdered – garam masala, cumin, cinnamon, paprika **Restricted – no dairy, meat or meat derivatives, fish or fish derivatives, gluten, wheat or bread products, no alcohol, no nuts or nut products** Chilled goods – garlic purée **Restricted – no wheat or wheat derivatives, no meat, fish or dairy products, no alcohol, no additives E223, no nuts or nut products**

Figure 12.2 Supplier restricted or unrestricted denotation for a standard special meal product.

The restricted and non-restricted supplier classification allows identification at a glance of any material crossover in the source of supply, and determination of where each raw material needs to come from. This process is in addition to all the usual supplier audit protocols determined by the prerequisite programmes. If the potential raw material supplier base is limited, it is advisable to continue to classify suppliers in the manner illustrated in Figure 12.2 so even if a restricted classification criteria cannot be met, it can be written into the HACCP plan to be monitored and controlled. It is essential at the raw material outsourcing stage that the supply chain integrity is established so any potential cross-contaminants in production and transit are identified from the outset.

For meal items that are to be bought in as an individual item and not manufactured from raw materials in-house, the supplier classification criteria can be less restrictive. As long as the supplier audit and product specifications are in place and they clearly document control of raw materials and process flow which are appropriate to restricted claims being made, then the classification can remain unrestricted. This classification procedure is particularly appropriate to ambient products such as bread, confectionery and bakery products. If the manufactured item is a chilled or frozen meal or a salad or fruit component, then the product specifications must meet retailer branded audit standards.

So it would seem then that most often it would be reasonable to expect that the contents and components which go towards making up the meal will constitute a combination of both products manufactured at the catering unit, in-house, and items which bought in and therefore manufactured elsewhere. This being the case, manufacturing protocols which need consideration are primarily those that operate in-house over which the airline's catering partner has jurisdiction and control; and secondly, those that are outside their parameters, at other catering units or manufacturers from which components are outsourced. Figure 12.3 illustrates this.

Post menu development and approval, it would make sense to devise a simple diagram as in Figure 12.3 for every meal in every sector, clearly documenting which items on the menu are manufactured and which are procured elsewhere and brought in as finished goods. This system will also assist security procedures (see Chapter 13) in terms of known and unknown stores classification, and make a contribution to the definition process of restricted and unrestricted supplier profiles.

Whilst it is possible that all or none of the components are made in-house, for the most part the illustration in Figure 12.3 is fairly representative of the manufacturing versus procurement divide and ratio that exists in the special meal arena and, indeed, very often in the standard meal environment also.

Tray set type Components inventory	Item	Source
Example economy breakfast	Special diet fruit	In-house
	Special diet bread item	Bought in
	Special diet yoghurt	Bought in
	Special diet milk jigger	Bought in
	Special diet butter/spread portion	Bought in
	Special diet breakfast hot meal	In-house
	Special diet juice/drink item	Bought in
	Special diet condiments/cutlery	Bought in
Example economy dinner	Special diet salad	In-house
	Special diet bread item	Bought in
	Special diet dessert item	In-house
	Special diet milk jigger	Bought in
	Special diet butter/spread portion	Bought in
	Special diet dinner hot meal	In-house
	Special diet condiments cutlery	Bought in

Figure 12.3 Denotation of supply chain sources for a special meal tray set.

So what does this mean for manufacturing protocols? Well, having established a system for raw material and finished goods outsourcing as previously described, the next step is to focus attention on manufacturing procedures that need to be employed in the production of components in-house.

Production paperwork such as that illustrated in Figure 12.4 should be in place for all manufactured components. This is very much in the same vein as standard production paperwork but benefits from having the additional supplier control information as an integral part of the process. Approved process flow information should also form part of the production paperwork, as a constant reminder of the attributable risks and controls, so that the risks are managed as part of the process and the staff involved are routinely aware. Assembly paperwork for the tray set assembly is essential and should draw together all previous information into one document that charts the full process of all finished products and raw materials.

Developing technical specifications

A full technical specification based on manufacturing protocols should exist for every meal, detailing all the necessary restricted practices and protocols pertinent to every meal category and its components. In the absence of industry directives, it is difficult without an advanced technical knowledge to know how to bring such a specification together, but there are several golden rules that must be followed so that the complete and necessary

BULK PRODUCTION RECORD

PRODUCT...................Chix veg pot   main meal   Category Rotation 1.

%	PRODUCT	USE BY	BATCH CODE	WEIGHT	RAW TEMP.	BATCH	FOOD ITEM	START TIME	COOK TEMP.	FINISH TIME
	Chix breast			100 × no.		×1	Chix			
	Potatoes diced			3 kg		×1	Pots			
	Carrots			1 kg		×1	Veg			
	Beans			1 kg		×1				
	Asparagus			1 kg		×1	Sauce			
	Tom purée			2 × 800 g		×1				
	Tom chopped			1×		×1				
	Onion 6 mm			1 kg		×1				
	Mixed herbs			50 g		×1				

BATCH SIZE 100

BATCH CODE

PRODUCTION AREA

METHOD

1. Verify clean-down activity as documented on schedule for production area and equipment to ensure allergen regimes have been applied   CCP1
2. Decant and weigh out recipe raw ingredients into 'blue' coded dedicated GFML containers. Document confirmed usage CCP2
3. Cover and transfer into production area   CCP3
4. Blanch vegetables in boiling water in dedicated allergen-defined equipment (code blue) CCP4
5. Steam potatoes in Allergen-only oven   CCP4
6. Sweat off onions in blast pan 1, add tomatoes, herbs, tomato purée and add water   CCP4
7. Steam chicken breasts in CCP4
   Oven (code blue) probe with probe 2 > 75°C and document
8. Transfer all cooled products into high risk and decant into high-risk allergen-coded containers 'pale blue' and document
9. Transfer into blast chiller CCP5 and chill to < 5°C within 90 minutes
10. Transfer into high-risk pan
11. Lay out GFML code blue foils and assemble, lid and label
    Transfer into chilled storage

SPECIAL NOTES

Ensure all 'high risk' and 'low risk' staff are wearing code blue PPE throughout all

DATE MADE

PRODUCE BY

CHECKED BY

USE BY

Temperatures	ASSEMBLY		
Holding Random 5	<5°C LIMIT	10 MINS	20 MINS	FINISHED TEMP.

USE CAPITAL LETTERS ONLY

Figure 12.4 Standard production paperwork for manufacture of special diet main meal.

technical picture is drawn accurately. Bearing in mind that most technical specs are about the information attributable to just one meal item, and that the meal in the case of airline set up is made up of as many as eight items, it is easy to see why it is so difficult to achieve the level genuinely required to assure full product safety.

The golden rules of specification development:

- Full traceability and raw material outsourcing information must be in place for every component that forms the meal.
- Full nutritional information must be documented for every component that forms the meal.
- Specifications must contain all component information from components that are manufactured in-house in the same format as those components that are bought in.
- HACCP and process flow attributable to the production of in-house items must be product-specific not process-generic.
- HACCP and process flow information of bought-in components must be product-specific in terms of manufacturing protocols undertaken by the manufacturer, but process-generic in terms of how they are handled once they arrive in-house.
- Allergen control in-house for in-house products must be process and product-specific. Allergen control in-house for bought-in goods may be process-generic.
- The technical spec in terms of ingredients detail should comprise a breakdown of named components, ingredients listing and country of origin, recipe illustration and percentage presence of each raw material, allergen identification, nutritional data and packaging information.
- The technical spec in terms of process application should comprise a breakdown of in-house and bought-in components, classification of restricted or unrestricted supply, process flow in manufacture, handling and assembly and allergen control.
- Labelling information should be attached in example format to the spec.

The protocols that need to be employed during the development and manufacture of SPMLs are immense if total product safety is to be assured. The burden is no greater than standard airline meal production if the development and food safety management systems work in tandem from the outset. The complexities arise when SPML development and standard meal development start to overlap and the requirement is to adopt aspects of the standard meal into the SPML. Even taking one component from a standard menu can cause all manner of technical issues, not least of which is trying to control the throughput of a standard item through a restricted diet manufacture and assembly area.

Special meal labelling

It has long been debated whether airline caterers should be subject to the same labelling standards as retail manufacturers. The simple fact is that any product which makes a restrictive claim, be it medically or preference-based, must back up the claim with sound nutritional data and analysis. Whether it is considered necessary to label the meal components appropriately and individually with the attributable nutritional information and data is also a matter for debate, but to avoid doubt and for peace of mind and improved perceptions of the client, I believe that it is essential.

If we refer back to Figure 12.3, we can identify how much labelling work would be required by the airline catering provider themselves. We can see from the figure that at least half of the average tray set will comprise bought-in items that, if procured from manufacturers, will carry all the necessary labelling information generically. If it does not appear on every product packet, then certainly it will be on a box or tertiary carton label. For bought-in items, the airline and their catering providers need to insist that full ingredients, allergen and nutritional data be applied to all products.

With regard to bought-in tray set components, e.g. fruit salads and salads, on the airline's own equipment the labelling information should form part of the generic tertiary packaging, albeit crate or box labels, as well as part of the documented technical specification. In this way the labelling information should correspond with the information held on the technical specification and is much easier to identify.

I am aware that the concept of full product labelling on food service-type products is alien and I would not necessarily suggest that it is appropriate or necessary when the airline meal is of a standard format. However, in terms of SPML provision and the range of dietary claims that are attributed, it is an absolute must to ensure full product safety and acceptability.

Applying international labelling standards

If one takes just an isolated example of a commonly ordered and provided airline meal such as gluten-free, it is interesting to refer to Codex Standard 118-1981 (amended 1983). In the industry's haste to cite Codex compliance in its IFCA/IFSA World Food Safety Guidelines, it has omitted reference to Codex standards on labelling and claims for 'foods for special medical purposes' (Codex STAN 180-1991). Both 118 and 180 dictate that meals which fall into categories such as

DBML, GFML and LFML should be subject to full nutritional labelling criteria and defined strictly in terms of their nutritional make-up, the parameters of which are described in both Codex documents. In both cases the labelling and nutritional criteria are arguably deemed necessary to be applied in the case of pre-packaged foods, but will the consumer notice the discernable difference between a pre-packaged airline offer and those adequately labelled by the major retailers?

In theory, if all the other development and manufacturing protocols are in place, the issue of labelling will become a minor one. With the technical specifications complete and the supplier outsourcing controls established, the devolvement of technical, ingredients and allergen information onto a product label should not be a difficult task.

In premium cabins, where the tray set components are likely to be presented on non-rotable equipment, and the tray sets themselves may contain fewer bought-in items, it is ideal to place all the necessary ingredients, allergen and nutritional data for the complete meal onto a menu card which can be placed on the tray and accompany the meal. In this way the passenger is afforded the same standard of information about their meal as the retail-style product.

There are conflicting opinions about whether passengers expect this type of detailed information bearing in mind the food service nature of the airline meal concept. However, one cannot escape the dietary claims aspects of the product which render it totally outside catering and food service standard protocols in every sense. What the customer expects is less relevant than what the customer requires.

Figure 12.5 shows a typical standard airline product in an SPML group that makes a dietary claim without applying the appropriate labelling.

The tray label carries the specific information required to ensure that the correct meal reaches the passenger for whom it is intended. This label should denote the following information:

- Passenger name.
- Flight number and class of travel.
- Meal type by IATA coding.
- Date of travel.
- Meal type, i.e. breakfast or lunch, main meal or snack.

Figure 12.6 illustrates some typical SPML label formats.

Whilst much of the above information is also dictated by the type of equipment used, it is still critical to apply all of the above data to the meal tray in order to defend against a suggestion that the wrong meal reached the passenger in error.

Figure 12.5 Typical standard airline product for special meal group, without full appropriate labelling.

In this regard it is also necessary to inaugurate in the systems management protocols some way of denoting which meal category every meal component belongs to. For example, if all of the meal categories are colour-coded and then in turn the attributable meal colour code was applied to every component on the tray, this would guard against a situation where, if a component should fall from a tray in transit post assembly, it would be immediately obvious to anyone handling it to which meal group the component belonged. In large SPML operations this coloured code could be applied post-production and pre-assembly to help to underwrite the protocols that control the risks.

Figure 12.7 illustrates a typical SPML tray set with colour-coded components denoted by a dot system.

So far we have considered the labelling aspects of bought-in items, be they prepacked or in the airline's own equipment. For those products which are made in-house by the caterer themselves, the same issues apply. The product label will need to be applied and the information devolved from the product specification. The alternatives to labelling every component individually are to apply all relevant ingredient, nutritional and allergen information for every component onto a menu card, which can then be added to the tray set. This will save labour costs and give a more food-service look to the tray set.

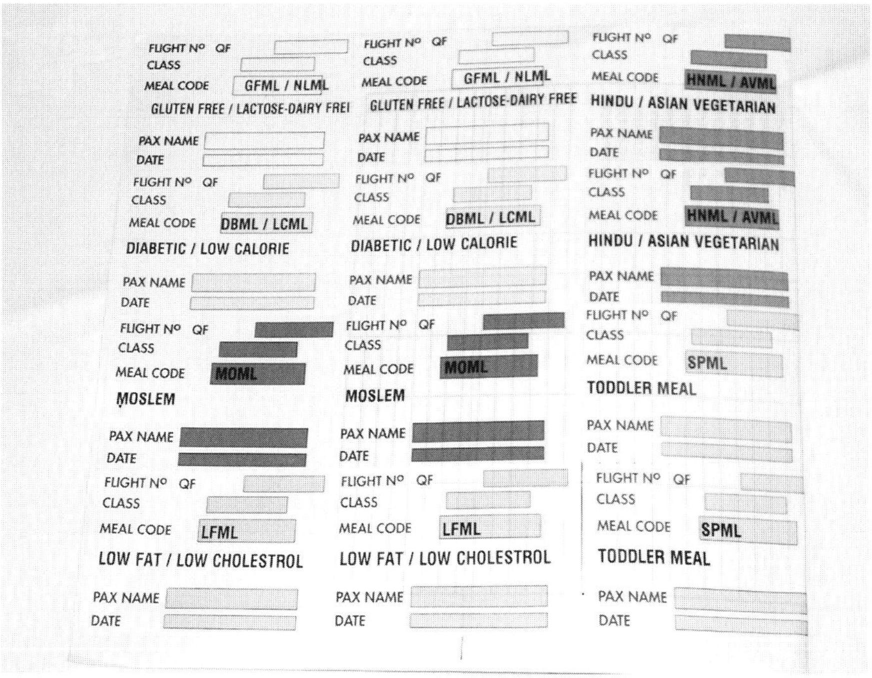

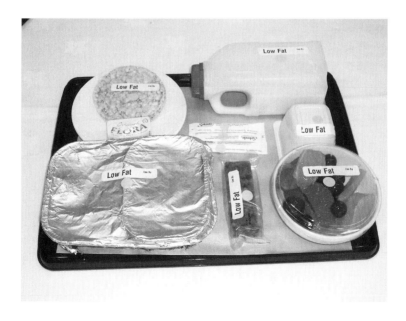

Figure 12.6 Typical special meal label formats.

Figure 12.7 Typical special meal tray set with colour-coded components denoted by a dot system.

Role of technical specifications in product labelling

In the event that full technical specifications are not in place, the job of labelling effectively will be extremely arduous and difficult. It is critical that full product specifications are completed for every meal component that comes together to make the tray set up meal. In terms of the specialist diet, medical and 'free from' meal categories, an even greater degree of technical data will need to be applied. For example, meals carrying a Codex or similar standard classification like gluten-free will require the specification and verification analysis to illustrate that the total nitrogen content of the gluten-containing cereals used in the product does not exceed 0.05 g per 100 g of these grains on a dry matter basis.

In the light of the burgeoning trends among the low cost providers to offer buy on board concepts, the labelling criteria described throughout this chapter will apply to all food service products retailed on board, whether special meal format or not. If a sandwich is retailed directly from the aircraft crew to consumer, all of the mandatory labelling issues that apply in retail environments on the ground will apply here. For those airlines considering even a partial rollover to buy on board concepts for some previously standard items, serious consideration of the labelling implications needs to be made.

So in summary, the following issues will impact on the effectiveness and correctness of SPML labelling:

- Ensure that all nutritional, allergen and ingredients declarations are in place for every component whether bought-in or made in-house.
- Ensure that all banned and restricted food groups are identified on each spec as a cross-reference against the labelling claim being made.
- Apply a coding system, be it colours or otherwise, to each component to avoid products being mismatched during assembly. Include condiments and cutlery packs in this.
- As well as product production paperwork, an assembly paperwork sheet must be in place to ensure that the meal assembly aspect is as much about temperature control as it is about correct product and component assembly. This is particularly critical in respect of components such as milk, sweetener and salt portions.
- Employ an end product checksheet that includes final product checks for correct labelling.
- Ensure that supplier specification requirements mandate the attribution of labelling information to every component if prepacked, and if bulk packed ensure that the information can be found on the outer or tertiary pack.

Labelling issues should be viewed as the visible manifestation of all centrally held and researched technical product information. A correctly applied label does much to enhance the consumer's perception that the meal attributes have been correctly considered, not just in terms of product quality and safety but also in terms of the overall suitability of the meal to fit the restricted diet profile of that consumer.

Nutritional analysis and data

Having spent the majority of this chapter focusing on the absolute necessity for nutritional analysis and data to be attributed to every component of every special meal, it is now to time to break down the requirements in terms of what type of data should be attributed to what type of meal.

The general rules are:

- Preference meals: Stage 1 Nutritional analysis.
- Religious meals: Stage 1 Nutritional analysis.
- Medical meals: Stage 2 Nutritional analysis.

For those not familiar with the differentials between the two stages, stage 1 is based on theoretical data derived from standard industry software or publications[116]; stage 2, however, involves full scientifically derived data based on full nutritional analysis.

Without this type of information the labelling claims can be challenged easily and due diligence will be effectively impossible to prove. By laying down GHPs and GMPs in production and process flow, such as those previously outlined in this chapter, it is possible to comply effectively in the preference and religious meal sectors. However, in the case of making medical claims, the only defence has to be stage 2 analysis as the verification aspect, in conjunction with all the protocols previously described.

The data should form part of the end tray set product technical specification and be centrally held. The necessity for retesting at stage 2 level should be negated by systems management protocols, outside of a recipe or menu change. In order to ensure that stage two testing funds are not mismanaged, it would make sense first to apply stage 1 assessments to the medical meal recipes to ensure that the theoretical data stack up, before proceeding down the stage 2 route at considerable expense. This allows an opportunity for recipe or even component modifications before the final product specification is formalised.

At this point it is interesting to consider the fiscal impact of carrying out nutritional analysis in-house on in-house products. It would be far better in the medical meal arena to buy in all components from manufacturers who have already undertaken the nutritional work and are in a position to provide all relevant data. The only work left in-house is then a compilation of the full meal specification.

Bearing in mind the export nature of the airline special meal, it is essential that nutritional data also contain information on possible ingredients which may be banned under the food legislation of countries into which the airline is travelling. For example, in the EU, cyclamates (artificial sweeteners) used to be permitted in some EC states such as Denmark but banned in the UK on safety grounds. Following a major safety review carried out by the Scientific Committee for Food, a directive on sweeteners for use in foodstuffs (94/35EC, now amended) was passed and integrated into UK legislation as the Sweeteners in Food Regulations 1995, No. 3123 as amended 1996, No.1477 and 1997, No. 81. The supremacy of EC law over UK law meant that the UK government now had to permit the use of cyclamates in foods sold in the UK.

This being the case, it is still essential that ingredients such as cyclamates are identified by way of specific labelling and technical data on the specification. Artificial sweeteners are an example I have focused on in particular because of their likely presence on every LCML and DBML tray set derived around the world, but there are numerous others.

The other major issue affected in the supply chain, for which data will have to be accrued, is the presence of genetically-modified organisms (GMOs). Bearing in mind the exhaustive use of soya products in the manufacture of special meals in the NLML, VGML, AVML and often GFML categories, the GM status of all soya derived raw materials and components needs to be established at audit level and translated onto the end product specification. It is usual practice for airlines to make a generic statement about their use of GM products globally. In the UK and most of Europe that declaration must be made as food service businesses are not exempt. However, with the potentially huge utilisation of GMOs around the world, the airline catering supply chain is littered with possibilities. It is for the consumers of special meals then to demand a specific statement attributable to their meal choices port-to-port in the absence of the airlines clarifying their statements.

Overall, the nutritional analysis and data issues should be a joining together of all aspects of the audit and supply chain onto one document. This provides at a glance all the legal compliance and suitability for purpose issues expected by the consumer and demanded by legislation.

Healthy options – healthy claims

Increasingly, on both airline menus and enclosed within the accompanying airline literature or publicity material, there are references to items on the standard menu that claim to be a healthy option/choice or more specifically may claim to be any one of the following: rich in fibre, reduced fat/cholesterol, no added salt, no added sugar or a lighter bite. Having spent the chapter coming to terms with the degree to which the food safety management and labelling protocols need to extend in order to fully satisfy the SPML criteria, I am always extremely alarmed to discover these kinds of descriptions, which basically constitute SPML classification being attributed to standard menu items.

To deliberately attempt to avoid the true meal type classification, hence falling subject to all the outlined procedures and protocols, and ostensibly 'candy coat' what is in effect a claim about a menu selection, is extremely misleading and ultimately unhelpful to the consumer. The passenger could not be blamed for assuming that the claim stands in the same manner that it has to on a pre-packaged, retailed product, and is unlikely to understand why his or her perceptions are not entirely satisfied by these healthy claims.

In many of the premium cabin menus that I have been exposed to, an asterix or star denotes the 'healthy' choices often without any explanation as to why these might be perceivably healthy. Other scenarios I have witnessed simply list the selections on the menu that are deemed 'healthy' or 'light' without any further specific information being given.

On one recent trip across the Atlantic it was suggested to me that New England clam chowder was the 'healthy' option, denoted on the menu as such with an appropriate symbol. When I enquired as to the basis on which this classification had been attributed, when clearly it was full of both calorie and fat-laden ingredients, I was met by a wall of silence. My perception as the consumer was that in the absence of any hard and fast nutritional data or product ingredient information from the crew, I could only assume that in actual fact the chowder was anything but 'healthy' or 'light'. The key ingredients were obviously cream, potatoes and butter in relatively equal quantities with a smattering of 'tinned in brine' clams thrown in for good measure. Upon returning home I wrote to the airline requesting a nutritional breakdown of the product and a clarification of the nutritional composition of the chowder. Needless to say I am still awaiting a reply.

The increasing trend among airlines, particularly in front-end cabins, to make this kind of loose and invalid claim is extremely misguided. Ultimately any type of 'healthy' or 'light' claim must be subject to the same product development and technical standards as any of the meals or meal components sited in the SPML arena.

To attempt to avoid SPML compliance by littering menus with the types of vagaries I have just described, can only add more weight to the increasing legislative demand for full product labelling on all products whether prepacked or otherwise.

EU legislative impact on airline SPML labelling claims

With the advent of new EU directives on allergen labelling, enforceable from November 2004, European-based aviation caterers engaged in the development, production and procurement of special meals have to meet stringent new labelling criteria in this regard where labelling 'claims' are made.

All the production and systems management protocols and procedures we have examined in this chapter have been suggested in the light of such legislative compliance requirements. It is clear that with a legislative climate aimed at considering allergens and their process control and labelling requirements, it is essential that the whole issue of SPML manufacture and supply is proactively dealt with in the context of not only legislative compliance issues but also within the context of best practice and quality assurance perspectives.

The specified allergens currently under consideration are:

- cereals containing gluten
- crustaceans
- sesame seeds
- soybeans
- celery
- nuts
- fish
- peanuts
- mustard
- milk
- sulphite at more than 10 mg/kg

as well as all associated finished products in which any of the above may form an integral aspect or compound aspect.

All food safety management systems are only as good as those who manage them and therefore it is essential that staff engaged in the manufacture of special meal products have a full and given knowledge of the issues appertaining to special diet product manufacture and allergen handling.

I hope that throughout this chapter I have managed to bring some kind of structure to the operational and safety management issues associated with the specific production of special meals for the aviation catering sector. In an increasingly more allergen sensitive and consumer aware society, it is vital that the aviation sector does not get left behind in its ability and willingness to handle special meal manufacture and a burgeoning increase in the amount of special meals being requested by the flying, fare-paying public.

13 Aviation food safety versus aviation food security

I consider security to be the issue above all others that perhaps best illustrates with pertinence and relevance the necessity for the aviation catering industry to adopt food manufacturing protocols. Here at last is a topic where the indisputable connections between the quality and integrity of the supply chain, manufacturing systems management and distribution logistics are intrinsically linked at every stage by security risk factors. In this way, the nature of the processes and procedural compliances employed should automatically be determined by the attributable risk factors, and managed accordingly.

Throughout this chapter we look at the real security issues that impact aviation catering supplies and stores, on a step-by-step basis, and examine how the implementation of food manufacturing protocols and procedures in aviation catering production and provision would do much to enhance the security integrity of the on-board service product. We also examine the plethora of industry guidance material on the subject and make distinctions between operational and physical security measures, which have formed most of the international basis for legislative mandates, and the more pertinent product outsourcing, manufacturing and supply aspects which have until now remained part of the best practice guidance debate.

The interrelated issues of food safety and food security systems implementation and management are examined in detail in this chapter, and the requirement to ensure that safety and security issues are viewed with the same perspective and in tandem is demonstrated by the linking of the strategies together in one cohesive plan, which is then managed by multidisciplined teams.

In terms of food manufacturing procedures and protocols the content of this chapter is the status quo. What this means for the airline catering industry is that with systems and strategies already proven effective and manageable in the manufacturing sector, the only stumbling block to successful implementation is the required culture and industry commitment for change.

The industry perspective has always been to view food safety and food security as operationally juxtaposed in terms of the systems mechanisms required to assure both in tandem. It is this entrenched perspective that has led to a situation where, despite regulation and guidance, the production process control security procedures that require implementation, either by mandate or otherwise, have been widely condemned by the airline caterers and virtually ignored as a result.

Later in the chapter when we look at mandatory security-related regulation and where it applies to aviation catering, it will become clear just how opposed to systems management review and implementation some of those charged with the greatest responsibility in this area really are, and what this means for the long-term safety and security integrity of the supply chain.

Even when faced with mandatory regulation the industry is desperately unwilling to comply, citing fiscal prohibition and systems overhaul requirements as their rationale whilst cowering behind catering or even restaurant denotation as a way of underwriting their argument. The reality is that the systems management styles required to meet appropriate operational catering security protocol, run concurrently with good manufacturing practice, and to admit defeat in the security arena is to acknowledge that airline catering provision does not comply with good manufacturing practices either.

The entire focus of this book has been a desire to highlight the necessity to align aviation catering production with good manufacturing process, and to attempt to explain not only why it is essential that operational transformation occurs to this end but also to illustrate the systems management protocols that need to be established in order to make the transition possible. How ironic then, that it is in fact operational food security as opposed to food safety mandates and regulation, that may ultimately pave the way for this to happen.

Aviation food safety and its relationship with bio-terrorism

Even before the devastating events of 9/11 in 2001, there had been a long-established threat to the security of civil aviation by those wishing to use aviation and aviation mechanisms to perform acts of 'unlawful interference'.

The 1970s bore witness to an era of aviation history blighted by the terrorist activities of hijackers. The 1980s followed with a series of bomb threats and explosions on board aircraft as the industry became highly sensitised to the huge potential afforded terrorists by aviation and aviation-associated industries. To this end and in response to the diversity and evolution of the types of threats posed to aircraft operations, the industry has seen fit to generate a vast range of both regulated and guidance-based security initiatives in an attempt to deflect the potential for terrorist activity to occur.

The indisputable damage to consumer confidence caused by any association with lapses in aviation security may prove terminal in terms of the fiscal future viability of the operators implicated. To this end it is both necessary and obvious that any investment in time or resources dedicated to the development, training and implementation of security systems and procedures is a proactive investment in business futures rather than an unnecessary, onerous burden, and it should be viewed as such by those charged with the responsibility of assuring aviation security.

The startling similarities between attitudes towards catering security systems implementation and advances in food safety protocols are reflected in the reluctance to evolve in tandem with emerging technologies, displayed by the airline operators and their catering partners. A bold statement, I know, but one I give evidence to support throughout the chapter.

In order to assess the vulnerability of catering operations to terrorist activities, it is necessary to seek clarification on the official category of risk designated to airline catering operation by the industry itself. Acts of unlawful interference as defined in the industry security guides[117] are:

'acts or attempted acts such as to jeopardise the safety of civil aviation and air transport.'

They can be broken down into several categories:

(1) Unlawful seizure of aircraft in flight.
(2) Unlawful seizure of aircraft on the ground.
(3) Hostage-taking on board aircraft or on aerodromes.
(4) Forcible intrusion on aircraft, at an airport or on the premises of an aeronautical facility.
(5) Introduction on board an aircraft or at an airport of a weapon or hazardous device or material intended for criminal purposes.
(6) Communication of false information such as to jeopardise the safety of an aircraft in flight or on the ground, or of passengers, crew, ground personnel or the general public, at an airport or on the premises of a civil aviation facility.

The aspects that apply to the supply and manufacture of aviation catering can be illustrated by points (4) and (5). Here we witness the divide in security risk factors between operational aspects and process control aspects. Forcible intrusion via catering supplies can occur through security breaches inherent in the recruitment of personnel operating on behalf of the airline caterer and afforded access to restricted areas via their catering supplier status.

Introduction on board of a weapon or hazardous device or material relates to the process control opportunity for food to be maliciously and deliberately contaminated during process and/or production, storage and transportation. Historically, this type of food tampering has been linked to the manufacturing food sector and has affected such products as baby food and dairy products. The manufacturing industry has developed systems and control measures to deal with malicious tampering and in some cases, where the products are designated as a high security risk, such risk factors have been amalgamated into the HACCP specific to the product and process.

Whilst the aviation industry has for a long time accepted the requirement to deal with the operational security aspects of the aviation catering supply chain, the process control aspects have been regarded as unnecessary and overly onerous in their potential application. Whilst the industry accepts certain aspects of its vulnerability, it is reluctant to accept the potential threats to security afforded by the process application for fear of having to amend the process control mechanisms within catering operations. As I have said repeatedly, if aviation catering production protocols were established in a manufacturing format, then the additional security-based control measures required would merge effortlessly with the overall production schematic because the culture already exists for constant referencing and cross-referencing of procedures with a defined chain of responsibility and non-conformance reporting.

As we will see in this chapter, superimposing process-controlled catering security management systems onto a catering production framework is almost impossible as it represents a clash of food safety and security management cultures. Therefore it is essential that the bio-terrorist threats to the aviation catering production supply chain are first acknowledged and then the production culture transformed to reflect the true nature of food manufacturing protocols.

Whilst I would not go so far as to suggest that aviation catering supply is at a higher risk from bio-terrorist activity than other areas of food supply and/or manufacture, the obvious connection with interstate or global transit and the enhanced capacity to affect the export supply chain do indeed prove attractive to any potential saboteur wishing to make an impact on the international stage. The added attraction of the aviation catering industry as a tool for terrorist activity is that the widely publicised reaction to any formal production process regulation in this regard has been fierce, suggesting that the industry is not only unwilling but unable to genuinely deal with the risks posed to catering processes by bio-terrorist activity.

Meanwhile in the manufacturing sectors, food sabotage protocols are an established accepted mechanism within the supply chain, that amalgamates with operational security systems management. Whilst the manufacturing sector works hard to maintain its security awareness, it is not viewed as a cultural management issue in isolation and is certainly not singled out for protest.

The nature of contamination threats to food during process can take many forms but mainly falls into the following categories:

- chemical
- physical
- microbiological.

310   *Aviation Food Safety*

Having established the basic route causes, the category of causation has to be determined from the following:

- natural
- accidental
- deliberate.

Most of what we will tackle in this chapter is of course concerned with deliberate or malicious causation, but it should be noted that the systems management protocols designed to tackle deliberate causation can be effective also in tackling the likelihood of natural and accidental causation also. Painful though the evolution of systems management may be, it is an essential tool in the quality and safety assurances of any food business and must continue as an organic process in consideration of all interrelated aspects of the business make-up.

Since 9/11 the indisputable link has been made between the aviation industry and compromises to security assurance. It is a dangerous strategy for the catering aspects of the industry to seek to challenge the extent of process-related security obligations and to contest the necessity for regulation. Bio-terrorist threats to the food and water supply are here to stay and the challenge for the industry is how to embrace them and act upon them with the same level of responsibility and systems assurance as is viewed in the food quality and safety arenas.

Aviation catering security rules, regulation and guidance strategies

As with many aspects of the aviation industry, the parameters by which aviation security mechanisms need to be established and operated are defined by a combination of industry guidance and legislative compliance. Whilst the industry guidance is designed to bring an international cohesion to standards and systems implementation, the mandatory regulation operates on a national level allowing governments to establish their own enforceable security directives which may or may not be based on the international industry guidance, depending on how the security threats to aviation are perceived.

The relevant industry guidance material with regard to aviation catering security, is broken down into offerings by the following organisations:

ICAO – Annex 17 International Standards and Recommended Practices
IATA – Security manual
NASP – National Aviation Security Programme
NASA – National Aviation Security Authority

WHO – Bio-Terrorist Threats To Food Guidance Notes

ECAC (European Civil Aviation Conference) – Document 30, Policy Statement In The Field Of Aviation Security

EFSA (European Food Safety Authority) – food risk management and risk communication strategies.

The regulatory and mandatory legislation is represented by the following national and international mandates, some of which are aviation-specific and some of which are exclusively food safety and security focused:

International conventions

- The Tokyo Convention 1963
- The Hague Convention 1970
- The Montreal Convention 1971 and Montreal Protocol 1988.

Bilateral agreements

Air Services Agreements – negotiated between international states that share scheduled air services. Many of these agreements contain a clause which provides a commitment to fulfil the standards and obligations set down in Annex 17 to the Chicago Convention.

National legislation UK and Europe

- The Civil Aviation Act 1982
- The Aviation Security Act 1982
- The Aviation and Maritime Security Act 1990
- EU Regulation (EC) No 2320/2002 (based on ECAC document 30)
- EU Food Directives.

National legislation USA

- The Aviation and Transportation Security Act 2001
- The Homeland Security Act 2002
- Public Health Security and Bioterrorism Preparedness Response Act 2002.

International Civil Aviation Organization (ICAO)

ICAO is a specialist agency of the United Nations constituted under the Chicago Convention to promote the safe and orderly development of civil aviation. It comprises a General Assembly that meets every three years and a Council that conducts the day-to-day business of the organisation. The council is supported by a number of standing committees, including the Committee for Unlawful Interference established in 1969. This committee reviews standards and recommendations for the safeguarding of civil aviation and submits proposals on them to the Council.

Standards and recommended practices developed by ICAO in respect of aviation security are published in Annex 17 of the Chicago Convention, and guidance on their implementation is published in a Security Manual for Safeguarding Civil Aviation Against Acts of Unlawful Interference[118].

Security aspects considered under Annex 17 include the following:

- General aspects.
- Organisation of operational security.
- Preventive measures – aircraft, passengers and cabin baggage, hold baggage, cargo, access control.
- Management of response to acts of unlawful interference.

International Air Transport Association (IATA)

Current IATA Recommended Security Standards were accepted and approved by the 175th meeting of the Board of Governors on 2 June 2002. The 58th AGM that followed adopted a Security Resolution that called on all member airlines to ensure that effective airline security programmes are in place, commensurate with ICAO Annex 17 requirements and the IATA Recommended Security Standards.

The IATA security standards give detailed clarification and operational guidance as to how to undertake the implementation of the recommendations laid down under ICAO Annex 17. Whilst compliance for member airlines is not arbitrary, it is suggested that legislative compliance cannot be achieved without the industry-specific recommendations being observed.

National Aviation Security Programme (NASP)

One of the standards in Annex 17 of the Chicago Convention requires contracting states to establish a National Aviation Security Programme (NASP). In addition there are recommended practices that each member state is urged to establish as a means by which to co-ordinate activities between departments, agencies and other organisations involved in its NASP, and to keep under constant review the level of threat within its territory whilst considering the national situation in the context of the international security situation.

Although the basis for the national programme is contained within the provisions of Annex 17, the state has wider responsibilities than those contained within the national programme. It also has the overall responsibility for maintaining law and order within its territory and for the discharge of its international legal obligations as they appertain to aviation security.

Different governments employ different methods of discharging this responsibility, but whatever the policy delegation dynamics, they must be published in the NASP. The development, application and implementation of a NASP constitutes the basis of civil aviation security and all specific airline, airport and cargo handling programmes will emanate from the state NASP framework.

In the UK the NASP is implemented through the Department for Transport (DfT) via the Transport Security Directive (TRANSEC) and directives issued by TRANSEC apply to the following groups:

- Aerodrome managers.
- Operators of passenger aircraft with a maximum take-off weight of 10 tonnes or more than 19 passenger seats.
- UK operators of passenger or cargo only aircraft operating outside the UK.
- Operators of cargo only aircraft.
- Regulated cargo agents.
- Directed catering companies.
- Operators of an aircraft where the aircraft is to be used for a flight which has been assessed by the DfT as being under substantial or higher threat, regardless of its size.

Within the general directives issued by TRANSEC, section 19 applies specifically to the supply of in-flight catering and stores.

National Aviation Security Authority

Under the recommendations laid down in ICAO Annex 17, Standard 3.1.6[118]:

'Each contracting State shall establish a national aviation security committee or similar arrangements for the purpose of coordinating security activities between the departments, agencies and other organisations of the State, airport and aircraft operators and other entities concerned with or responsible for the implementation of various aspects of the national civil aviation security programme.'

To this end, each contracting state must establish an organisation, develop plans and implement procedures to achieve the aim and secure the objectives of aviation security within the national framework. In the UK such authorities are coordinated through the Secretary of State for Transport who carries out these functions through a Director of Transport Security who heads the Transport Security Directorate within the DfT.

World Health Organization (WHO)

In 2002 the WHO drafted a report on the potential threats posed to food and water supplies by terrorist activities[119]:

> 'Threats from terrorists, criminals and other anti-social groups who target the safety of the food supply chain, are already a reality. During the past two decades, WHO member States have experienced concern about the possibility that chemical and biological agents and radio-nuclear materials might deliberately be used to harm civilian populations. In recent months, the health ministries of several countries have increased their state of alert for international malevolent use of agents that may be spread through air, water or food.'

Whilst this guidance was not solely dedicated to the aviation catering sector, more a general set of guidance principles that can be applied to any food manufacturing or catering environment, its pertinence in terms of the overall framework of national and international aviation security measures is immense.

Whilst the security guidance contained in ICAO Annex 17 and its subsequent translations into IATA and NASP guidelines in relation to the supply of catering goods and stores, focus on the operational aspects of food security, policy documents such as the WHO guidelines focus far more attentively on the procedural and process-specific aspects and risk attribution. There is little of the production and process security aspects of the WHO guidance incorporated into the catering security recommendations of the aviation security guidance documents. This is a shame but is not surprising bearing in mind that much of the consultation on the content of such documentation is carried out directly with the aviation catering industry itself. I believe that successful security risk management strategies are far better defined by subjective, independent bodies that hold no pecuniary interest in fiscal considerations and implications to their businesses.

European Civil Aviation Conference (ECAC)

The ECAC was constituted in 1953 to review the development of European air transport in order to promote the coordination, better utilisation and orderly development of air transport and to consider any special problems which may arise in the field of air transport.

The ECAC working group on security problems reviews aviation security and drafts recommendations on security that the ECAC issues for the guidance of member states. These are promulgated in ECAC Document 30, ECAC Policy Statement in the Field of Civil Aviation Security.

European Food Safety Authority (EFSA)

The European Food Safety Authority was established in 2002 following an EU council regulation. Its primary mandate is to provide scientific advice on all matters relating to food and feed safety. EFSA risk assessments now provide the European Commission, the European Parliament and the European Council with a sound scientific basis on which to base legislation and policies related to food safety. EFSA's core business is the provision of risk assessments and independent scientific advice and the communication of these assessments to all interested stakeholders including consumers. Whilst risk management is not part of EFSA's brief, consumer protection most certainly is. As consumer confidence is dependent on how successfully the risks are seen to be managed, attempting to achieve consistent risk management across 25 member states is no easy task.

Centralising food safety and security risk assessments at EFSA ensures that member states are taking action on the basis of the same conclusions. With the establishment of the EFSA Advisory Forum for representatives of the national food safety agencies across Europe, there is now a focal point for discussions and dialogue between the risk managers. Relationships developed in this Forum are having an impact outside the remit of EFSA, with food safety agencies sharing information and consulting each other regarding approaches to dealing with problems. This is aiding the enhancement of the day-to-day efficiency of the public authorities charged with consumer protection. The Forum is currently addressing the issue of how national agencies, the European Commission and EFSA work together in dealing with emerging incidents and crisis. The environment in which EFSA operates is continually evolving, bringing new challenges into the food safety arena. The expansion of the EU and the global trade in food creates challenges for the harmonisation of controls and enforcement. Free trade has to be safe trade. The developing relationship between EFSA and the risk assessment bodies outside the EU will introduce consistency into risk assessment and increase the likelihood that the same scientific information will form the basis for risk management decisions internationally.

It is important to recognise that the role of EFSA is limited to risk assessment and risk communication, not including risk management, and as such it is critical that expectations as to what it can deliver in the assurance of European food safety and security practices are realistic.

Tokyo Convention 1963

At the Tokyo Convention, contracting states recognised offences committed on board aircraft on international flights, acknowledged powers accorded to aircraft commanders and undertook to restore an aircraft subject to unlawful interference to its lawful commander; they also accepted a range of procedures for bringing offenders to justice.

Hague Convention 1970

At the Hague Convention, contracting states agreed to make the seizure of an aircraft by force or intimidation, i.e. hijacking, an offence punishable by severe penalties, and to make offenders subject to prosecution or extradition.

Montreal Convention 1971 and Montreal Protocol 1988

At the Montreal Convention, contracting states agreed to make an increased range of offences relative to the safety of aircraft (acts of violence against persons on board, destruction of or damage to an aircraft or navigation facility, communication of false information, etc) punishable by severe penalties, with offenders subject to prosecution and/or extradition. This was supplemented by the Montreal Protocol 1988 which commits signatory states to make offences under national law: armed attacks at international airports and the causing of damage to facilities or the disruption of services from such an airport, where these have endangered safety at such an airport.

National ratification of international legislation

The three international conventions detailed above were consolidated in the UK by the following three Acts of Parliament:

- The Civil Aviation Act 1982, which consolidated the provisions of the Tokyo Convention Act 1967.

- The Aviation Security Act 1982, which consolidated the provisions of the Hijacking Act 1971 and the Protection of Aircraft Act 1973.
- The Aviation and Maritime Security Act 1990, which ratified the Montreal Protocol 1988, amended the Aviation Security Act 1982 and added provisions for air cargo and maritime security.

The Aviation Security Act 1982 provides for the issue of directions to aerodrome managers and aircraft operators on a range of matters in the interests of aviation security, whilst the Aviation and Maritime Security Act 1990 further provides for directions to be issued to air cargo agents and others. These directions provide the legal basis for the UK NASP that is promulgated in this document.

In the UK the NASP, as mentioned earlier, is instigated via the DoT and is enforced by the transport security arm of the DoT, TRANSEC.

European Parliament and Council of the European Union

In the wake of the terrorist incidents that occurred on 9/11 in 2001 in the USA, the European Parliament decided to legislate on aviation security to ensure common minimum standards throughout member states. Accordingly Regulation (EC) No 2320/2002 came into force. The regulation is based on Document 30 of the ECAC.

Aviation and Transportation Security Act 2001

On 19 November 2001, the Aviation and Transportation Security Act (ATSA) was enacted. ATSA created the Transportation Security Administration (TSA) and as a result transferred authority for enforcement of civil aviation and security requirements from the Federal Aviation Administration (FAA) to the TSA. The TSA has operated its civil aviation enforcement programme utilising many of the FAA procedures and policies already in place.

Homeland Security Act 2002

On 25 November 2002 the Homeland Security Act increased the statutory maximum penalty amounts for civil violations of the TSA's security regulations. The increased civil penalty amounts became effective on 25 January 2003.

Impact of legislative directives

Having established definitions for most of the relevant legislative and guidance directives appertaining to the security of aviation catering, we now need to focus on the two most pertinent directives that have had or will have an immediate impact on how airline catering operations are mandated with regard to the establishment and nature of their GMPs and SOPs.

The two pieces of legislation that will impact on airline catering operational intercourse more than anything else are the EU Food Directives 2006 and the Public Health Security and Bioterrorism Preparedness and Response Act 2002. They were developed in response to two very different levels of crisis and were instrumental in forcing a requirement for governments to legislate in order to safeguard the quality, integrity and security of the food supply chain.

The EU Food Directives have been devolved from an EU white paper commissioned in 2000, as a direct response to a host of food safety crises that permeated the European supply chain during the late 1990s and culminated in the BSE debacle in the UK. The Directives set out key objectives and have seen the homogenising of European legislation with regard to food safety, based on a 'farm to fork' principle of effectiveness. In less than four years the Directives culminated in the passing of 80 new pieces of legislation, which have directly impacted on the operational activities of every level of food business across the EU.

Meanwhile in the USA, the Federal Government was taking legislative steps to assure the safety and security of the supply chain for a host of different reasons. The Public Health Security and Bioterrorism Preparedness and Response Act 2002[120] set out dramatic new objectives to attempt to prevent terrorist activity from impacting on the food chain in the wake of 9/11.

As we will witness in the following few paragraphs, the measures considered necessary to assure the safety of the supply chain under EU Food Directives, and the measures considered necessary to assure the security of the supply chain under The Bioterrorism Preparedness and Response Act, are startlingly similar. It is interesting to note also that it was essential under the conditions of the World Tourism Organization (WTO) that both continents, whilst considering and implementing measures to deal with the integrity of their supply chain, simultaneously considered them in the context of any possible impact on or unfair preclusion to international trade.

The operational and cultural impact that these two crucial pieces of legislation have on the production and supply of aviation catering will be examined later, as will a conceptual overview of where and how operational aviation catering security mechanisms and food safety legislative requirements come together. First, we need to understand the provisions contained under both pieces of legislation.

EU Food Directives 2000

The European Commission adopted its white paper on Food Safety on 12 January 2000 and set out a 'farm to table' legislative programme. Headed by Commissioner David Byrne, responsible for Health and Consumer Protection, it set out three key objectives:

(1) The establishment of the EFSA.
(2) A total revision and consolidation of all existing hygiene directives.
(3) A legislative programme with over 80 new regulations within a time-frame of four years.

These key objectives were devolved into several more specific and detailed aims:

- The EFSA to be established to deal with risk identification and assessment and risk communication.
- With the introduction of a 'farm to fork' principle comes the requirement for hygiene rules within primary production.
- With the identification of the need to identify hazards at source come the following requirements:
    - full traceability of all food ingredients
    - compulsory registration of all food businesses
    - adequate records
    - transfer of the responsibility of safe food production from officials to food producers
    - mandatory HACCP systems.

The new legislative proposals covered a whole host of topics for regulation and looked something like the following in terms of content headings:

- general food law
- the primary food chain
- feeding stuffs and animal health
- BSE/TSE
- contaminants
- food additives
- food contact materials
- novel foods and GM
- irradiation
- dietetic foods and food supplements
- food labelling
- pesticides
- nutrition

- seeds
- food policy.

Whilst all the directives under this new legislation have had an effect on the nature of all food business operations, Regulation (EC) No 178/2002 of the European Parliament and of the Council – The General Food Law Regulation has had the greatest impact on the requirements for operational transformation and SOPs in catering establishments.

The law came into force on 21 February 2002 with the key provisions being enforceable on 1 January 2005. The primary objective of Directive 178/2002 was to lay down common principles underlying European food safety legislation, particularly the scientific basis, the responsibility of food producers and suppliers, traceability, product recall, effective controls and enforcement to improve the transparency, consistency and legal security of foodstuffs. The key principles of the Directive are defined in several Articles within the Directive that underwrite the impact of the overall proposal:

Article 14 – Food Safety Requirements

This prohibits food being placed on the market in any type of food business if it is unsafe, and specifies what this means. Food shall be deemed to be unsafe if it is considered to be:

- injurious to health
- unfit for human consumption.

In determining whether any food is injurious to health, regard shall be had:

- not only to the probable and/or immediate short-term and/or long-term effects of that food on the health of a person consuming it, but also on subsequent generations
- to the probable cumulative toxic effects
- to the particular health sensitivities of a specific category of consumers where the food is intended for that category of consumers.

In determining whether any food is unfit for human consumption consideration shall be given to:

- its intended use
- reasons of contamination whether by extraneous matter or otherwise, or through putrefaction, deterioration or decay.

As far as possible food business operators are to ensure that primary products are protected against contamination, having regard to any processing that primary products will subsequently undergo.

Article 16 – Presentation

This stipulates that the labelling, advertising and presentation of food shall not mislead customers.

Article 18 – Traceability

This requires food businesses to keep records of their suppliers and businesses that they supply to and to make sure that such records are available to the competent authorities on demand.

Traceability is defined in Article 2 as:

'the ability to trace and follow a food, food producing animal or substance intended to be incorporated into a food or feed through all stages of production, processing and distribution.'

The traceability systems are designed under this legislation to demonstrate a greater level of product control in terms of product history and must demonstrate a greater degree of traceability than 'one step forward and one step back'.

Article 19 – Product Recall/Withdrawal

This places obligations on food businesses to recall and/or withdraw food from the market if it is not in compliance with food safety requirements, and to notify competent authorities.

Article 5 – HACCP

All food business operators shall put in place, implement and maintain a permanent procedure or procedures based on the principles of hazard analysis and critical control points (HACCP). All seven principles of HACCP as defined in the Codex Alimentarius must apply to all food businesses.

Overall the focus of the new Directives is to place an emphasis on the connections between safety and quality and the requirement to consider all applicable hazards in the food chain before formulating a HACCP plan. Interestingly, the traceability issues are focused on strongly with an obvious connection with the ability to trace the supply chain in both directions in order to facilitate prompt and effective action in the event of having to instigate a product recall. Whilst the emphasis in the regulations on mandatory HACCP underwrites an ongoing commitment and culture for 'getting it right first time', the traceability focus suggests a direct connection with effective safety measures necessary to deal swiftly with a situation in which problems in the food chain have occurred.

In the US legislation, whilst the focus is on food security, the intentions appear to be the same as in Europe. Whilst the requirement for HACCP is not mandatory, it is difficult to see how compliance with the rest of the legislation could be possible without an operational HACCP, and the emphasis on full product traceability both forward and back is deemed crucial in the assurance of food security mechanisms. We will look later at what all this means to aviation catering operations.

Public Health Security and Bioterrorism Preparedness and Response Act 2002

The events of 9/11 reinforced the perceived requirement for the security of the US food supply chain to be enhanced. Congress responded by passing the Public Health Security and Bioterrorism Preparedness and Response Act 2002 and divided it into five titles:

Title 1 – National Preparedness for Bioterrorism and Other Public Health Emergencies
Title 2 – Enhancing Controls on Dangerous Biological Agents and Toxins
Title 3 – Protecting Safety and Security of Food and Drug Supply
Title 4 – Drinking Water Security and Safety
Title 5 – Additional Provisions

Having established a heightened awareness that nations could become the target for biological or chemical terrorism, the FDA took steps to improve its ability to prevent, prepare for and respond to incidents of food sabotage. Whilst initially motivated by concerns about deliberate contamination, those activities built upon and expanded the agency's continuing efforts and resolve to protect consumers from foods that have been unintentionally contaminated either via process failures or handling errors.

The most pertinent aspects of the Act to affect suppliers of aviation catering were contained within Title 3 and made provision for several key requirements:

- The registration of food facilities.
- The prior notice of food imports into the USA.
- The administrative detention of suspect foods.
- The establishment and maintenance of records.

The registration of food facilities required that all domestic and foreign facilities that manufacture, process, pack or hold food as defined in the regulations, for human or animal consumption in the USA, must have registered with the FDA by 12 December 2003. Included food groups classified under this legislation included:

- Dietary supplements and dietary ingredients.
- Infant formula.
- Beverages including alcohol and bottled water.
- Fruits and vegetables.
- Fish and seafood.
- Dairy products and shell eggs.
- Raw agricultural commodities for use as food or components of food.
- Canned and frozen products.
- Bakery goods, snack foods and confectionary including chewing gum.
- Live food animals.
- Animal feeds and pet foods.

Exemptions included the following:

- Private residences of individuals, even though food may be manufactured/ processed, packed or held in them.
- Non-bottled drinking water collection and distribution systems.
- Transport vehicles that hold food only in the usual course of their business as carriers.
- Farms.
- Restaurants, i.e. facilities that prepare and sell food directly to the consumer for immediate consumption. Facilities that provide food to interstate conveyances such as commercial aircraft, or central kitchens that do not prepare and serve food directly to consumers, are not classified as restaurants for the purposes of this rule.
- Retail food establishments that sell food directly to consumers as their primary function, meaning that the annual food sales directly to consumers are of a greater dollar value than annual sales to other buyers.
- Non-profit food facilities, which are charitable entities and prepare and serve food direct to the consumer.
- Fishing vessels that harvest and transport fish.
- Facilities currently regulated by the US Department of Agriculture, i.e. facilities handling meat, poultry or egg products.

The prior notice of foods into the USA required importers to provide the FDA with advance notice of human and animal food shipments imported or offered for import on or after 12 December 2003. The purpose of this ruling was to allow the FDA to know in advance when specific food shipments would be arriving into US ports of entry and what those shipments would contain. The idea was to allow the FDA, working with US Customs and

Border Protection (CBP), to more effectively target inspections and ensure the safety of imported foods.

The ruling on administrative detention of suspect foods allows the FDA to detain an article of food:

'on the strength of credible evidence or information resulting from an inspection, examination or investigation, that the article of food presents a threat of serious adverse health consequences or death to humans or animals.'

The rule requires a detention order to be approved by the FDA Director of the District where the detailed article of food is located, or by a higher official. A copy of the detention order will be given to the owner, operator and/or agent in charge of the place where the article of food is located and to the owner of the food, provided the owner's identity can be determined readily. If the FDA issues a detention order for an article of food located in a vehicle or other carrier, the agency also must provide a copy of the detention order to the shipper of record and the owner and operator of the vehicle or other carrier, provided the owner's identity can be determined readily. The requirement also is that foods be detained in secure locations and may not be transferred from the place where they have been ordered to be detained or from the place where the detained article has been removed, without seeking prior FDA approval or until the termination of the detention order. A detention may not exceed 30 days.

The final ruling requires the establishment and maintenance of records and requires that manufacturers, processors, packers, distributors, receivers, holders and importers of food must establish and maintain records, for no longer than two years, that would demonstrate the following:

- The identity of the immediate non-transporter previous sources, whether foreign or domestic, of all foods received, including the name of the company and the responsible individual. This would be satisfied by the following information being available:
  - company name
  - contact name, address, telephone number, fax number, email address
  - specifics of type of food supplied
  - date received
  - batch number, quantity and packaging specification
  - all contact details for supplier who delivered the product also.

The records must include all information that is reasonably available to identify the specific source of each ingredient that was used to make every lot of finished product.

- The identity of the immediate non-transporter subsequent recipients of all foods released, including the name of the company and the responsible individual. This would be satisfied by the following information being available:

  – company name
  – contact name, address, telephone number, fax number, email address
  – specifics of products supplied including brand name and variety
  – date released
  – batch number, quantity and type of packaging
  – all contact details for supplier who delivered the product.

For companies engaged in the transportation of goods, all of the above rules would apply to all food products transported.

Whilst the specifics of the type of information that must be kept is determined by the regulations, the form in which the information must be maintained is not specified. The records therefore may be kept in any form, paper or electronic, as long as they contain all the required information.

Industry response to legislative directives and guidance

Having completed an overview of both USA and European legislation, it is clear that whether the mandates are formulated in the name of food safety or food security, the principles by which compliance is necessary are the same. The principles of full product traceability have been at the essence of food manufacturing product safety protocols for over 30 years and the establishment and implementation of all seven HACCP principles have to happen to ensure that all of the attributable risks have been considered and are being managed effectively. In terms of the US model, this must include hazards posed to food security also, which in essence is the same as food safety in the European model.

So what does this plethora of legislation mean for the aviation catering industry? In essence, much of this legislation should have had no impact at all if the operational standards proffered by the industry are to be proved effective. The aviation catering industry best practice has advocated HACCP for over 15 years and the IFCA/IFSA World Food Safety Guidelines also make full product traceability an audit requirement. It is startling to note, therefore, that in response to the proposal and then implementation of the USA legislation, the aviation catering industry backlash was significant and prolific.

Industry heavyweights, including representatives from all of the major internationally-based catering companies as well as the trade organisations including IFSA and IATA, were quick to condemn the aspect of the US legislation appertaining to 'establishment and maintenance of records', which basically mandated full product traceability. They demanded in a series of letters to the FDA to be included in the exemption list and classified alongside restaurants and other service environments where the food products served were 'consumed directly by the customer'.

In one letter, the quality assurance corporate director of one major international aviation catering company states:

> '... one flight is likely to include hundreds of individual foods from scores of different sources, representing many suppliers and there are thousands of flights every day from hundreds of airports across the country. Tomorrow's meal service will be quite different, as will the food on the return flight.
>
> What this proposal does is create a nightmare for our industry in the name of making food more secure. [We] believe this proposal must be modified to be significantly less burdensome or to exempt the catering business altogether.'

Another representative from a major airline catering company writes in support of the IFSA request for exemption from the regulations on the basis that airline caterers are not 'food processors' but 'in reality we are more similar to a large restaurant or hotel kitchen, producing a wide variety of meals within a matter of hours.'

Having examined in minute detail the provisions contained within the industry-offending section of the US legislation, I can find no unreasonable request that should faze the industry if systems management protocols were established in the manufacturing genre, and even complied with industry best practice. To suggest that a catering facility with several different locations across the world, supplying millions of meals to travel vectors flying all over the globe, should be exempt from mandates that demand full finished product and raw material traceability, tray set by tray set and component by component, is in my view insane!

I would strongly challenge the view also that the systems management protocols required to be established to assure total product safety, traceability and legal compliance, not just in the USA but also in the EU, would be overly onerous and cost prohibitive. In my company, a supplier of catered products to the business and commercial aviation industry, as well as retailer branded recipe dishes to the supermarket sector, we have indeed established exactly these types of systems with huge success. We can indeed trace every component on every tray flight by flight. We can indeed verify the very essence of our supply chain and produce detailed food technology end product specifications for every product, be it a single-unit component or a multi-component tray.

A simple reference to Codex Standards on mass catering will underwrite my belief that far from running operations akin to a restaurant or a hotel, airline catering operations are massive producers of prepared meals and therefore cannot possibly operate under catering guidelines and legislative enforcement standards. The multimillion unit replication, the export status of the products and the associated industry security risks all serve to build the strictest possible case for mandatory regulation to require the toughest possible standards of product safety, integrity and traceability.

As I mentioned earlier in this chapter, the irony for me is that it has been food security not food safety legislation that has finally proved that the industry has to evolve into the 21st century, embrace contemporary standards of food manufacture and establish a new culture for change in food safety systems management and protocol. If full product traceability in aviation catering supply is to become more than an idle boast and start to become a reality, as the legislative requirements on both sides of the Atlantic are now demanding, product and process-specific HACCP plans must also become integrated with hazard analysis, taking account of security issues also, in the overall spectrum of risk management.

In the following sections we look at the steps necessary and already adopted by the manufacturing sector to assure process control security in food processing, and then at a direct comparison with the operationally-based directives adopted by the transport security industry that have been amalgamated into airline catering operations all over the world.

Security assurance in the food manufacturing sector

In May 2002, the 55th World Health Assembly expressed serious concerns about the real and current threat posed to the international food chain by the malicious contamination of food for terrorist purposes. The potential for chemical, biological or radio nuclear agents to be used to deliberately harm civilian populations, using food as a vehicle to disseminate such agents, suddenly became immense.

It was recognised and acknowledged that the most effective way to deal with such threats was proactive systems management via the establishment or enhancement of existing food safety management systems within all food businesses. Food businesses of every size and genre were encouraged to accept responsibility for the requirement to assess the security risks posed to their business, and through the development of a cooperative approach with government and enforcement agencies to amalgamate new security-based protocols into existing food safety management programmes. Whilst the responsibility for the prevention of a terrorist atrocity using food as the vehicle was acknowledged as a shared one, between central government and the food industry itself, the primary means for minimising food risks lies indisputably with the food industry.

As in all considerations of health and safety-related matters, prevention of an incident is by far the most desirable option. In the context of food terrorism it relies on protecting the food from contamination during production, processing, distribution and preparation. As the capacity to prevent deliberate sabotage lies primarily with the food industry, it is crucial that the potential security threats are redressed throughout the supply chain. It is true to say that not all businesses are as vulnerable as others; however, the proactive nature of the food manufacturing sector has seen many of the manufacturing industry codes of practice interface long-established systems designed to prevent accidental contamination of foods with those now required to be considered, to prevent deliberate and malicious contamination of foods also.

The production process, and therefore the steps within the process considered to be vulnerable to attack, will vary for each type of product and each type of process, which is why it is essential that any food security strategies are a homogenisation of existing procedure designed to do as much to enhance successful systems as to prevent food security violations. In addition, strategies adopted and resources dedicated to the development of security-based systems need to be considered in the overall assessment of risk posed to the product, the process or the business.

Having established that prevention is the best culture to adopt in the industry's attempt to deal with potential acts of food terrorism, and having established also that amalgamation of specific food security considerations into GMPs should not cause too much difficulty post risk assessment, it is essential for the food industry to be realistic about its capacity to completely prevent an act of food sabotage. It is essential then that the second wave of strategies required are surveillance, preparedness and ultimately response.

The management and effect of such strategies on a national or international basis are quite different to the localised application of them in the food manufacturing environment itself. The types of surveillance approaches undertaken will be directly defined by the relative risks associated with the business activity itself. However, they may include a revision to the verification end product and incoming raw material testing regimes by placing products where possible on a positive release basis. Where the potential risk contaminants remain undetectable through conventional testing, more operationally-based surveillance strategies need to be put in place.

Preparedness and the food business's capacity to cope in the event of a terrorist emergency are linked directly to the historical effectiveness of eliminating accidental contamination and avoiding product recall situations. The more robust the GMPs and the more proficient and well resourced the make-up of the technical team, the more prepared for any potential crisis management situation the food business will be.

The essence of food manufacture and food processing is the culture of self-preservation and the overwhelming requirement to get it right first time. The fiscal implications of a breakdown of procedure or protocol are immense and the ensuing publicity disastrous. Unlike the catering sector, the ability to embrace any aspect of food security in systems management and procedure is made possible by the existence of compatible food safety management strategies: systems that assure full product traceability and are linked to the requirement to demonstrate the success of product recall on a regular basis. To inaugurate a security-based risk management strategy into already well established and robust controls designed to assure quality, safety and product legality consistency, is no huge mountain to climb for this sector, assuming the culture for security assurance is there.

As part of the required national and international strategies on food surveillance capabilities following an act of terrorism involving food, consideration has to be given, when awarding valuable public health resources, to the types of food businesses most susceptible to attack. Broadly speaking the most vulnerable foods and food processes can be generalised as follows[121]:

- The most readily accessible food processes.
- Foods that are most vulnerable to undetected tampering.
- Foods that are most widely disseminated and spread.
- The least supervised food production areas and processes.

In each case four broad-based factors will need to be considered for security assessment to be undertaken:

(1) The personnel involved in the food business with a direct or indirect connection with product or product data.
(2) The design and fabric of the manufacturing and processing facility.
(3) The nature of the production process from goods receipt to dispatch.
(4) The nature of the transportation requirements of the product through to final purchase or consumption by the end consumer.

In any overview and assessment of security risk, the capacity for food business activity to be compromised by the shortfalls in the security assurances of the supply or distribution chains, needs to be considered in the scope of the business security schematic. For example, raw material imports from countries where there may be uncontrolled access to toxic chemicals – including pesticides, heavy metals and industrial chemicals as well as a whole host of naturally occurring microbiological pathogens that could be used as agents in terrorist threats to food – are likely to warrant greater consideration than those from a domestic or known source.

Security assurances to the supply chain need to be applied throughout and should include the following industries:

- farming
- fisheries and aquaculture
- distribution including transportation and storage
- food and beverage manufacture
- food and drink contact packaging
- wholesale and retail points of sale.

As part of the food industry's commitment to assuring not only the safety and integrity but also the security of the supply chain, the Food Industry Security Assurance (FISA) scheme[122] was developed and provided as a model for companies to follow to introduce measures to safeguard their operations, employees and products from deliberate acts of tampering and/or malicious harm.

FISA

FISA is based on the recommendations contained within the WHO document and seeks to provide assurance on systems amalgamation between food safety, quality and legality control measures and security-based control measures. The standard deals with every aspect of process and production security protocol, not least the establishment of a management and fiscal commitment for resources to be allocated specifically to food security assurances.

FISA applies exactly the same food safety management ethics to enhanced security systems development and uses assessment of risk in the broadest possible sense and in consideration of the wider supply and distribution chain. Every product is considered in isolation within the scope of the scheme, with security controls being assessed on a product-by-product and process-by-process basis in much the same way that food safety risk assessment is product and process-specific in food manufacturing.

As we will see in the next section, the product and process focus throughout the security assessment criteria in food manufacture and processing is in stark contrast to the operational and logistical focus witnessed in the security guidelines inherent in aviation catering. Instead of an holistic approach to aviation-specific security issues in tandem with product security assurance, a marginalised and secular standardisation of catering security protocols has emerged which does little if anything to assure the security of the food products themselves, let alone deal with tampering post production.

Operational security directives in the supply of aviation catering

The security of in-flight catering and stores directives, as laid down under the provisions of the state NSAP's requirements, are designed to ensure that no prohibited article is taken on board a qualifying aircraft such as may endanger the safety of the aircraft, its passengers or crew.

As I have said from the beginning of this chapter, the industry guidance is based purely on the security measures that need to be applied to prevent the infiltration of catering goods and supplies by the smuggling of a prohibited item on board. These operational security measures do not deal with the security integrity of the products themselves in terms of bio-terrorist activity during manufacture or production, where the fabric and composition of the supplies may be compromised; rather they concern themselves with the prevention of the corruption of the vehicle of supply mechanisms.

The directives are devolved from ICAO guidelines, which state[117]:

> 'Each contracting State shall establish measures to ensure that catering supplies and operators stores and supplies intended for carriage on passenger flights are subjected to appropriate security controls.'

The aircraft operator and their contracted caterer have a joint obligation for the responsibility of the implementation of the measures described and are required to assure that the measures are carried out. Generally the directives are divided into the measures required to deal with products manufactured from raw materials in-house, known stores, and those designed to be applied to products that are bought in and merely transit through the catering facility as part of their logistics function, unknown stores. The measures to be considered can be broken down into several different sections, as follows.

General security considerations

- The aircraft caterer shall develop an approved security programme under the state NASP that allows for the application of in-house security measures for all stores and supplies taken on board a qualifying aircraft, during preparation, storage and transportation.
- The caterer shall nominate a suitable person or persons to be responsible for the security functions relating to in-flight catering and stores.
- The supplies and stores that have had all necessary security measures applied to them may be classified as known stores and as such, so long as their security status has been maintained throughout the supply chain, may be

placed on board qualifying aircraft with a minimum of further security controls.
- Items from sources other than the airline caterer should be regarded as unknown stores and should not be taken on board an aircraft until they have had the necessary security measures applied to them such as would make them known stores.

Security measures at the catering premises

- Premises utilised for the preparation or storage of in-flight catering supplies should be secured and access controlled.
- Pre-employment security screening of personnel employed in the preparation and delivery of in-flight catering supplies.
- Staff engaged in the preparation and delivery of catering supplies should be trained to understand their security responsibilities as they relate to the state NASP. This training should be carried out before staff are allowed access to any supplies or stores which are to be despatched to the aircraft as known stores.

Preparation of in-flight catering supplies and stores

- All deliveries of raw materials received by the airline caterer for use in the preparation of known stores shall be decanted to ensure that they do not contain any prohibited article.
- Deliveries of products that are not decanted (exemptions may include special meals bought in from other contractors, sealed amenity kits, first-aid kits, blankets where these have been sealed by the laundry, sealed mineral waters, dry goods and bulk deliveries of frozen meals) shall be accompanied by appropriate security certification from the supplier.
- Caterers should only accept bulk deliveries of precooked frozen meals that cannot be searched or broken down providing they have been sealed or made tamper evident.
- Staff engaged in the preparation of known stores should be supervised to ensure that they cannot place a prohibited article within such stores.
- Catering carts of containers carrying known stores should be searched prior to being sealed for despatch by a supervisor. The record of such a check having been carried out should be placed on the delivery documentation and signed against.
- Carts and/or containers carrying known stores on board should be designed so that access to the interiors is not possible once they have been sealed.

- Chilled storage areas used for the holding of prepared meals, tray sets and other known stores should be secured and access-controlled.
- Tamper-evident seals should be used for the sealing of known stores containers and access to the seals should be security controlled.
- Documentation detailing the content and nature of the delivered consignment should accompany each delivery and supply of known stores.
- Appropriate security certification shall be issued by the catering contractor once they are satisfied that all security criteria have been met. These certificates shall be issued and signed by those within the business charged with a security responsibility. This documentation should be handed over to an appropriate representative of the aircraft operator.

Transportation and delivery

- The security of known stores should be maintained throughout the transfer of goods from the catering facility to the aircraft itself by an approved haulier operating approved procedures and security measures. These may include the caterer and the aircraft operator themselves.
- All vehicles used for the transportation of known stores must be searched prior to loading and sealed with tamper-evident means. This includes all containers, doors and load entry points. The seals and accompanying documentation should be checked at airport points of entry and by airport authority representatives at vehicle control posts.

Application of operational security measures

The operational security directives, whilst devolved from aviation industry guidance, are often formulated after consultation with the airline catering industry itself. The generality of the directives means that compliance standards are catering industry generic as opposed to catering operation specific and this is symptomatic of the fact that the evolution of the content has not been based on any formalised risk assessed criteria.

It is obvious that transport security, whilst having to account for a whole host of operational and supply-based activity, cannot be supplier specific in terms of what it requires; however, within the scope of the general guidelines I believe there should be a requirement for operational security measures to be applied following in-depth security risk assessment of each individual supplier. Whilst the general heading of aviation caterer may suffice in terms of the general application of the title, what actually constitutes an airline caterer must be considered in the wider context of those operations that supply either directly or indirectly to aircraft operators and their catering partners. With the advent of buy on board profile products and with the modern day requirement for many airline caterers to act merely in a logistical capacity, they may well produce little if anything that can ultimately be classed as unknown stores.

With the supplier base constantly growing and evolving, I would suggest that it should be a mandatory aspect of any general NSAP guidance to caterers that both the raw material and finished goods, supplier audit criteria, should include security-based auditing as well as food quality and safety-based assurances. In this way, supplier assurances need to be graded in terms of the security risks posed to known stores by the products supplied as unknown, which ultimately the airline caterer has the responsibility to apply.

In any consideration of the overall impact that the supply of catering goods and stores has on aviation security, the full gamut of possibilities have to be considered. It is impossible for blanket, operationally-based directives to do this with any measure of success unless such broad-based directives are backed up by a catering security assurance scheme that deals with the specifics of the security supply chain as well as the logistical issues in an operation-by-operation basis.

The future success of aviation catering security rests with the industry's ability and willingness to accept the possibility of process and production-based terrorist activities, as well as operational and logistical infiltration-based ones. To assume that the only directives required to be applied to aviation catering operations are the ones designed to inhibit the capabilities of terrorists to get a prohibited item on board a qualifying aircraft, is to assume that the potential for bio-terrorist activities at the supply and production end of the chain poses no risk. It is essential that any aviation catering security assurance schematic inaugurates both the process control attributes and the operational and logistical aspects as one, whilst both must be based on specific, independent, operational risk assessment.

Crew food protocols

Throughout other chapters in this book we have looked at the risks posed to aviation safety by the quality and integrity of the crew catering provision. In Chapter 8, 'Fitness to fly', we looked at the links between pilot incapacitation and in-flight food poisoning, and in Chapter 9, 'Cabin crew – the missing link', we looked at the implications on passenger food safety of crew flying while sick. In this chapter I feel it is important to focus some degree of attention on the nature and level of security assurance that should be paid to the provision of in-flight crew meals.

Food safety versus food security 335

When we looked at the mandates and industry best practice in place to ensure that the two flight-deck crew do not consume the same in-flight meals, we looked at the inherent safety shortcomings of such a blanket directive policy if, even though the meals were different, they had been supplied by the producer. In the same way, we now look at some of the potential security shortcomings inherent in the manufacture and the supply of crew meal options.

First, it is essential to consider the source of supply, in security terms, of all crew food products where they are specific to the crew. It is essential to remember that all of the industry security assurances are based on the prevention of prohibited article transmission on board, not malicious contamination of products at source, therefore crew food provision risk assessment criteria will be dominated by source of supply. If crew food products are to be manufactured under known stores classification in-house, then separate, secure, access-controlled areas should be designated and only security-cleared personnel should be engaged in crew food manufacture. Where such measures are not practicable an external supply source should be sought, preferably one engaged in the mass production and supply of products so that the specifics of their intended usage may remain undisclosed.

If crew food products are to be outsourced to another catering-based supplier, it is essential that their crew food status remains undisclosed, particularly if as a result of the make-up of the products they are classified under unknown stores exemption.

I can see no necessity to disclose the nature of the recipients of products to external suppliers when the nature and make-up of most of the crew food inventory is likely to closely resemble that of passenger food applications.

Once produced, crew food should be not be placed into separate containers or identified in any other way as specific to the intended crew members' consumption, especially by labelling 'crew'. Whilst this may sound obvious to those in the food manufacturing sector who have been subject to such undisclosed recipient practices in order to protect the intellectual property data of a competing client base, in the world of aviation catering supply the labelling and packing of crew food as such is the status quo.

In consideration of the process-specific risks attributable to the supply chain, any detailed subjective risk-based security assessment of the supply chain would render this type of practice dangerous and in direct conflict with risk minimisation strategy. The vulnerabilities of crew food to malicious contamination, if unlawful interference of aircraft is the aim, are immense and it is essential that operational activities surrounding crew food provision are devised with both food safety and food security in mind.

One of the simplest and most effective ways to assure crew food product safety at source for hot meals is to despatch them on a batch-specific positive release basis. This can only be done using frozen products unless the chilled products have defined extended life capacity and the results turn-round is no more than 48 hours. However, with frozen meals positive release mechanisms will work very well as long as batch integrity validation is assured.

Invariably, the decisions about the specifics of the security measures that need to be applied to the process aspect of crew catering provision, should be determined by the aircraft operator in collaboration with the catering provider at each individual outstation. In tandem with this, strict production or process control parameters need to be established based on whether the products are bought in or manufactured in-house. During the menu development stages of crew food options, consideration needs to be given also to where the products are going to be made and what supply chain issues may be thrown up by enhanced security considerations.

I am amazed that the operational security aspects of crew food provision have not been predetermined before, alongside other aspects of catering security and bearing in mind the obvious connections between what crew, particularly technical crew, eat in-flight and the capacity it has to compromise their abilities to carry out their duties. Further consideration needs to be given also to crew special meal requests, if indeed special meal supply is exempt from known stores classification under the NASP transport security directives of the member states involved. This is a great example of exactly the types of product and supplier-specific factors that need to be considered when developing risk-assessment criteria for crew food provision.

14 Food safety in the business aviation environment

Outside the mainstream vision and perspective of where catering provision to the aviation sector begins and ends, lies a somewhat diverse and self-regulating area of the aviation catering industry which is, for the most part, overlooked and ignored in terms of standard operating procedures.

The provision of catering to business or general aviation aircraft is also a multimillion pound business globally. The diversity of aircraft type, design and configuration, the non-existence of airline-style menus and rotations, the absence of in-flight schedules and the requirement to cater absolutely anything at a moment's notice, render the business aviation catering industry naturally predisposed to a food safety nightmare!

What is business aviation?

The utilisation of business aircraft globally has grown hugely over the past 50 years. The USA is the home of general aviation (GA); the National Business Aviation Association (NBAA), based in Washington DC, recently reported a considerable upturn in the use of GA since 9/11 (NBAA update September 2003). Historically business aircraft have been utilised to maximise a corporation's two greatest assets: its people and its time.

A major corporation owning and operating its own aircraft has the capacity to get its people anywhere they want to be in the quickest possible time. Most major corporations will own and operate any of a number of types of business aircraft and use them to constantly transit their executives to the four corners of the globe.

The environments are as eclectic as the aircraft types, everything from hotel suite splendour to the stark functionality of the office environment. The technology is staggering, 21st century specification; satcom telephones, fax machines, computers, DVDs and televisions are the mainstream. The environments are owned and operated by the company to which they belong in order to control both the personal safety and security of their most high profile and critical personnel.

In tandem with advancing technologies comes the need to protect company data. What better way to conduct confidential company business than on board the company-owned and operated business jet?

The use of GA is not exclusive to business activity. Many highly successful and high profile individuals fly their own aircraft for reasons of convenience, privacy and security. For many, a corporate jet is merely an extension of the family home or car. If owning your own aircraft is not an option – the $20–$100 million price tag may be cost prohibitive for some – then the opportunity to charter (hire) a multitude of aircraft is available worldwide. Since the early 1990s it has also been possible to buy an eighth or sixteenth share of a business aircraft. Fractional ownership, as it is called, is a burgeoning industry with the number of fractional shares sold nearly doubling every year. For an annual fee this time share-type scheme buys owners a certain number of flying hours each year and the flexibility and assurance that at less than eight hours' notice they can be on their way to any destination of their choice.

History of business aviation catering

GA accounts for some 77% of all flights in the USA, involving over 200 000 aircraft, 650 000 pilots, and over 19 000 airports and landing strips. The GA industry is reported to provide jobs and opportunities for thousands of people who work at airports and run small businesses providing goods and services sustained by GA activity. General aviation-related economic activity in the USA alone is estimated to be in excess of $100 billion annually and is related to approximately 638 000 jobs[123].

In the 21st century, the type of business aircraft available varies hugely in size, range and interior design. All the major manufacturers, Gulfstream, Bombardier, Embraer, Falcon, Boeing, Airbus, offer a variety of aircraft types and styles to meet the demands of even the most discerning clientele. The most commonly held misconception about privately owned and operated aircraft is that they are, for the most part, small five-seat Learjets used for short-haul travel. The reality is that the flying ranges of some of the most popular aircraft can be in excess of 15 hours and 8000 miles, flying at altitudes of 42 000 feet.

So what does this mean to those who are dedicated to providing catering for such a high profile and perceivably discerning client base? How exactly does the provision of catering for this exclusive sector of the flying public vary from the service to commercial airlines?

Fifty years ago, the catering requirements of the business aviation passenger were vastly different from those of today. In much the same way as airline catering provision has evolved from the simplistic cold tray meal or sandwich in a box, the diversity and range of foods available to business aircraft is nowadays limitless. Bizarrely, the sources of food supply have remained the same over that time. This is where and how we begin to gain an understanding of why the nature of business aviation catering and its historical evolution have resulted in the potential for a modern day food safety crisis.

A further misconception about business aircraft is that they operate primarily from major hubs at major international airports. This was certainly not the case 50 years ago and is not today. The USA alone boasts nearly 20 000 airports servicing business aircraft, whilst the UK has at least ten that fulfil this requirement exclusively.

Historically, what this meant in regard to outsourcing catering was that the nearest restaurant or coffee shop to the airfield was initially asked to provide the cockpit crew (early GA aircraft did not have the capacity to carry a third crew member) with whatever repast they required for both themselves and their passengers. Later, with the evolution of turboprop aircraft into jet aircraft, the existence of a flight attendant or stewardess became more prolific and the catering service demands and advanced food service demands emerged accordingly. The crucial aspect of the story, which did not evolve or change, was the critical source of food supply.

As menu requirements evolved to include not just one but several meal choices, and cold menu requests developed into hot menu selections in addition, the local coffee shop and café remained the single source of food supply. There are several reasons for this but they focus on a simple economic principle of demand/pull.

Even busy business aircraft-dedicated airfields in the early days would not service many aircraft that required catering. The requirement was ad hoc and inconsistent, in some areas resulting in seasonal pockets of activity and then nothing for weeks on end. It made sense to utilise a local restaurant or coffee shop. Another pervading factor, which remains consistent with catering provision for business aircraft today, is that even when an upturn in the volume of aircraft orders was witnessed, the resulting passenger loads were still small.

It wasn't until about 30 years ago that the requirement for dedicated business aviation caterers at major business aircraft hubs emerged. New York and Chicago realised this opportunity first, yet still this dedicated new breed of business aviation caterers were spawned from the coffee shop and restaurant environments which had been the original source of supply.

The catering requirements and logistics continued to vary hugely from those of the commercial airlines. There were no set menu rotations and no standard meal requests, the demand came and went as and when the need arose, and food preparation and production did not happen in line with a set airline schedule, rather at the whim of those who decided they wished to travel. The short-order restaurant ethic translated into short-order catering provision, with the added complication of chilling, packaging and transportation to the aircraft.

At this time a take-away food service ethic came into play as a reaction to the inherent difficulties of preparing food in this way. This happened in tandem with the emerging fast food industry in the USA in the 1950s and 1960s. It made logistical sense to prepare and cook the food to order and then have the crew come and pick it up warm and run it out to the aircraft. The food safety considerations of further reheating of precooked foods and the absence of temperature-controlled storage on board the aircraft were not a concern, in much the same way that they weren't for the airlines during this period.

The dedicated commercial airline caterers were not known for their successes in providing catering for the business aviation community, for all the reasons we have already witnessed. Absence of schedules, no set menus, no place for tray set up meal presentation, the requirement to provide anything the passengers required at a moment's notice, the location of the airfields away from major airports; the list goes on. The only two things these emerging industries had in common it seems were food and aircraft!

The comparison between a typical airline-style menu versus a business aviation catering order illustrates the point beautifully (Figures 14.1 and 14.2).

What is obvious from the menu illustrations are the enormous quality, quantity and packaging differentials. Much of the catering for business aircraft was and still is packed in bulk, i.e. menu items are broken down into components and then packaged as such in small bags and containers so that the flight attendant can assemble the food on board. It is only recently that the commercial airlines have begun to lend themselves to this idea in the front-end cabins but the logistical problems for the airline caterers of catering short-notice, small orders in this fashion, is immense.

Only in the very early days of business aviation catering did the tray meal ethic have any place and much of that was born out of the absence of a flight attendant and the relative size and stowage constraints of the galleys. As business aircraft utilisation increased and the ergonomics evolution progressed, the galley design issue came into play. The recognition by the aircraft manufacturers that the food service aspect of a trip on a business aircraft was crucial to its overall perceived success, was instrumental in moving galley design forward. Business aircraft were now equipped with ovens and coffee makers, stowage areas and work tops, inventories of the finest china and glassware, but still no refrigeration for safe food storage.

Throughout the history of the development of business aircraft design, little or no attention was paid to the increasing urgency to integrate dedicated chilled food storage areas into galley design. Today the picture looks pretty much the same. Despite the capacity and requirement of the long range business

Breakfast Menu

Selection of breakfast cereals
Cornflakes, Weetabix, Sultana Bran, Special K
oOo
Warm bagels with cream cheese
Butter croissants and blueberry scones served with a
selection of preserves
oOo
Seasonal fresh fruit served on its own or with natural yoghurt
oOo
Bacon roll served with either tomato ketchup or brown sauce
oOo
Scrambled egg with chives served on grilled bruschetta with pork centre loin,
potato wedges and Roma tomato

oOoOoOo

Parsnip soup with stem ginger
Fresh salad leaves with your choice of balsamic vinaigrette,
Caesar dressing or extra virgin olive oil

oOoOoOo

Fresh pasta with your choice of creamy mushroom sauce
or cherry tomato and oregano sauce
South-western Tamale casserole served with roasted corn tortilla sauce

oOoOoOo

Fresh fruit
Cox's apple pie with clotted cream

Figure 14.1 An example airline menu for first and business class.

jet to circle the globe with upward of 12 passengers and crew, all requiring several eclectic meal services, there is no mandatory requirement to equip these aircraft with dedicated chilled food storage areas. The reasons for this are simple. The assumption is that if the aircraft is privately owned and operated, then it is at the discretion of the owner whether refrigerated storage areas are included in the specification. They are available, but despite the advanced technological status of the rest of the aircraft equipment, they comprise a 'stone-age' style ice drawer or a 'gasper air' style system which draws in cold air from outside whilst the aircraft is in flight, but does not operate on the ground even with ground power. The added complication of no integral methods of temperature calibration renders the situation on board far from ideal.

We look at issues appertaining to food service safety in this environment later, but it was crucial to highlight the point here in terms of what restrictive chilled storage facilities on board mean to those who provide catering to this sector. The startling reality is that there is immense capacity for a plethora of

Qty	Description	Qty	Description
1	Medium foil tomato bread	7 kg	Ice
1	Platter tomato mozzarella and basil	5 pkt	Dry ice
1	Medium foil stir fry spicy beef	Assort	USA and United Kingdom papers
1	Medium foil mixed rice	Assort	Magazines
1	Medium foil stir fry vegetables	1	Medium foil of garnish
4	Grilled chicken with BBQ sauce	6	1.5 L of water
1	Medium fussilli Arrabiata	6	0.5 L of water
3	Passengers – sole with lemon butter	1	Litre of orange juice
4	Passengers – veal scallopine	1	Litre grape fruit juice
1	Medium foil rosti potatoes	1	Small foil sliced lemons
1	Medium foil coq au vin	5	Half lemons in muslin
1	Medium foil mixed vegetable	1	Large foil crudités including fennel
1	Medium foil beef stew	4	Assorted Italian breads
1	Medium foil farfalle in tomato sauce	4	Naan bread
1	Medium foil rice	4	Assorted rolls
1	Medium foil mash	1	Garlic bread
1	Medium foil sweet and sour chicken	4	Packets of Swiss chocolate biscuits
2	Beef fillets with pepper sauce	1	Large foil fruit slices
6	Cans of Perrier	6	Petit fours
3	Cans of Diet 7 Up	6	Pastries
2	Cans of Diet Coke	1	Large chocolate cake
3	Assorted ice-cream bars	6	Chicken kebabs
1	Cheese and ham sandwich	4	Beef kebabs
1	Beef and Dijon sandwich	6	Vegetable tartes – mini
1	Turkey and cheese sandwich	1	Large zip-lock romaine
2	Medium foil humous – no garnish		Croutons
			Caesar
			Parmesan
Customer's signature		**Representative**	

Figure 14.2 Typical catering order for a business aircraft.

high risk foods to remain outside temperature control on board for extended periods of time, whilst subjected to the same food service demands of a top class restaurant. Any business aviation caterer must operate with a full and given knowledge that the ultimate food service environment (the galley) is far from ideal in terms of food safety standards commensurate with the requirements. It would make sense for these caterers to make provision for this, not only at the menu development stage of the arrangement but throughout the aspects of the extended chill chain over which they have jurisdiction.

It is at this point that I take you on a journey back down the history of business aviation catering. We discuss the risks inherent in providing catering to this high profile sector of the aviation industry and focus on the standard operating procedures required to fulfil this incredibly complex requirement safely and effectively, with a full and given knowledge of the products, processes, logistical and regulatory issues.

Who are business aviation caterers?

From what we have learned about the evolution of GA and the simultaneous evolution of the catering requirement for business aircraft, it now makes sense that those that cater these types of aircraft are as diverse and eclectic a mix of business types as the aircraft themselves.

The short notice yet high specification requirement of GA catering renders GA flight crews incapable of any type of standardised ordering practices and even less of standard sources of catering supply. The critical source of supply can be anything from a hotel or restaurant kitchen to a hot food take-away, from a supermarket or grocery store to a mainstream business aviation caterer, or a combination of all of the above.

A fundamental lack of understanding of the on-board requirement, from all except the mainstream business aviation caterer, often leaves the crew with little choice but to handle the bulk of the food preparation and presentation themselves. In terms of food safety practices and procedures, this situation leaves the crew wide open to a variety of risk management issues of which, for the most part, they have little or no knowledge.

Bearing in mind what we have already learned about the evolution of the business aviation catering industry and the historical roots of its dedicated providers, it is very rare to find caterers who hold anything like the kind of food hygiene and safety qualifications commensurate with the risks inherent in the activity. The unique nature of operating a short-notice cook chill unit alongside chilled food preparation, fresh squeezed juice, sandwich and salads manufacture, with a little chilled distribution thrown in for good measure, is an extraordinary prospect.

It is catering like no other, for a group of end users whose profile is of the highest calibre. Whilst there is no suggestion that any particular consumer's health and safety is more important than anybody else's, it is undeniable that impacting upon the integrity of the food supplied to those who have the capacity to fly their own aircraft, the so called movers and shakers of the world, has a remarkable burden of responsibility attached to it.

As with the airline industry, the statistical data suggest that less than 10% of corporate flight crews have any formal food hygiene background or training (Castle Kitchens Ltd Survey 1999), a remarkable statistic bearing in mind the incredibly risky and diverse nature of the job and the profile of the consumer group. This, coupled with an unconventional global supplier base that also exhibits little or nothing in the way of food safety management protocols, and any attempt to accommodate the safety shortfall at the food service end of the spectrum seems unlikely.

Having taken a brief look at the types of businesses involved in the provision of catering to GA, the safety picture looks rather shaky. The lengthy supply chain, high risk and diverse nature of the products, global outsourcing and extensive transit requirements all combine to create a very unstable framework of activities, underwritten by a widespread ignorance of the risks.

To illustrate the point still further, the next logical step is to examine the typical types of foodstuffs required and supplied to GA aircraft in tandem with the packaging requirements and likely source of supply.

Catering supplied to general aviation aircraft

Having gained an insight previously into the short notice and high specification requirements of the customer base, one may be forgiven for wondering why a provider to the industry would indeed bother with a printed menu or brochure at all! By logging onto the website www.castle-kitchens.com you can view a typical example of a GA dedicated menu.

So having established that the requirement of the client is to expect the caterer to provide whatever they require whenever they require it 24/7, caterers use their menus as a means by which to guide the client towards some of their most traditionally successful and regionally prepared dishes. If the caterer is a dual-operation facility, covering other markets, it may well include items from function, restaurant, retail or sandwich menus they are running. Many caterers will assure you that the menus are used as a guide by the flight crew to assess flexibility and an essence of the capability of the catering provider, and the crew in turn will judge many a business aviation caterer on the aesthetic strengths and weaknesses of their literature. Whilst this may be a common perspective in any industry, it is a particularly prevalent one within the business aviation community and it contributes hugely to influencing dubious catering supply decisions among those who are ultimately responsible for passenger safety and security, namely the crew.

They can of course be forgiven for this. Typically a GA crew member will be faced with outsourcing catering fit for a king or queen, from an unknown source, in an unfamiliar country where language barriers will most certainly play a part, at potentially short notice (24 hours is typical but less than 12 hours not uncommon). Assessing the reliability of said provider, as well as the aesthetic appeal and quality of the food, will be of primary importance as will geographical location to the airfield.

As we have already discussed, most GA aircraft arrive and depart from small regional airports, making access to supply crucial if the catering fails to arrive or the caterers forget something critical. Given all these pervading issues, the small matters of food safety and security slide way down the agenda compared with achieving something vaguely palatable and aesthetically appealing, packaged suitably and arriving at the airport in time for departure!

So what exactly does a typical catering order for a private aircraft consist of? In reality there is no typical order, but for the purposes of highlighting the safety issues inherent in this type of catering provision we will assume there is!

All GA crews have to consider several things before deciding on a menu and the manner in which the packaging of the food should be handled. Aircraft type and size will dictate stowage availability and size constraints. Number of passengers travelling will also have a major bearing on what and how catering is ordered, as will the duration of the flight, take-off time, and finally any passenger profile considerations such as special diets or young or elderly passengers. In this way much about the manner in which menu decisions are made is similar to that of the airlines, the major difference being that for the most part, it is the flight attendant (FA) who will devise the menu and dictate the packaging specifications and not the aircraft operator or caterer.

Getting back to our 'typical' catering example, if one took a flight from London Stansted to the USA, travelling on a Gulfstream IV carrying six passengers and three crew, take-off time 0700 local and arrival at 1200 local, how would a typical menu look?

Breakfast would likely consist of a combination of continental-style breakfast breads, cereals, fruit and yoghurt, along with a hot option. Eggs would most certainly feature in some way, either scrambled or omelettes, for which the FA may well choose to order the raw materials and cook from scratch on board, or alternatively ask the caterer to supply ready for reheat.

To accompany the eggs would be either bacon and/or sausage with mushrooms and tomatoes or breakfast potatoes. In addition a selection of freshly squeezed fruit juices would be requested, mainly orange and grapefruit but often others such as mango, apple, peach or pineapple. All juices will be expected to be freshly squeezed by the purveyor and not pasteurised or shop bought (Table 14.1).

Having dealt with breakfast, the next meal service would be mid-flight snacks. The plethora of combinations here is potentially endless. Everything from airline-style crisps, peanuts and pretzels to incredibly high specification canapés and hors d'oeuvres, sliced cheeses and fruit to tea sandwiches and mini patisserie. The packaging dictates of any of the above could be a combination of preplated ready-to-serve and bulk packaged raw materials made and assembled by the flight attendant on board. Sandwiches assembled by the crew in flight, having been supplied in component form from the caterer, are a very

Table 14.1 Potential sources for breakfast items on a typical general aviation (GA) aircraft menu

Menu item	Required state	Potential source of supply	Potential state
Breakfast breads	Ambient	GA caterer/grocery store bakery/hotel	Ambient
Breakfast cereals	Ambient	GA caterer/grocery store hotel	Ambient
Milk/yoghurt	Fresh chilled	GA caterer/grocery store hotel	Chilled/ambient
Raw eggs	Fresh shell	GA caterer/grocery store hotel	Ambient/chilled
Cooked eggs poached/omelette	Cooked/chilled	GA caterer/hotel restaurant	Chilled/hot/warm
Raw meat bacon/sausage	Fresh chilled	GA caterer/hotel restaurant	Chilled/hot/warm
Freshly squeezed juice – orange/apple	Fresh chilled	GA caterer/grocery store hotel/restaurant	Chilled/ambient

Table 14.2 Potential sources for main meal items on a typical general aviation (GA) aircraft menu

Menu item	Required state	Potential source of supply	Potential state
Ambient snacks crisps/nuts/pretzels	Ambient	GA caterer	Ambient
		Grocery store	Ambient
		Hotel	Ambient
Cold canapés prawn/lobster/	Fresh chilled	GA caterer	Chilled/ambient
		Grocery store (raw materials only)	Chilled/ambient
Caviar		Hotel	Chilled/ambient
		Restaurant	Chilled/ambient
Cheese and fruit	Fresh chilled	GA caterer	Chilled/ambient
		Grocery store	Chilled/ambient
		Hotel	Chilled/ambient
		Restaurant	Chilled/ambient
Sandwiches	Fresh chilled	GA caterer	Chilled/ambient
		Grocery store	Chilled/ambient
		Hotel	Chilled/ambient
		Restaurant	Chilled/ambient

common request in order to ensure that presentation standards are kept as high as possible (Table 14.2).

The final meal service would most certainly comprise a hot lunch/supper option. The list is limitless but let us suppose for the sake of this example that it is a cold seafood starter, a hot fillet steak with a choice of sauces accompanied by vegetables and potatoes/rice, followed by ice cream and fruit (Table 14.3).

Table 14.3 Potential sources for dinner items on a typical general aviation (GA) aircraft menu

Menu item	Required state	Potential source of supply	Potential state
Seafood salad Lettuce/prawns/ lobster	Fresh chilled <5°C	GA caterer Grocery store (raw materials only) Hotel Restaurant	Chilled/ambient Chilled/ambient Chilled/ambient Chilled/ambient
Dressing	Ambient	GA caterer Grocery store (raw materials only) Hotel Restaurant	Ambient Ambient Ambient Ambient
Fillet steak (raw)	Fresh chilled <5°C	GA caterer Butcher Grocery store Hotel	Chilled/ambient Chilled/ambient Chilled/ambient Chilled/ambient
Fillet steak (cooked)	Cooked/chilled From 75°C to <5°C	GA caterer Grocery store Hotel Restaurant	Chilled Chilled Chilled Chilled
Vegetables Potatoes Rice	Cooked and chilled Raw	GA caterer Grocery store Hotel Restaurant	Chilled/ambient Chilled/ambient Chilled/ambient Chilled/ambient
Ice-cream	Frozen	GA caterer Grocery store Hotel Restaurant	Frozen on dry ice Frozen on dry ice Frozen on dry ice Frozen on dry ice

Tables 14.1–14.3 illustrate the potential sources of food supply for each item included on the 'typical' GA menu. The purpose of these examples is not to illustrate the risks attributable to each product at source of supply, but to help gain a firm understanding of the nature of the product and the broadest spectrum of the potential supply chain. Later in the chapter when we look at GA catering systems management utopia, the magnitude of the risk management issues in GA catering will become clearer.

The general aviation food chain explained

Figure 14.3 (adapted from *In-flight Food Safety Passenger Health and You* Erica Sheward 2000) illustrates the GA food chain from an operational perspective. Figure 14.4 (adapted from *In-flight Food Safety Passenger Health and You* 2000 Erica Sheward) illustrates the GA food chain from a purveyor's perspective.

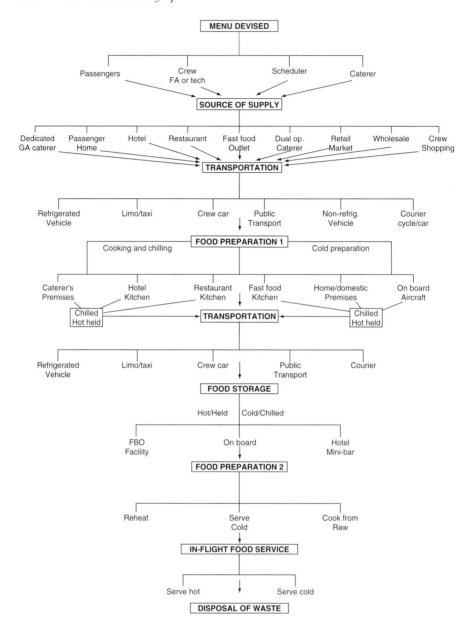

Figure 14.3 General aviation food chain from operational view.

Food safety in business aviation 349

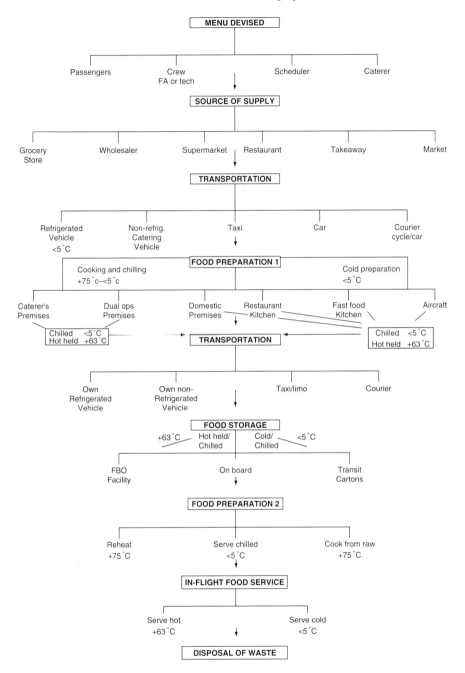

Figure 14.4 General aviation food chain from caterer's/purveyor's view.

Both these figures clearly illustrate the remarkable similarity between the supply, transportation and preparation components in each case, a situation I believe is almost unique to GA catering provision. To find such critical supply chain issues shared by both the aircraft operator and the food manufacturer is fascinating and crucial in establishing the appropriate basis for the risk management protocols required to assure safe catering parameters for both.

It is interesting to compare the GA food chain at this stage with that established within the same relationship framework in commercial aviation (Figures 14.5 and 14.6). The picture is vastly different and this is attributable to several key factors around which the difficulties of assuring product safety in GA hang.

First, there is no defined product specification against which to set quality and safety parameters; secondly, the supply chain is vast and diverse and established supplier relationships even at the home domicile of GA aircraft are rare; thirdly, those who contribute to the GA supplier network are small, at best dual operation and at worst not aviation-related businesses; finally, there is no industry standard or code of conduct by which GA aircraft operators and caterers abide, despite the incredibly high profile nature of the end users.

By combining the information represented by all the figures so far, a picture begins to emerge as to the extremely diverse nature and broad spectrum of the potential supplier base. Obviously an understanding of this is critical to accepting the huge potential for a variable level of product safety being delivered to the end user. Add to that the global outsourcing and replication requirement and the non-standard menu elements of the product, and the ability to dictate quality and safety parameters by the operator to the extended supplier base becomes seemingly impossible.

Flight crew impact on the general aviation food chain

By this part of the chapter, I would hope that even a complete newcomer to the concept of GA catering provision would have noticed that the major focus of all food safety risk management issues is the individuals who are charged with the responsibility for making crucial decisions as to what, where, how and by whom the catering will be provided.

For the most part this remains the role of the third crew member. I choose this term to describe them, rather than flight attendant or stewardess, because in the world of GA many of those who fulfil this incredibly responsible and

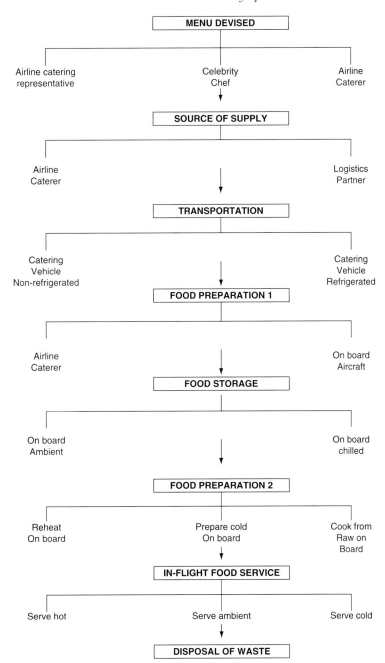

Figure 14.5 Airline food chain from operational view.

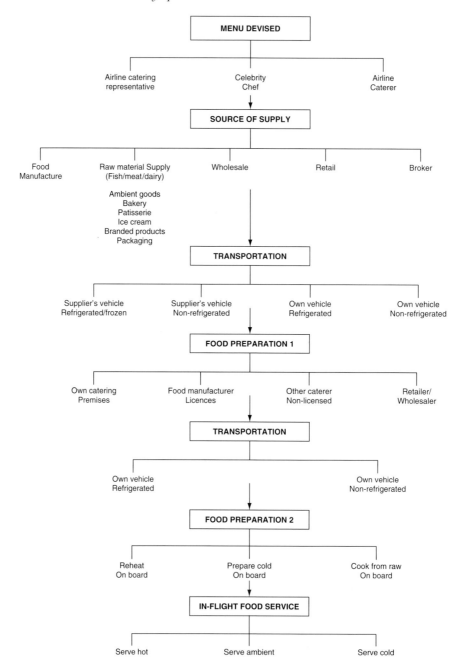

Figure 14.6 Airline food chain from catering/logistics view.

technical function are neither flight attendants nor stewards/stewardesses. A large proportion of those who procure catering services to GA are pilots or flight mechanics.

This startling fact is an acceptable normality in the GA world, as historically when GA was at its inception, aircraft types could not accommodate a third crew member of any description whether they were flight tech or hostess. Having become established as the acceptable norm early on, the industry has seen no reason to challenge such a protocol. Even in the wake of safety advancements in other areas of the cabin, and the legislative demand that no crew member involved in the technical operation of the aircraft leave the cockpit in-flight to attend anything other than a call of nature, the number of GA pilots and flight mechanics involved in in-flight food service is huge.

Quite apart from the food safety considerations of having a non-food professional taking care of catering procurement and provision, one has to consider the general aviation safety implications of technical crew members performing this function when they should be undertaking their duties in the cockpit. Our focus, however, is to look at the impact of industry practice on food safety management.

Having examined closely the manner in which technical crew make catering decisions versus the manner in which cabin crew make the same decisions, it is clear that the focus is very different. Tech crew view catering provision and procurement as not their primary function and as a result will leave many more decisions to third parties, albeit the handling agent, flight scheduler or the catering provider themselves. In this arena, never has the connection between critical source of catering supply and safety been less obvious to those with passenger safety in every other area squarely at the forefront of their minds!

The reasons for using technical as opposed to cabin crew to fulfil a catering function on board are many, but for the most part they focus on owner/operator preference, or the requirement of some large corporations which fly business aircraft to have flight mechanics on board at all times in case of a maintenance failure down route. If they are there, they may as well serve the food!

The likelihood of a flight technician or pilot possessing the food safety qualifications or expertise required to perform the task illustrated by the GA food chain effectively and safely, is very slim. In the absence of any formal requirement, the practice will continue to jeopardise the safety and integrity of the food service environment on board GA aircraft.

In any GA environment, the requirement of those fulfilling any part of the catering function, but especially those connected directly with procurement and/or service, must obtain the necessary knowledge and risk management experience to understand the impact on passenger and crew safety posed by the catering product.

General aviation food safety management utopia

With seemingly so many constraints on the industry's ability to apply fundamental food safety management protocols to the GA environment, it is not surprising that the industry goes unchecked and unregulated. Even those dedicated GA catering providers who have expressed an interest in doing so, are barely equipped with the resources or expertise to instigate the levels of product safety required. Any advancement and attempt at safety management systems standardisation is met with suspicion and fear by both the catering providers and the aircraft operators themselves. The perception is that with standardised operating procedures will come an unfavourable impact on the aesthetic quality of the product, resulting in airline-style standardisation and product downgrading which has no place in GA.

Having spent eight years as a service provider and advisor to the industry, I have witnessed at first hand the inherent difficulties in applying GMPs and GHPs within the GA framework of operation. What I have learned, however, is that by applying adaptations of mainstream food safety, quality assurance and systems management protocols, it is possible to deliver the same standard of food quality and safety without compromising GA customer demands.

For an industry which is so demanding in terms of the standard of aesthetic quality and service that it insists on, it is ironic that the level of demand for product safety is so unequivocal. It is left, therefore, to the dedicated GA catering service providers to set the safety bar, and where there are none, for the aircraft operators to consider the safety implications of the catering decisions that they make.

In the next section we look at the best practice scenarios required by both GA caterers and catering providers, as well as those required by aircraft operators to ensure product safety globally.

Having previously noted the extraordinary similarities between the aspects of the food chain over which both catering providers and aircraft operators have jurisdiction, (Figures 14.3 and 14.4), one would expect the safety picture to be similar. However, once the source of supply issues have been dictated by the operator, the divergence of risk management topics attributable to both catering providers and operators begins to look very different.

Catering providers – general factors influencing product safety

The product

- Absence of any defined product and product specification.
- Safety and quality parameters not defined owing to lack of specification.
- Short notice provision requirement and no definitive product list resulting in inconsistent raw material supply chain.
- Requirement for the product to fall within undefined food production parameters, e.g. product can be ambient/chilled/raw/high risk/low risk.
- Capacity for both the product itself and raw material outsourcing to be dictated by the customer. Items include branded products and finished goods produced by third parties outside quality and safety jurisdiction of the caterer.
- Inherent difficulties in product traceability to standards required to exercise appropriate levels of due diligence. Diverse product range and inconsistent supply chain contribute to this.
- Labelling techniques conducive to product traceability and product safety protocols, i.e. storage conditions, etc., non-existent.
- Requirement for food brokering by the customer without appropriate controls in place for the caterer.
- Requirement for water and ice to be included in the possible product requests, requiring the same level of safety management in the potential production and provision.

Premises

- Range of premises fulfilling the requirement of GA catering provision diverse and 'invisible'.
- Most premises perform dual operation function, i.e. they are also hotel kitchens, restaurant kitchens, outside catering or function caterers, can even be retail, wholesale, domestic or take-away food premises.
- At any given time the range of premises used to provide catering to a single aircraft within a single order is wide.
- Premises design and specification may not lend itself to the safe action of all food production methods required to accommodate the safe manufacture of the product.
- Location of premises may impact on other areas of safety in the supply chain, in terms of transportation issues or raw material supply issues.
- Capability of premises to conform to related prerequisite programmes, which would class as GMPs or GHPs.

- Size of catering premises dictated by the level of business activity, which is inconsistent and unpredictable in most markets, results in overburdening in many cases at times of high activity.

Personnel

- Product requirement dictates that food production personnel are multifaceted and engaged in both low and high-risk applications simultaneously.
- GA catering personnel are most commonly recruited from the restaurant and hotel sector with little experience of manufacturing standard GMPs.
- Medical and criminal background checks are essential to ensure ultimate product safety to the high profile clientele.
- Packaging dictates are complex and labelling requirements critical to ensuring product safety. Personnel have the capacity to compromise these if a full and comprehensive understanding of the end user is not apparent.
- The global outsourcing requirement allows for language barriers to play a part in product interpretation and production.
- Qualifications in establishing the required food safety and quality management protocols must be apparent in key members of personnel and those in a supervisory capacity.

Transportation

- Requirement for extended levels of transportation to fall under caterer's jurisdiction.
- Multifaceted nature of the product requiring ambient/chilled/frozen transportation capabilities.
- Inconsistent raw material supply chain resulting in varying levels of raw material transportation, often dictating that standards of supply fall outside critical limits.
- Unpredictable nature of business activity and no set schedules resulting in appropriate transport not being available and secondary or third party sources of transportation being utilised.

Food preparation

- The diverse nature of the product requires the premises and personnel to be capable of a number of food preparation processes simultaneously: hot prep, cook chill, cold prep, fresh juice and ice manufacture, sandwich and salad preparation.

- Temperature controlled and segregated preparation processes are a requirement but not likely to be evident among the typical supplier base.
- The diverse nature of the product dictates that the preparation specification can vary from plated and prepared to unprepared and unpackaged.
- Packaging requirements range from disposable, rotable, plastic and ceramic to fine china and glassware.
- Much of the packaging used will not be of food grade quality. Knowledge of what does or doesn't constitute food grade packaging unlikely to be found among GA catering providers.

Food storage

- In the same way that size of premises considerations are dictated by the activity levels of the individual providers, storage capabilities are subject to the same conditions.
- High levels of activity result in storage levels being overburdened and secondary storage facilities being used, whether appropriate or not.
- Chilled food storage areas often contain cooked and raw materials side by side.
- Raw material storage equipment may not be of industrial quality, particularly among small providers.
- Storing food in third-party facilities renders the product subject to tampering and abuse. Safety management protocols should eliminate interim storage processes, e.g. fixed base operator (FBO) facilities.

Refuse disposal and dish-wash

- GA catering providers involved in a de-cater function must consider waste food disposal protocols, particularly if disposing of waste food from aircraft arrived from outside the home domicile.
- The nature of the operation requires that any dish-wash activities meet required safety standards and that the action can be validated effectively.
- Inappropriate regard for waste food disposal protocols will potentially impact the wider food chain, whilst ineffective dish-wash procedures have the capacity to impact pax safety.
- Agreeing to hold over foods from incoming aircraft must be viewed in terms of the impact on the wider food chain and the transgression of any Port Health requirements.

Consideration must be given to the impact on the caterer's in-house food chain of storing unknown high risk foods and stores in the same areas as their own raw materials.

Conclusions

From an analysis of the GA food chain comes a heightened appreciation of the advanced level of food safety management issues facing GA caterers. In understanding what constitutes GA food safety management utopia, an extended and in-depth process flow (Figure 14.3) illustrates a given knowledge of factors influencing the end user, which the caterer must consider when attempting to build in safety features at the production/preparation stages of the supply chain.

By acknowledging an association and connection with the food service environments of the client base, product safety considerations can be made in the same way as aesthetic quality and packaging considerations are made in tandem with the clients' demands.

Aircraft operators – general factors influencing product safety

The product

- Absence of any defined product and product specification.
- Safety and quality parameters not defined due to lack of specification.
- Requirement for global replication of product despite absence of specification.
- Short notice production requirement and no definitive product list resulting in unknown and inconsistent supply chain.
- Diversity of what constitutes the product resulting in a supply chain as eclectic as the product itself. Supply chain may consist of ten different contributors to one order.
- Requirement for the product to fall within undefined production parameters leading to a broad supply chain which may not be equipped to carry out the function safely, e.g. cook chill product requirement outsourced from a hot food take-away.
- Requirement for the product to be a branded component not suitable for the purpose, e.g. Kentucky Fried Chicken.
- Water and ice supply included in the product, requiring the same levels of safety consideration and management.

The product provider

- No defined group of dedicated GA catering providers with attributable, appropriate and recognised food safety management protocols commensurate with the product risks.

- Diverse nature of the product dictates providers can be numerous and multitasked, or brokers.
- Lack of industry guidelines allows for catering providers to remain invisible and unregulated in terms of appropriate expertise and standards.
- Even dedicated or dual operation providers have no minimum requirements for food safety management systems analysis and expertise.
- Source of supply often unknown owing to orders being placed through third parties.
- Global replication of product requirement allows for inappropriate components, e.g. hot food service branded goods, to compromise safety in the GA food service environment.

The crew

- Capacity for in-flight service personnel to have no hygiene qualifications commensurate with the risks inherent in the role they play in the GA food chain.
- Crew jurisdiction over sources of catering supply.
- Crew jurisdiction over menu development and delivery in-flight without the qualifications necessary to understand the impact on pax safety and security.
- Fitness to fly – does the nature of the job render the crew naturally predisposed to heightened food poisoning risk and therefore render them unfit food handlers?
- The propensity for catering procurement decisions to be made without taking adequate food safety protocols into account due to lack of knowledge and safety awareness.
- Tech crew propensity to place catering decisions in the hands of third parties.
- Capacity for poor food safety management decisions to impact not just on the pax but directly on the crew themselves, rendering them unfit to fly.

Transportation

- Requirement for the multifaceted nature of the product to be transported under various conditions – chilled, ambient, frozen, hot held, etc.
- Jurisdiction for catering transportation divided between catering provider and, where wholesale, retail or take-away components have been sourced by the crew, the operator.
- Capacity to transport components under the correct storage conditions unlikely.
- Global requirement for outsourcing of the product to impact on transportation issues, particularly taking account of the likely geographical location of the airfield.

- Where branded hot food take-away components are required to form part or whole of the product, the capacity to transport and hot hold until service is unlikely to be realised.

In-flight storage

- The diverse nature of the product requires that in-flight storage capabilities should include chilled (0–5 °C), reheating capacity (+75 °C), hot holding capability even with no ground power (+63 °C), segregated storage for high risk foods, ice holding capacity for both clean and dirty ice and ambient storage facilities that do not require the product to held in situ on the floors or lavatories.
- Diverse and potentially high risk nature of the product is not conducive to safe storage considering the likely quantities of food required to be stored and the chilled space available.
- Potentially ineffective methods of chilled storage available.
- Potentially non-existent methods of chilled storage space available.
- Likely overspill storage areas designated the lavatories or floors.
- Catering stored in interim facilities, i.e. FBOs, pre- and post-flight, render the high risk, high profile nature of the product open to issues of tampering and temperature abuse.

In-flight service/preparation

- Diverse nature of the product requiring a restaurant-style service delivery renders space on board overburdened and ill equipped for the purpose.
- Majority of heating equipment uncalibrated and directly linked to ground power requirements.
- No effective surface preparation segregation.
- Service requirement for fine glassware and china poses breakage issues in both the cabin and the cockpit, with the capacity to physically contaminate.
- Requirement for crew to be food service not food safety professionals first and foremost, renders hygiene awareness limited among those preparing, heating and serving the food.
- Where technical crew are engaged in food service, additional training will be required.
- Unavailability of dedicated hand-washing facilities in the galley requires risk management protocols to be established, particularly where the bathroom hand-wash is located in the main cabin.

Refuse disposal and dish-wash

- Diverse and mainly high risk nature of the product renders it unfit for consumption at the conclusion of the flight. Consideration of the storage conditions pre- and post-flight must be taken.
- Absence of product definition and specification suggests the existence of scientifically applied 'use by' attached to the product is unlikely.
- Absence of product definition and specification leaves purveyors open to product life abuse and compromise.
- Absence of adequate in-flight waste storage may result in food waste being stored in food service receptacles, e.g. ice bins/drawers and buckets.
- Holding over catering supplies, even in transit, can result in violations of certain international Port Health Regulations. The safety of the food chain may be compromised by unauthorised disposal of waste food post-flight.
- High risk, diverse and untraceable nature of the product renders it a danger to the food chain. Consideration of the interstate transit of food regulations has to be given for GA travel within the USA.
- Engaging non-regulated catering suppliers in the de-catering function may pose a risk to the food chain.
- On-board dish-wash will require the on-board provision of water in excess of 83°C.
- If outsourcing dish-wash to catering purveyor or FBO, consideration of the hygienic practices involved has to be given to avoid dish contamination.

Conclusions

It is extremely unusual to witness such similar connections made to the food chain between both the catering purveyor and the aircraft operator. In the case of the airlines and the equivalent relationship they share with their catering partners, the picture is completely different. Food chain issues impacting on each vary markedly (Figures 14.5 and 14.6) whilst safety protocols should be akin.

In the world of GA, food chain issues are almost identical whilst the safety protocols required are at opposite ends of the spectrum. The safety principle are the same but the differing SOPs of the provider and operator are what cause the safety management divergence.

Having looked at the generalised factors influencing product safety in the GA environment, we now turn to the specific systems management framework upon which catering and operational safety in the GA arena will hang.

Catering HACCP in the general aviation environment

In advance of any detailed look at the manner in which appropriate HACCP protocols can be established in GA catering operations, our focus must begin with the prerequisite programmes and quality management systems required for effective operation and process control.

Previously, having gained an understanding of the nature of both the likely product and likely provider, the basis of the HACCP plan must first be underwritten by the prerequisite programmes. Historically, in the GA catering network arena, these have been shaky to begin with.

> 'The World Health Organization defines prerequisite programmes as, "Practices and conditions needed prior to and during the implementation of HACCP and which are essential for food safety." Many would class these as good GMPs (good manufacturing practices) or GHPs (good hygiene practices).'[124]

In short, prerequisite programmes provide an essential support framework upon which to hang a HACCP system and are fundamental to the success that hazard management and process controls achieve, through HACCP.

Prerequisite issues

The following constitutes a general list of prerequisite programme considerations attributable to GA catering providers.

Facilities

- The location, layout and structure of the food premises, internal and external – easy to clean, located so that drainage and water supply is of hygienic quality and direct to mains, ventilation should not be compromised by 'dirty' surrounding air, i.e. adjacent to an airport without appropriate filtration.
- The provision of adequate hand-wash, changing facilities and toilets.
- Design of equipment – hygienic and easy to clean.
- Drainage and waste management system in place for storage and disposal. Drainage should be kept clean and flow should be such that cross-contamination from clean to dirty areas does not occur.
- Air quality and ventilation – meets food industry standards of air change and flow to preserve food quality and ensure personnel health and safety.
- Appropriate lighting.
- Water supply – meets microbiological standards.
- Pest control – contract in place to control all pests.

- Temperature controlled storage, preparation and transportation.
- Food container design – food grade quality.

Operational

- Time and temperature control – during preparation, storage and transit.
- Raw material supply controls – audit and product specification required.
- Cross-contamination – physical, microbiological, chemical.
- Packaging material control – food grade quality.
- Water quality when used as an ingredient and for cleaning purposes.
- Personnel – hygiene and training.

All of the above operational and facilities considerations have to be made to ensure the safe and effective running of any GA food premises. In the absence of any one of the above conditions, the realisation of an effective HACCP system is unlikely. Managing hazards without the benefit of sound prerequisite standards, facilities or operation will be extremely difficult. Even when hazards are identified, the ability to monitor and control is severely compromised by the unhygienic foundation of the business.

By referring back to the issues raised in the earlier section, 'Catering Providers – General Factors Influencing Product Safety', it is clear that many prerequisite programme fundamentals are missing from the procedures of a large proportion of potential providers of GA catering. What this means for the effective implementation of HACCP in such environments is that hazards, even if identified, will not be able to be controlled. It is essential therefore that GA catering providers consider prerequisite issues as paramount before embarking on the implementation of a HACCP system. The benefit of focusing attention on general rules for hygienic operation first, before embarking on any further quality management steps, is that GHPs will begin to come as second nature, making the next steps into hazard awareness, identification monitoring and process controls much easier to take.

Quality management systems

Quality management systems are primarily focused on ensuring that customer expectations and legal compliance standards are met. In tandem with HACCP they aim to prevent product non-conformities, with the focus being corrective action as opposed to hazard analysis and an attempt to get it right first time.

It is important to be clear at this stage that a QMS is *not* a prerequisite programme, but may prove useful in managing both the HACCP and prerequisite systems in terms of issues raised to ensure quality, and how they impact on hazard identification and GHPs.

If one imagines creating a HACCP plan as a series of blocks, which put together form the foundations of a situation where total food safety is achievable, then the blocks would consist of:

- prerequisite programmes
- quality management system
- HACCP.

The quality management system in any business, but particularly in GA catering operations, can be used to draw a clear distinction between CPs and CCPs in any product process. In this way quality assurance, aesthetic conformity and legal compliance standards are identified in the QMS as CPs, whilst those aspects critical to product safety are identified as CCPs.

Figure 14.7 illustrates the relationship between CPs and CCPs within the QMS framework.

Before commencing, therefore, with the development of an in-depth HACCP plan, the status of both the prerequisite systems and the QMS within the business must be established by the GA caterer. Let us not forget the likely heritage of these catering providers and the chances that their expertise in this arena is likely to be limited. Knowing what we do about the nature of the product and its processes, these too will also have a bearing on the ability of each purveyor to establish the status of the HACCP plan which needs to evolve, and their potential ability to achieve it.

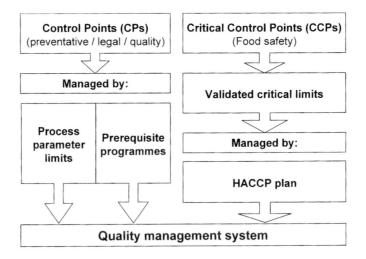

Figure 14.7 Control point differentiation (from Mortimore & Wallace[125]).

Suitability of HACCP approaches to general aviation catering

The next step is to understand the type of HACCP systems which need to be built, alongside the QMS and prerequisite programmes. This decision will be based entirely on the operational nature of the business type:

- Dedicated GA caterers.
- Restaurants.
- Hotels.
- Dual operation businesses – retail/GA caterer; restaurant/GA caterer; outside caterer/GA caterer.

At this stage we are leaving out operational sources of supply, i.e. flight attendant, flight department, pilots and flight mechanics, as we look at those in the last section in this chapter.

My experience in assisting caterers involved in GA catering provision to implement both a QMS and HACCP plan, has led me to witness the fear and confusion that many experience as they attempt to formally qualify and quantify their product processes and protocols in a structured, managed system of operation which has previously been totally alien to them.

Deciding on the correct type of HACCP plan is critical. In the GA provision area of my own business we developed a strategy for a combination of plans which met the diverse needs of the product requirement and production methods. By looking at the options first, we can see that many of the traditional approaches cannot accommodate the business needs entirely, and indeed I believe this is true of many catering-based operations. What becomes clear later is that any deviation from the manufacturing-based, systematic approach protocols results in an incomplete identification of all the processes and hazards involved. It is by an amalgam of plan protocols that the optimum plan is achieved. More of that later in the chapter.

Plan 1: The linear approach

The linear approach is based on applying HACCP ideals to each individual product or process, i.e. beginning with raw material receipt and ending with the finished product. This would work well if the operation was focused simply, the range of products was small and similar in type and the production process flow involved relatively few steps.

Conclusion

This simple approach is unlikely to be effective in a GA environment in isolation, unless the purveyor was a specialist provider of one type of product or of several types of product undergoing similar processes, i.e. morning goods, Danish, bakery or cookies and cakes. The chances are that this approach might work well applied to the product process flow, but would likely not account for the transit and extended storage activities typical of GA catering provision.

Attempting to adopt the linear approach in isolation, in an environment with such an undefined range of products, would be difficult.

Plan 2: The modular approach

The modular approach is most effective in environments where a variety of basic processes are utilised to produce a number of different products. A good example is sandwich-making. The basic process will be the same, however much the types of filling and the manner in which the fillings are manufactured will vary. The assembly of each different type of sandwich will have a combination of different processes or modules which, when combined, will form the complete picture. In this case several different processes will have HACCP principles applied to them and together they will combine to form the whole picture (Figure 14.8).

Conclusion

The modular approach could work well in a GA catering environment which was dedicated to that activity alone. The work involved would be great, however, in order to incorporate all the basic processes. A modular plan for hot/cold and hot hold activities would have to be formulated independently, underwritten by advanced standards of prerequisite programmes. Adopting this approach, however, would satisfy the need to include the extended nature of the supply chain and identify the transportation and extended storage risks inherent in all GA catering provision.

Plan 3: The generic approach

The generic approach is the one with which GA catering providers are probably most familiar, but it is the one which, if used in isolation, is likely to be the least effective. It involves one basic plan being formulated, which is effective only if there are one or two primary processing techniques being undertaken. The generic nature of the plan is designed to allow for similar activities to be undertaken at a variety of locations. Clearly this is not the case in the provision of GA catering unless one is a single product provider, which is unlikely. For this plan to be most effective, both raw materials and processes would have to be the same, again very unlikely in a GA catering provider.

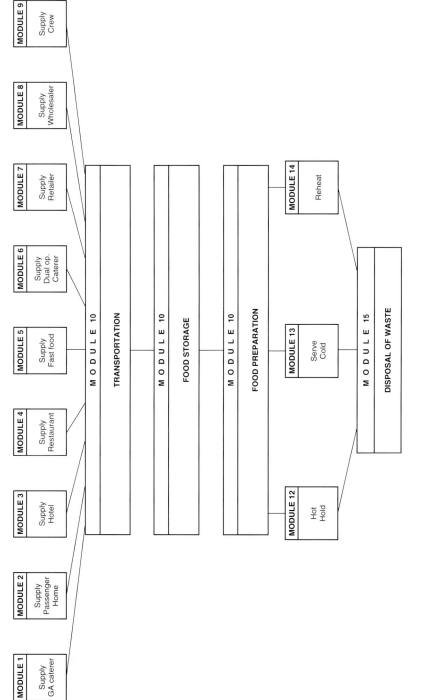

Figure 14.8 General aviation supply chain process flow.

Conclusion

Generic plans are arguably better than nothing and among GA catering providers whom I have worked with who claim to have a HACCP system in place, most commonly this is what I have found. However, in isolation and with the prerequisite programmes for the most part unestablished, they will not prove effective in such a multifaceted, multitasked business environment.

It is foolish to underestimate the hard work and commitment involved in establishing a HACCP system which *really* does what it needs to do in any catering environment. This is why systems for caterers like Assured Safe Catering (ASH)[126] were introduced in the UK in the early 1990s, to try to bridge the gap between basic risk assessment techniques and the more advanced qualitative risk assessment techniques established through HACCP.

However, my feeling is that this job is made so much easier by taking on board manufacturing standard HACCP techniques from the outset. Even in this small-volume catering-based environment, the amount of manufacturing process controls is overwhelming. How many catering businesses that are true catering businesses are involved in such a vast array of preparation techniques and are involved in such a proliferation of raw material outsourcing? As we have seen with the airline manufacturing activities, the only way to proceed effectively and safely in the GA environment is to adopt the advanced levels of prerequisite systems and HACCP programmes typical of a large manufacturer, and adapt them.

The next section illustrates some examples of basic paperwork templates that could be employed by any GA catering provider to verify the systems management and process controls.

Operational HACCP in the general aviation environment

To attempt to apply HACCP protocols to an ostensibly non-food group seems a strange concept until one reminds oneself of the direct connection with the food chain that GA operators have (Figure 14.3).

We have already examined the situation where GA operators and crews make critical decisions about the supply chain, which impact directly on their responsibilities and liabilities in this arena. By choosing to undertake the catering function themselves, either by catering directly or shopping for raw materials which are then prepared and served in-flight, they unknowingly become the catering provider and as such liable for food safety deficiencies in this regard.

The food chain diagram also illustrates the propensity of GA operators to use restaurants and hot food take-aways as sources of food supply, a practice which, in the absence of any advance knowledge of the food safety risks inherent in these activities when one utilises the products outside of the areas for which they were developed, can prove fraught with risk.

The question at this point is how does one develop a HACCP plan for GA operators, who have little or no knowledge of basic food safety principles and concepts to begin with? The answer is that a simple risk assessment structure needs to be established which can be utilised by a variety of personnel wherever they are in the world.

If one looks at the supply chain, a modular format can be used to demonstrate the risks attributable to the supply processes, and then a building of separate plans to illustrate the specific risks associated with the product group attributes, can occur.

When I have been involved in the training of tech and operational crew in this regard, we have begun by looking at the basic principles of risk assessment and applied them to the purchasing process (see Table 14.4 taken from 'In-Flight Food Safety Passenger Health and You 2001') with which they are involved.

Table 14.4 Basic risk assessment for hazards associated with purchasing catering for a general aviation aircraft

Source of supply	Potential hazards	Action/Control CCP	Monitoring
Commercial airline caterers			
Bespoke bizav caterers			
Department stores/ supermarkets/food halls/street markets			
Restaurants/hot food takeaways/ hotels			
FBO in-house catering facilities			
Pax bringing on food			

Whilst the source of supply is not the only aspect of operational activity that needs to be considered, it is by far the most critical and forms the foundation upon which all other GA food safety protocols will be based.

The next aspect of the operational supply chain, which needs a separate management system, is the in-flight preparation and service aspect. The extent of the in-flight meal service will dictate the complexity of the systems required. However, temperature control monitoring and paperwork are critical in any scenario, given the 'stone-age' nature of the chilled storage space and limited capacity of the preparation environments. Figure 14.8 also illustrates the service aspects of the process, which require systems attributable to them.

The following list shows examples of the sort of protocols that need to be established by aircraft operators to bring the risk assessment techniques to fruition:

- Supplier audit questionnaire where the supplier is known and unknown.
- Goods receipt paperwork for all raw materials at any point of receipt.
- In-flight temperature monitoring paperwork.
- Equipment calibration paperwork.
- Crew health audit.
- Microbiological verification of water supply.
- Passenger profiling documents to record special dietary requirements and allergies.
- Supplier security declarations.

Finally, it is crucial for GA aircraft operators to acknowledge and embrace the huge range of food safety and security issues over which they have jurisdiction. Whatever the catering supply decisions made, the responsibility for in-flight service remains theirs alone. A few simple steps and appropriate training of all those who have a connection with catering procurement and service is a must if passenger and crew safety in the GA arena is to be assured.

References

1. United Kingdom Parliament (10 July 2003) *Transport Sixth Report*. Transport Committee Publications. www.publications.parliament.uk
2. World Tourism Organization (WTO) (2000) *Tourism Market Trends 2000*. www.world-tourism.org
3. The late Captain Glen Stewart (1991) Personal letter to Erica Sheward.
4. Kaferstein, F. (2000) Food-borne disease as related to travellers – a public health challenge. First NSF International Conference and Exhibition, *Food Safety In Travel And Tourism*, 12–14 April 2000. Conference Proceedings, pp. 1–8.
5. Jones, P. & Kipps, M. (eds) (1995) *Flight Catering*, pp. 105–19. Longman, London.
6. Codex Committee On Food Hygiene (1993) Code of Hygenic Practice For Precooked and Cooked Foods in Mass Catering, CAC/RCP 39-1993, section 1.1 p. 3. In: Codex Alimentarius Commission *Food Hygiene Basic Texts*, Food & Agriculture Organization of the United Nations, World Health Organization, Rome.
7. Pakkala, P. (1989) Ruokamyrkytysten Aiheuttamat Kustannukset. (Costs caused by food poisoning outbreaks.) In: *Proceedings of Annual Meeting of Finnish Veterinary Association*, Helsinki, pp. 125–33.
8. Williams, R. B., Morley, L. A. & Kohler, M. (1950) Food-borne typhoid outbreak with rapid dissemination of cases through air transportation. *Northwest Med.*, 49, 686–9.
9. Munce, B. A. (1978) Microbiological hazards of airline catering. *Food Technol. Aust.*, pp. 470–76.
10. Munce, B. A. (1986) *Salmonella* serotypes from international hotel and airline food. In: *Proceedings of 2nd World Congress on Foodborne Infections and Intoxications*, pp. 705–8. Institute of Veterinary Medicine, Robert von Ostertag Institute, Berlin, Germany.
11. Tauxe, R. V., Tormey, M. P., Mascola, L, Hargrett-Bean, N. T. & Blake, P. A. (1987) Salmonellis outbreak on transatlantic flights: food-borne illness on aircraft 1948–1984. *Am. J. Epidemiol.*, 125, 150–57.
12. Svensson, C. (1998) Matforgiftningar Pa Flygplan. (Food poisoning on passenger aircrafts.) *Svensk Vet. tidning.*, 50, 745–52.
13. Bottiger, M. & Romanus, V. (1977) Salmonellautbrott Bland Flygresenarer Fran Paris. (Outbreak of *S. brandenburg* in air passengers from Paris). *Lakartidningen.*, 74, 2507–8.
14. Burslem, C. D., Kelly, M. J. & Preston, F. S. (1990) Food poisoning – a major threat to airline operations. *J. Soc. Occup. Med.*, 40, 97–100.
15. WHO (1999) Worldwide increase in cholera cases in 1998. *Weekly Epidemiol. Rec.*, 74, 1–3.
16. Hatakka, M. (2000) *Hygienic quality of foods served on aircraft*. Academic dissertation, Department of Food and Environmental Hygiene, Faculty of Veterinary Medicine, University of Helsinki, Finland.
17. Jahkola, M. (1989) *Salmonella enteritidis* outbreak traced to airline food. WHO Surveillance Programme for Control of Food-borne Infections and Intoxications in Europe. *Newsletter*, No. 22, p. 3. Institute of Veterinary Medicine, Robert von Ostertag Institute, Berlin, Germany.
18. Lambiri, M., Mavridou, A. & Papadakis, J. A. (1995) The application of hazard analysis critical control point (HACCP) in a flight catering establishment improved the bacteriological quality of meals. *J. Roy. Soc. Health*, February, 26–30.
19. De Jong, B. (1998) Salmonellautbrott Bland Hemvavnande Turister Fran Kanarieoarna. *Smittskydd*, 1, 6–7.

20. Centers for Disease Control and Prevention (CDC) (1961) Staphylococcal food poisoning on a trans-Pacific airline – Hawaii. *Morbidity and Mortality Weekly Report (MMWR)*, 10, 8.
21. CDC (1973) Staphylococcal food poisoning aloft. *MMWR*, 22, 381–2.
22. Eisenberg, M. S., Gaarslev, K., Brown, W., Horwitz, M. & Hill, D. (1975) Staphylococcal food poisoning aboard a commercial aircraft. *Lancet*, 595–9.
23. CDC (1976) Outbreak of staphylococcal food poisoning aboard an aircraft. *MMWR*, 25, 317–18.
24. Socket, P., Ries, A. & Wieneke, A. (1993) Food poisoning associated with in-flight meals. *Communicable Dis. Rep*, 3, 103–4.
25. Sutton, R. G. A. (1974) An outbreak of cholera in Australia due to food served in flight on an international aircraft. *J. Hyg. Camb.*, 72, 441–51.
26. Dakin, W. P. H., Howell, D. J., Sutton, R. G. A., O'Keefe, M. F. & Thomas, P. (1974) Gastroenteritis due to non-agglutinable (non-cholera) vibrios. *Med. J. Aust.*, 2, 487–90.
27. Desmarchelier, P. (1978) Vibrio outbreaks from airline food and water. *Food Technnol. Aust.*, December, 477–81.
28. Eberhart-Phillips, J., Besser, R. E., Tormey, M. P., Feikin, D., Araneta, M. R., Wells, J., Kihlman, L., Rutherford, G. W., Griffin, P. M., Baron, R. & Mascola, L. (1996) An outbreak of cholera from food served on an international aircraft. *Epidemiol. Infect.*, 116, 9–13.
29. Oden-Johanson, B. & Bottiger, M. (1972) Erfarenheter Fran Utbrotten av Shigella Sonnei Dysenteri Hosten 1971. (The outbreaks of Shigella sonnei dysentery in the fall of 1971.) *Lakartidningen*, 69, 3815–17.
30. CDC (1971) Gastroenteritis aboard planes. *MMWR*, 20, 149.
31. Hedberg, C. W., Levine, W. C., White, K. E., Carlson, R. H., Winsor, D. K., Cameron, D. N., MacDonald, K. L. & Osterholm, M. T. (1992) An international food-borne outbreak of Shigellosis associated with a commercial airline. *JAMA*, 268, 3208–12.
32. CDC (1969) Acute gastroenteritis among tour groups to the Orient – United States. *MMWR*, 18, 301–2.
33. Lester, R., Stewart, T., Carnie, J., Ng, S. & Taylor, R. (1991) Air travel-associated gastroenteritis outbreak, August 1991. *Communicable Dis. Intelligence*, 15, 292–3.
34. CDC (1994) Foodborne outbreaks of enterotoxigenic *Escherichia coli* – Rhode Island and New Hampshire, 1993. *MMWR*, 43, 81–9.
35. Preston, F. S. (1968). An outbreak of gastroenteritis in aircrew. *Aerospace Med.*, 39, 519–21.
36. Mossel, D. A. & Hoogendoorn, J. (1971) Prevention of food-borne diseases in civil aviation. *Ind. Med.*, 40, 25–26.
37. CDC (1991) Cholera – Peru, 1991. *MMWR*, 40, 108–10.
38. CDC (1992) Cholera – Western Hemisphere, 1992. *MMWR*, 41, 667–8.
39. Chalmers, J. W. & McMillan, J. H. (1995) An outbreak of viral gastroenteritis associated with adequately prepared oysters. *Epidemiol. Infect.*, 115, 163–7.
40. Leeds, D. N., Henshilwood, K., Green, J., Gallimore, C. I. & Brown, D. W. G. (1995) Detection of small round viruses in shellfish by reverse transcription-PCR. *Appl. Environ. Microbiol.*, 61, 4418–24.
41. Taylor, M. (2004) Passengers grounded by food poisoning outbreak. *The Royal Gazette, Bermuda* (www.royalgazette.com), 26 October.
42. Hardiman, M. (2004) Senior Advisor, Communicable Disease Surveillance and Response (CSR), World Health Organization (WHO). IATA Cabin Health Conference, Geneva, June 2004.
43. WHO (1974) International Health Regulations, Article 14.2 (http://policy.who.int) second annotated edition. WHO, Geneva.
44. WHO (1974) International Health Regulations, Article 14.3 (http://policy.who.int) second annotated edition. WHO, Geneva.
45. IATA (2002–2003) *Inflight Management Manual* 2nd edition, section 7, pp. 65–92. Ref No. 9347-02. IATA, Montreal and Geneva.
46. Chicago Convention on International Civil Aviation (Chicago Convention) 7 December 1944, Article 44, 15 UNTS 295; ICAO Doc. 7300/5.

47. Chicago Convention on International Civil Aviation (Chicago Convention) 7 December 1944, Article 47, 15 UNTS 295; ICAO Doc. 7300/5.
48. Center for Science in the Public Interest, Codex Alimentarus Commission (CAC)/RPC 39-1993 Code of Hygiene Practice for Pre-cooked Foods in Mass Catering. CSPI, Washington.
49. International Air Transport Association (IATA) *Inflight Management Manual*, Standard Catering Services Agreement, 2nd edition, effective 1 July 2002 – 30 June 2003, Article 8.1, p. 79.
50. Statutory Instrument 2002/1817. The Food for Particular Nutritional Uses (Addition of Substances for Specific Nutritional Purposes) (England) Regulations 2002, Section 2-1a and 2-1b.
51. WHO (1999) Strategies for Implementing the Hazard Analysis and Critical Control Point (HACCP) in Small and/or Less Developed Businesses. WHO/SDE/FOS/99.7. WHO, Geneva.
52. Mortimore, S. & Wallace, C. (2001) *HACCP*, p. 26. Food Industry Briefing Series, Blackwell Publishing, Oxford.
53. Mortimore, S. & Wallace, C. (2001) *HACCP*, Fig. 8, p. 26. Food Industry Briefing Series, Blackwell Publishing, Oxford.
54. Mortimore, S. & Wallace, C. (2001) *HACCP*, Fig. 11, p. 32. Food Industry Briefing Series, Blackwell Publishing, Oxford.
55. Mortimore, S. & Wallace, C. (2001) *HACCP*, p. 39, Fig. 14. Food Industry Briefing Series, Blackwell Publishing, Oxford.
56. Mortimore, S. & Wallace, C. (2001) *HACCP*, p. 41. Food Industry Briefing Series, Blackwell Publishing, Oxford.
57. Mortimore, S. & Wallace, C. (2001) *HACCP*, pp.44–5. Food Industry Briefing Series, Blackwell Publishing, Oxford.
58. Mortimore, S. & Wallace, C. (2001) *HACCP*, pp. 59–61. Food Industry Briefing Series, Blackwell Publishing, Oxford.
59. Codex Committee on Food and Hygiene (1997) HACCP System and Guidelines For Its Application, Annex to CAC/RCP 1-1969, Rev 3 in Codex Alimentarius Commission *Food Hygiene Basic Texts*, Food and Agriculture Organization of the United Nations, World Health Organization, Rome.
60. Mortimore, S. E. & Wallace, C. A. (1998) *HACCP*, p. 83, Fig. 17. Food Industry Briefing Series, Blackwell Publishing, Oxford.
61. Mortimore, S. & Wallace, C. (2001) *HACCP*, pp. 82–3. Food Industry Briefing Series, Blackwell Publishing, Oxford.
62. IFCA/IFSA World Food Guidelines (2002) http://www.ifcanet.com/teams/food safety/
63. Proceedings of the Convention for the Unification of Certain Rules Relating to International Carriage by Air (Warsaw Convention), signed at Warsaw on 12 October 1929; ICAO Doc. 7838, 9201.
64. Proceedings of the Protocol Signed at The Hague on 28 September 1955 to amend the Convention for the Unification of Certain Rules Relating to International Carriage by Air signed at Warsaw on 12 October 1929; ICAO Doc. 7632.
65. Guadalajara Convention Supplementary to the Warsaw Convention for the Unification of Certain Rules Relating to International Carriage by Air performed by a person other than the contracting carrier (Guadalajara Convention), 18 September 1961; 500 UNTS 31; ICAO Doc. 8181.
66. Montreal Agreement (1966) Relating to Liability Limitations of the Warsaw Convention and The Hague Protocol (Montreal Agreement), 4 May.
67. Guatemala City Protocol to amend the Warsaw Convention (Guatemala Protocol) 8 March 1971; ICAO Doc. 8932/2.
68. The Proceedings (1975) of the additional protocols Nos 1, 2, 3 and 4 signed at Montreal on 25 September to amend the Convention for the Unification of Certain Rules Relating to International carriage by Air signed at Warsaw on 12 October 1929. (Montreal Protocols). ICAO Docs 9145, 9146, 9147, 9148.

69. Agreement between the Government of the United Kingdom of Great Britain and Northern Ireland and the Government of the United States concerning air services (Bermuda II). *Air Law* (1977) Vol. II, p. 194.
70. Batra, J. C. (2003) *International Air Law*, p. 44. Reliance Publishing, New Delhi.
71. Proceedings of the Montreal International Diplomatic Conference on Air Law 1999 which adopted the Convention for the Unification of Certain Rules for International Carriage by Air on 28 May 1999. ICAO Doc.9740; DCW Doc. No 57 28/5/99.
72. Proceedings of the Convention for the Unification of Certain Rules Relating to International Carriage (Warsaw Convention) by Air, signed at Warsaw on 12 October 1929. ICAO Doc. 7838, 9201, Articles 17.
73. Bailey, J. (1977) *Guide to Hygiene and Sanitation in Aviation*, 2nd edition, p. 24. WHO, Geneva.
74. Buley, L. E. (1969) Incidence, causes and results of airline pilot incapacitation while on duty. *Aero. Med.*, 40 (1), 64–7.
75. Kulak, L. L., Wick, R. L. Jr & Billings, C. E. (1971) Epidemiological study of in-flight airline pilot incapacitation. *Aero. Med.*, 42 (6), 670–2.
76. Raboutet, J. & Raboutet, P. (1975) Sudden incapacitation encountered in-flight by professional pilots in French civil aviation 1948–1972. *Aviat. Space Environ. Med.*, 46 (1) 80–1.
77. Federal Aviation Administration (FAA) (2004) *In-flight incapacitation and impairment of US Airline Pilots, 1993–1998*. Office of Aerospace Medicine, Washington DC.
78. Lane, J. C. (1971) Risk of in-flight incapacitation of airline pilots. *Aero. Med.*, 42 (12), 1319–21.
79. James, M. & Green, R. (1991) Airline Pilot Incapacitation Survey. *Aviat. Space Environ. Med.*, 62, 1068–72.
80. Bailey, J. (1977) *Guide to Hygiene and Sanitation in Aviation*, 2nd edition, p. 56. WHO, Geneva.
81. Federal Aviation Administration, Regulations Part 91, section 533. www.gofir.com/fars/part91
82. ICAO Manual Operation of Aircraft (1998) International Civil Aviation Organization. AN 6-2-92-91-94-249-9. I Annex 6, Chapter 12, 12.4.
83. WHO (2004) *Guidelines For Drinking Water Quality*, 3rd edition, Vol. 1. WHO, Geneva.
84. Evins, C. (2004) WHO: Safe Piped Water. Managing Microbial Water Quality in Piped Distribution Systems. Chapter 6 – Small animals in drinking-water distribution systems, p. 100. IDA Publishing, London.
85. WHO (1996) *Guidelines for Drinking Water Quality*, 2nd edition, Vol. 2, Health Criteria and other supporting information, pp. 68–74. WHO, Geneva.
86. Chang, S. L., Berg, G, Clarke, N. A. & Kabler, P. W. (1960) Survival and protection against chlorination of human enteric pathogens in free living nematodes isolated from water supplies. *American Journal of Tropical Medicine and Hygiene*, 9, 136–42.
87. Smerda, S. M., Jensen, H. J. & Anderson, A. W. (1971) Escape of Salmonellae from chlorination during ingestion by Pristonchus ihertheri (Nematoda diplogasterinae). *Journal of Nematology*, 3, 201–4.
88. Levy, R. V., Cheetham, R. D., Davis, J., Winer, G. & Hart, F. L. (1984) Novel method for studying the public health significance of macroinvertebrates occurring in potable water. *Applied and Environmental Biology*, 47, 889–94.
89. Burfield, I. & Williams, D. N. (1975) Control of parthenogenetic chironomids with pyrethrins. *Water Treatment and Examination*, 24, 57–67.
90. Abram, F. S. H., Evans, C. & Hobson, J. A. (1980) *Permethrin for the control of animals in water mains*. Technical Report TR 145, Water Research Centre, Medmenham.
91. Mitcham, R. P. & Shelley, M. W. (1980) The control of animals in water mains using permethrin, a synthetic pyrethroid. *Journal of the Institution of Water Engineers and Scientists*, 34, 474–83.
92. Crowther, R. F. & Smith, P. B. (1982) Mains infestations control using permethrin. *Journal of the Institution of Water Engineers and Scientists*, 36 (3), 205–14.

93. Evins, C. (2004) WHO: *Safe Piped Water. Managing Microbial Water Quality in Piped Distribution Systems*. Chapter 6 – Small animals in drinking-water distribution systems, p. 112. IDA Publishing, London.
94. Evins, C. (2004) WHO: *Safe Piped Water. Managing Microbial Water Quality in Piped Distribution Systems*. Chapter 6 – Small animals in drinking-water distribution systems, p. 113. IDA Publishing, London.
95. Daszubshi, D. M. (2003) Detroit Food and Drug Administration. Detroit MI48207. http://www.fda.gov/foi/warning-letters/g4125dhtm
96. Keates, N. & Costello, J. (2002) How safe is airline water? *Weekend Journal of The Wall Street Journal*, USA Edition, 1 November.
97. Mattingly, P F. (1969) *The Biology of Mosquito Borne Disease*, pp. 53–6, 59, 252–5. George Allen & Unwin, London.
98. WHO. (1995) Report of the Informal Consultation on Aircraft Disinsection, pp. 6–10, 29. WHO, Geneva.
99. Collier, L. & Oxford, J. (1993) *Human Virology*, pp. 221–30. Oxford Medical Publications.
100. WHO (1969) International Health Regulations (IHR), 3rd annotated edition, pp. 6, 18, 31, 36–37. WHO, Geneva.
101. Chitty, N. (2004) Personal communication on advanced fumigation and pest control. Meeting with Erica Sheward at Washington, West Sussex,
102. Carnevale, P. (1995) *Potential spread of vector borne human diseases*. WHO Informal Consultation on Aircraft Disinsection, 6–10 November 1995, pp. 2–8.
103. Muentener, P., Schlagenhauf, P. & Steffen, R. (1999) Imported malaria (1985–1995). *Trends and Perspective Bulletin of WHO*, 77 (7), 560–66.
104. WHO (1984) *Ports designated in application of the International Health Regulations*, pp. 3–4, 29–30. WHO, Geneva.
105. Carter, W. (1973) *Insects in Relation to Plant Disease*, 2nd edition, pp. 9,115–119, 317, 329, 487–534. Wiley Interscience.
106. WHO (1995) *Report of the informal consultation on aircraft disinsection*, pp. 4–21. WHO/HQ, Geneva. 6–10 November, WHO/PCS/95.51.
107. Curdt-Christiansen C. (1995) Review of aircraft disinsection requirements and methods. WHO: *Informal consultation on aircraft disinsection*, 6–10 November.
108. Russell, R. & Paton, R. (1989) In-flight disinsection as an efficacious procedure for preventing international transport of insects of public health importance. *Bulletin of the WHO*, **67** (5), 543–7.
109. Australian Quarantine & Inspection Service (AQIS) (1999) www.dpie.gov.au
110. Bailey, J. (1977) *Guide to Hygiene and Sanitation in Aviation*, 2nd edition. WHO, Geneva.
111. WHO (1972) *Vector Control in International Health*, pp. 3–30, 25–34, 35–38, 44–55, 73–76 WHO, Geneva.
112. WHO (1985) Recommendations on the disinsecting of aircraft: procedure for disinsection of aircraft by weekly residual insecticide film. *Weekly Epidemiol. Rec.*, 8 November, 45, 60, 345–52.
113. Guillet, P., Germain, M., Giacommi, T., Chandre, F. & Akogbeto, M. (1998) Origin and prevention of airport malaria in France. *Tropical Medicine*, **3** (9) 700–705.
114. Manga, L. (1995) Other methods of aircraft disinsection besides those routinely used. WHO: *Informal consultation on aircraft disinsection*, 6–10 November, pp. 2–3.
115. Fairechild, D. (1992) *Jet Smart*, pp.90–93. Flyana Rhyme, Kauai, Hawaii.
116. McCance, R. A. & Widdowson, E. M. (2002) *The Composition of Foods*, sixth summary edition. Royal Society of Chemistry, Cambridge and Food Standards Agency.
117. ICAO (2004) Annex 17, Standard 4.5.4 Doc. 8973–6 Restricted. http;//www.icao.int/icao/en/sales.htm
118. ICAO (2004) Annex 17, Standard 3.1.6. Doc. 8973–6 Restricted. http;//www.icao.int/icao/en/sales.htm
119. WHO (2002) *Terrorist Threats to Food. Guidance for Establishing and Strengthening Prevention and Response Systems*, p.3. Food Safety Department, World Health Organization, Geneva.

120. FDA (2002) The Public Health Security and Bioterrorism Preparedness and Response Act of 2002. www.fda.gov./oc/bioterrorism
121. WHO (2002) *Terrorist Threats to Food. Guidance for Establishing and Strengthening Prevention and Response Systems.* Food Safety Department, World Health Organization, Geneva.
122. Product Authentication International (PAI) (2003) Food Industry Security Assurance Scheme (FISA), Edition 02 PAI – April 2003. www.food-standards.com
123. Transport Security Administration (TSA) (2004) Homelands Security. TSA Home Page www.tsa.gov
124. Mortimore, S. & Wallace, C. (2001) *HACCP*, p.22. Food Industry Briefing Series, Blackwell Publishing, Oxford.
125. Mortimore, S. & Wallace, C. (2001) *HACCP*, p.27, Fig.9. Food Industry Briefing Series, Blackwell Publishing, Oxford.
126. Department of Health (1996) H16/011 709 3P 220k.

Index

Note: All figures and tables are highlighted in **bold**.

allergens, 290–293
Australian Quarantine and Inspection Service (AQIS), 256, 259
aviation safety, 4–5, **11**

business aviation, 337–338
buy on board, 21–23
    caterers, 343–344
    catering, 17–19, 43–44, 338–342

cabin crew, 197–225
    auditors, 207–211
    food safety management, 223–225, 359
    food service, 204–205, 359
    health, 211–213, 334–336, 359
    responsibilities, 205–207, 359
    training, 197–199, 214–223, 359
Captain Glen Stewart, 8–9
Clostridium perfringens, **26**, 29
Codex Alimentarius Commission (CAC), 35–40, 51, 84, 86, 296, 300
critical control points (CCP), 75–76, 79, 82–86, 88, 90, 364–365

delayed flights, 45
deep vein thrombosis (DVT), 3–4, **11**, 195, 253
disinfestation, 267
disinsection, 47–48, 253–256, 258–260
    methods, 260–267, 269–270

Escherichia coli, **26**, 29
European civil aviation conference, 314–315
European Food Safety Authority (EFSA), 315–316, 319
European Union Food Directives, 319–322
evaluation
    product safety, 178, 179
    product, 177–178
    supplier, 178

fitness to fly, 190, 199–200
food brokers, 175–177
food hygiene training, 214
Food Industry Security Assurance Scheme (FISA), 330–331
food poisoning outbreaks, **25**, **26**, 27–30
    Clostridium perfringens, **26**, 29
    Escherichia coli, **26**, 29
    Norovirus, **26**, 29
    Salmonella spp., **25**, **26**, 27
    Shigella spp., **26**, 28, 29
    Staphylococcus aureus, **25**, **26**, 27
    Vibrio spp., **25**, 28
food safety legislation, 49–50, 54–58
foot and mouth disease, 5, 319
Four Montreal Protocols 1975, 157
fumigation, 267–268

378　*Index*

general aviation, 344–350
　food chain, 350–354
　food safety management, 354–362
good hygiene practice (GHP), 66–67, 71, 74, 301
good management practice, 67, 71, 74, 89, 301, 354–356
Guadalajara Convention 1961, 156
Guatamala City Protocol 1971, 156, 157
Guide to Hygiene and Sanitation in Aviation, 62–63, 170, 196–197, 233, 242, 258

Hague Protocol 1955, 156
Hazard Analysis and Critical Control Point (HACCP), 64–92, 99, 100, 365–370
Homelands Security Act 2002, 317

ice, 47, **234**
integrated aviation safety debate/agendas, 4
Inter Carrier Agreement 1977, 157
International Air Transport Association (IATA), 42–47, 52, 53, 271–273, 312
International Civil Aviation Organization (ICAO), 42, 49, 310–312, 331
International Flight Catering Association (IFCA)/ International Travel Catering Association (ITCA)/ International Food Service Association (IFSA) World Food Safety Guidelines, 51, 58–60, 271, 290, 296, 325–327

International Health Regulations (IHR), 33–35, 40–42, 232–233, 235, 255, 258–259
intrinsic factors, 77–79

labelling
　healthy claims, 303–304
　legislation, 304–305
　shelf-life, 145–147
　special meals, 288–289, 296–299, 300–301
liability, 154
logistics, 179–182

malaria, 256–258
micro-organisms
　Clostridium perfringens, **26**, 29
　Escherichia coli, **26**, 29
　Norovirus, **26**, 29
　Salmonella spp., **25**, 26, 27
　Shigella spp., **26**, 28, 29
　Staphylococcus aureus, **25**, 26, 27
　Vibrio spp, **25**, 28
Montreal Agreement 1966, 156
Montreal Convention, 155, 160–161, 311
Montreal Convention 1971 and Montreal Protocol 1988, 316

National Aviation Security Authority, 310, 313–314
National Aviation Security Programme, 310, 312–313
Norovirus, **26**, 29
nutritional analysis, 301–303

personal hygiene, 45
pilot factor, 7–8
pilot incapacitation, 1, 7, 190–194
prerequisite programmes, 66, **68**, 226–227, 362–363

product development, 94–101, 171, **172**, 173
production records, **129**, **130**, **131**, **132**, **133**, **134**, **135**, **136**, **140**, **141**
product recall, 148–150
product specifications, 104–120, 293–295, 300–301
Public Health Security and Bioterrorism Preparedness and Response Act 2002, 322–325

quality management systems (QMS), 66–67, **68**, 71, 74, 81, 87, 363–364
Quantitative Ingredient Declaration (QUID), 119

raw material procurement, 104–117

Salmonella spp., **25**, 26, 27
SARS, 3, **11**, 253
Shigella spp., **26**, 28, 29
special meals
   labelling, 54–58, 296–299, 300–301
   medical meals, 275–277
   menu development, 277–288
   preference meals, 273–274
   religious meals, 274–275
   standards, 46

standard operating procedure (SOP), 93–150, 199–200
Staphylococcus aureus, **25**, 26, 27
supplier outsourcing, 101–104

terrorism, 307–309
terrorists' threats to food, 60–62
traceability, 184–189

vector-borne disease, 253–256, 257–258
Vibrio spp., **25**, 28

Warsaw Convention, 155, 158–160, 162–164
water
   disinfection, 242, 247–248
   microbial risks, 239–240
   risk management, 244–249
   safety management, 223–236, 248
   safety plan, 41, 240–242
   safety, 249–252
   sampling, 243
   small animals in supply system, 236–239
   standards, 42, 47, 228–233
World Food Safety Guidelines, 58–60
World Health Organization (WHO), 35–39, 67, 270, 311, 314–315
World Trade Organization (WTO), 35–39, **40**

ERAU-PRESCOTT LIBRARY

BELFAST INSTITUTE LIBRARIES
346.730482
STRA
335877 ✓

STAFF LIBRARY

BELFAST INSTITUTE LIBRARIES

DATE DUE	DATE DUE	DATE DUE
B.I.F.H.E. LIBRARY MILLFIELD	WD 11/23	

APPLICATION FOR RENEWAL MAY BE MADE
IN PERSON, BY WRITING OR BY TELEPHONE